TAKING SIDES

Clashing Views on
Environmental Issues

TAKING SIDES

Clashing Views on

Environmental Issues

FIFTEENTH EDITION, EXPANDED

Selected, Edited, and with Introductions by

Thomas A. Easton
Thomas College

TAKING SIDES: CLASHING VIEWS ON ENVIRONMENTAL ISSUES, FIFTEENTH EDITION, EXPANDED

Published by McGraw-Hill Education, 2 Penn Plaza, New York, NY 10121. Copyright © 2014 by McGraw-Hill Education. All rights reserved. Printed in the United States of America. Previous editions © 2013, 2011, and 2009. No part of this publication may be reproduced or distributed in any form or by any means, or stored in a database or retrieval system, without the prior written consent of The McGraw-Hill Education, including, but not limited to, in any network or other electronic storage or transmission, or broadcast for distance learning.

Some ancillaries, including electronic and print components, may not be available to customers outside the United States.

This book is printed on acid-free paper.

Taking Sides® is a registered trademark of the McGraw-Hill Education.
Taking Sides is published by the **Contemporary Learning Series** group within the McGraw-Hill Education.

1 2 3 4 5 6 7 8 9 0 DOC/DOC 1 0 9 8 7 6 5 4 3

MHID: 0-07-351454-3
ISBN: 978-0-07-351454-3
ISSN: 1530-0757 (print)

Acquisitions Editor: *Joan L. McNamara*
Marketing Director: *Adam Kloza*
Marketing Manager: *Nathan Edwards*
Senior Developmental Editor: *Jade Benedict*
Project Manager: *Erin Melloy*
Cover Design: *Studio Montage, St. Louis, MO*
Buyer: *Jennifer Pickel*
Content Licensing Specialist: *Rita Hingtgen*
Media Project Manager: *Sridevi Palani*

Compositor: MPS Limited
Cover Image: *© Getty Images RF*

Editors/Academic Advisory Board

Members of the Academic Advisory Board are instrumental in the final selection of articles for each edition of TAKING SIDES. Their review of articles for content, level, and appropriateness provides critical direction to the editors and staff. We think that you will find their careful consideration well reflected in this volume.

TAKING SIDES: Clashing Views on ENVIRONMENTAL ISSUES
Fifteenth Edition, Expanded

EDITOR

Thomas Easton
Thomas College

ACADEMIC ADVISORY BOARD MEMBERS

Preface

Most fields of academic study evolve over time. Some evolve in turmoil, for they deal in issues of political, social, and economic concern. That is, they involve in controversy.

It is the mission of the *Taking Sides* series to capture current, ongoing controversies and make the opposing sides available to students. This book focuses on environmental issues, from the philosophical to the practical. It does not pretend to cover all such issues, for not all provoke controversy or provoke it in suitable fashion. But there is never any shortage of issues that can be expressed as pairs of opposing essays that make their positions clearly and understandably.

The basic technique—presenting an issue as a pair of opposing essays—has risks. Students often display a tendency to remember best those essays that agree with the attitudes they bring to the discussion. They also want to know what the "right" answers are, and it can be difficult for teachers to refrain from taking a side, or from revealing their own attitudes. Should teachers so refrain? Some do not, but of course they must still cover the spectrum of opinion if they wish to do justice to the scientific method and the complexity of an issue. Some do, though rarely so successfully that students cannot see through the attempt.

For any *Taking Sides* volume, the issues are always phrased as yes/no questions. Which answer—yes or no—is the correct answer? Perhaps neither. Perhaps both. Perhaps we will not be able to tell for another hundred years. Students should read, think about, and discuss the readings and then come to their own conclusions without letting my or their instructor's opinions dictate theirs. The additional readings mentioned in the introductions should prove helpful.

This edition of *Taking Sides: Clashing Views on Environmental Issues* contains 42 readings arranged in pro and con pairs to form 21 issues. For each issue, an *introduction* provides historical background and context, recent developments, and a brief description of the debate. The *postscript* after each pair of readings offers recent contributions to the debate, additional references, and sometimes a hint of future directions. Each part is preceded by an *Internet References* page that lists several links that are appropriate for further pursuing the issues in that part.

Changes to this edition   Over half of this book consists of new material. Two issues, *Is Global Warming a Catastrophe that Warrants Immediate Action?* (Issue 7) and *Does Commercial Fishing Have a Future?* (Issue 14), were added for the 2012 partial revision of the fourteenth edition. Because of the new interest in energy issues, we have added *Is Shale Gas the Solution to Our Energy Woes?* (Issue 9) and *Is Renewable Energy Really Green?* (Issue 10). We also added *Should Society Impose a*

Moratorium on the Use and Release of "Synthetic Biology" Organisms? (Issue 16). We renamed the precautionary principle issue *Should the Precautionary Principle Become Part of National and International Law?* (Issue 1), the ecosystem services issue *Do Ecosystem Services Have Economic Value?* (Issue 3), the biofuels issue *Are Biofuels a Reasonable Substitute for Fossil Fuels?* (Issue 11), the population issue *Do We Have a Population Problem?* (Issue 13), and the Superfund issue *Should the Superfund Tax Be Reinstated?* (Issue 18).

In addition, for five of the issues retained from the previous edition, one or both of the readings have been replaced. In all, 24 of the readings in this edition were not in the fourteenth edition.

A word to the instructor An *Instructor's Resource Guide with Test Questions* (multiple-choice and essay) is available through the publisher for the instructor using *Taking Sides* in the classroom. Also available is a general guidebook, *Using Taking Sides in the Classroom,* which offers suggestions for adapting the pro-con approach in any classroom setting. An online version of *Using Taking Sides in the Classroom* and a correspondence service for *Taking Sides* adopters can be found at http://www.mhhe.com/cls.

Taking Sides: Clashing Views on Environmental Issues is only one title in the *Taking Sides* series. If you are interested in seeing the table of contents for any of the other titles, please visit the *Taking Sides* website at http://www.mhhe .com/cls.

Thomas A. Easton
Thomas College

Contents in Brief

Contents

UNIT 1   ENVIRONMENTAL PHILOSOPHY   1

Issue 1.   Should the Precautionary Principle Become Part of National and International Law?   2

YES: **Agne Sirinskiene,** from "The Status of Precautionary Principle: Moving Towards a Rule of Customary Law," *Jurisprudence* (October 2009)   *6*

NO: **Ken Cussen,** from "Handle with Care: Assessing the Risks of the Precautionary Principle," *Australasian Journal of Environmental Management* (June 2009)   *14*

Agne Sirinskiene argues that the evidence from treaties, legislation, and court cases clearly indicates that the precautionary principle is becoming or has already become a rule of customary national and international law, and international applications of the principle are developing rapidly. Ken Cussen argues that the precautionary principle is so vague, ill-defined, and value-ridden that it is either vacuous or dangerous. Its underlying assumptions must be clarified before it can be used to guide public policy.

Issue 2.   Is Sustainable Development Compatible with Human Welfare?   21

YES: **Richard Heinberg,** from *The End of Growth: Adapting to Our New Economic Reality* (New Society Publishers, 2011)   *25*

NO: **Ronald Bailey,** from "Wilting Greens," *Reason* (December 2002)   *36*

Richard Heinberg argues that the era of economic growth as we have known it is over. A major cause of the world's recent (and continuing) economic crisis is depletion of resources such as oil and environmental degradation. We must learn to live sustainably, in "a healthy equilibrium economy." Ronald Bailey argues that sustainable development results in economic stagnation and threatens both the environment and the world's poor.

Issue 3.   Do Ecosystem Services Have Economic Value?   40

YES: **Rebecca L. Goldman,** from "Ecosystem Services: How People Benefit from Nature," *Environment* (September/October 2010)   *44*

Rebecca L. Goldman argues that ecosystem services are crucial to human well-being, both now and for the sustainable future. They are also affected by human behavior, at both the individual and the national levels. Assessing their economic value is difficult but essential to public decision making. Professors of applied ecology Marino Gatto and Giulio A. De Leo contend that the pricing approach to valuing nature's services is misleading because it falsely implies that only economic values matter.

UNIT 2   PRINCIPLES VERSUS POLITICS   65

C. Josh Donlan proposes that because the arrival of humans in the Americas some 13,000 years ago led to the extinction of numerous large animals (including camels, lions, and mammoths) with major effects on local ecosystems, restoring these animals (or their near-relatives from elsewhere in the world) holds the potential to restore health to these ecosystems. There would also be economic and cultural benefits. Dustin R. Rubenstein, Daniel I. Rubenstein, Paul W. Sherman, and Thomas A. Gavin argue that bringing African and Asian megafauna to North America is unlikely to restore prehuman ecosystem function and may threaten present species and ecosystems. It would be better to focus resources on restoring species where they were only recently extinguished.

Benedict S. Cohen argues that environmental regulations interfere with military training and other "readiness" activities, and that although the U.S. Department of Defense will continue "to provide exemplary stewardship of the lands and natural resources in our trust," those regulations must be revised to permit the military to do its job without interference. Jamie Clark argues that reducing the Department of Defense's environmental obligations is dangerous because both people and wildlife would be threatened with serious, irreversible, and unnecessary harm.

energy sources essential to maintaining its economy and standard of living. Mary Annette Rose argues that the environmental impacts of exploiting offshore oil—including toxic pollution, ocean acidification, and global warming—are so complex and far-reaching that any decision to expand U.S. oil drilling must be based on more than public opinion driven by consumer demands for cheap energy, economic trade imbalances, and politics.

Diane Katz argues that new technology has made it possible to release vast amounts of natural gas from shale far underground. As a result, we should stop spending massive sums of public money to develop renewable energy sources. The "knowledge and wisdom of private investors" are more likely to solve energy problems than government policymakers. Deborah Weisberg argues that the huge amounts of water and chemicals involved in "fracking"—hydraulic fracturing of shale beds to release natural gas—pose tremendous risks to both ground and surface water, and hence to public health. There is a need for stronger regulation of the industry.

Andrea Larson argues that "green" technologies include, among other things, renewable energy technologies and these technologies are essential to future U.S. domestic economic growth and to international competitiveness. Senator Lamar Alexander (R-TN) argues that the land use requirements of solar and wind power threaten the environment. We must therefore be very careful in how we implement these "green" energy technologies. He also believes the best way to address climate change (by cutting carbon emissions) is with nuclear power.

Keith Kline, Virginia H. Dale, Russell Lee, and Paul Leiby argue that the impact of biofuel production on food prices is much less than alarmists claim. If biofuel development focused on converting biowastes and fast-growing trees and grasses into fuels, the overall impact would be even better, with a host of benefits in reduced fossil fuel use and greenhouse gas emissions, increased employment, enhanced wildlife habitat, improved soil and water quality, and more stable land use. David Pimentel, Alison Marklein, Megan A. Toth, Marissa N. Karpoff, Gillian S. Paul, Robert McCormack, Joanna Kyriazis, and Tim Krueger argue that it is not possible to replace more than a small fraction of fossil fuels with biofuels. Furthermore, producing biofuels consumes more energy (as fossil fuels) than it makes available, and because biofuels compete with food production for land, water, fertilizer, and other resources, they necessarily drive up the price of food, which disproportionately harms the world's poor. It might also damage the environment in numerous ways.

Allison MacFarlane argues that although nuclear power poses serious problems to be overcome, it "offers a potential avenue to significantly mitigate carbon dioxide emissions while still providing baseload power required in today's world." However, it will take many years to build the necessary number of new nuclear power plants. Professor Kristin Shrader-Frechette argues that nuclear power is one of the most impractical and risky of energy sources. Renewable energy sources such as wind and solar are a sounder choice.

Sir David Attenborough argues that the environmental problems faced by the world are exacerbated by human numbers. Without population reduction, the problems will become ever more difficult—and ultimately impossible—to solve. Tom Bethell argues that population alarmists project their fears onto popular concerns, currently the environment, and every time their scare-mongering turns out to be based on faulty premises. Blaming environmental problems will be no different. Societies are sustained not by population control but by belief in God.

Carl Safina argues that despite an abundance of bad news about the state of the oceans and commercial fisheries, there are some signs that conservation and even restoration of fish stocks to a sustainable state are possible. The Food and Agriculture Organization of the United Nations argues that the proportion of marine fish stocks that are overexploited has increased tremendously since the 1970s. Despite some progress, there remains "cause for concern." The continuing need for fish as food means there will be continued growth in aquaculture.

Ed Hamer and Mark Anslow argue that organic agriculture can feed the world if people are willing to eat less meat. It would also use less energy and water, emit fewer greenhouse gases, provide better nutrition, protect ecosystems, and increase employment. D. J. Connor argues that a major report claiming that organic methods could produce enough food to sustain a global human population even larger than that of today has serious faults. At best organic methods could support a population less than half as large as today's (over 7 billion).

UNIT 5   TOXIC CHEMICALS   293

Jim Thomas, Eric Hoffman, and Jaydee Hanson, representing the Civil Society on the Environmental and Societal Implications of Synthetic Biology, argue that the risks posed by synthetic biology to human health, the environment, and natural ecosystems are so great that Congress should declare an immediate moratorium on releases to the environment and commercial uses of synthetic organisms and require comprehensive environmental and social impact reviews of all federally funded synthetic biology research. Gregory E. Kaebnick of the Hastings Center argues that although synthetic biology is surrounded by genuine ethical and moral concerns—including risks to health and environment—which warrant discussion, the potential benefits are too great to call for a general moratorium.

Professor of biological sciences Michele L. Trankina argues that a great many synthetic chemicals behave like estrogen, alter the reproductive functioning of wildlife, and may have serious health effects—including cancer—on humans. Michael Gough, a biologist and expert on risk assessment and environmental policy, argues that only "junk science" supports the hazards of environmental estrogens.

Stephen Lester and Anne Rabe argue that because toxic waste cleanup is complicated by extreme weather events, corporations dodge their cleanup and payment obligations, and the taxpayer is left with the bill, Congress must reinstate the "polluter pays" fees. J. Winston Porter argues that Superfund cleanup efforts can be made much more efficient and that "polluter pays" taxes are unfair. The primary funder of cleanup work should be the people responsible for the problems. Taxpayers should foot the bill as a matter of last resort.

Kate J. Dennis, Jason Rugolo, Lee T. Murray, and Justin Parrella argue that nuclear fuel reprocessing extracts more energy from nuclear fuel and reduces the amount of nuclear waste to be disposed of. "If the United States truly wants to proceed with nuclear energy as a viable, low-carbon emitting source of energy, it should pursue reprocessing in combination with the development of fast reactors. Once such a decision is made, the debate should turn to how best to develop cheaper and safer reprocessing options, rather than denying its general benefit." David M. Romps, Christopher D. Holmes, Kurt Z. House, Benjamin G. Lee, and Mark T. Winkler argue that reprocessing is both dangerous and unnecessary. "It is in the best interests of the United States—from the perspective of waste management, national security, economics, and environmental protection—to maintain its de facto moratorium on reprocessing and encourage other countries to follow suit."

Jessica Tsai argues that even though marketing is all about gaining attention in a very noisy environment, it is possible to improve brand image by being environmentally responsible without being guilty of greenwashing. Richard Dahl argues that consumers are reluctant to believe corporate claims of environmental responsibility because in the past such claims have been so overblown as to amount to "greenwashing."

Mike McKee, Uintah County Commissioner, argues that the government's new "Wild Lands" policy is illegal, contradicts previously approved land use plans for public lands, and will have dire effects on rural economies based on extractive industries. It should be repealed. Robert Abbey, director of the Bureau of Land Management, argues that the government's new "Wild Lands" policy is legal, restores balance and clarity to multiple-use public land management, and will be implemented in collaboration with the public. Destruction of local extractive economies is by no means a foregone conclusion.

Correlation Guide

The *Taking Sides* series presents current issues in a debate-style format designed to stimulate student interest and develop critical thinking skills. Each issue is thoughtfully framed with an issue summary, an issue introduction, and a postscript. The pro and con essays—selected for their liveliness and substance—represent the arguments of leading scholars and commentators in their fields.

Taking Sides: Clashing Views on Environmental Issues, 15/e, Expanded is an easy-to-use reader that presents issues on important topics such as *energy, environmental philosophy, and sustainability.* For more information on *Taking Sides* and other *McGraw-Hill Contemporary Learning Series* titles, visit http://www.mhhe.com/cls.

This convenient guide matches the issues in **Taking Sides: Environmental Issues** with the corresponding chapters in three of our best-selling McGraw-Hill Environmental Science textbooks by Enger/Smith and Cunningham/Cunningham.

Taking Sides: Clashing Views on Environmental Issues, 15/e, Expanded	Environmental Science, 13/e by Enger/Smith	Principles of Environmental Science, 6/e by Cunningham/Cunningham	Environmental Science: A Global Concern, 12/e by Cunningham/Cunningham
Issue 1: Should the Precautionary Principle Become Part of National and International Law?	**Chapter 3:** Environmental Risk: Economics, Assessment, and Management	**Chapter 1:** Understanding Our Environment **Chapter 15:** Environmental Policy and Sustainability	**Chapter 8:** Environmental Health and Toxicology **Chapter 24:** Environmental Policy, Law, and Planning
Issue 2: Is Sustainable Development Compatible with Human Welfare?	**Chapter 5:** Interactions: Environments and Organisms	**Chapter 2:** Environmental Systems: Connections, Cycles, Flows, and Feedback Loops **Chapter 3:** Evolution, Species Interactions, and Biological Communities	**Chapter 1:** Understanding Our Environment **Chapter 23:** Ecological Economics **Chapter 25:** What Then Shall We Do?
Issue 3: Do Ecosystem Services Have Economic Value?	**Chapter 6:** Kinds of Ecosystems and Communities	**Chapter 6:** Environmental Conservation: Forests, Grasslands, Parks, and Nature Preserves **Chapter 7:** Food and Agriculture **Chapter 11:** Environmental Geology and Earth Resources **Chapter 14:** Economics and Urbanization	**Chapter 23:** Ecological Economics

Taking Sides: Clashing Views on Environmental Issues, 15/e, Expanded	Environmental Science, 13/e by Enger/Smith	Principles of Environmental Science, 6/e by Cunningham/ Cunningham	Environmental Science: A Global Concern, 12/e by Cunningham/ Cunningham
Issue 4: Should North America's Landscape Be Restored to Its Prehuman State?	**Chapter 12:** Land-Use Planning	**Chapter 5:** Biomes and Biodiversity	**Chapter 12:** Biodiversity: Preserving Landscapes **Chapter 13:** Restoration Ecology
Issue 5: Should the Military Be Exempt from Environmental Regulations?	**Chapter 18:** Environmental Regulations: Hazardous Substances and Wastes **Chapter 19:** Environmental Policy and Decision Making	**Chapter 8:** Environmental Health and Toxicology **Chapter 15:** Environmental Policy and Sustainability	**Chapter 24:** Environmental Policy, Law, and Planning
Issue 6: Will Restricting Carbon Emissions Damage the Economy?	**Chapter 16:** Air Quality	**Chapter 9:** Air: Climate and Pollution **Chapter 14:** Economics and Urbanization	**Chapter 1:** Understanding Our Environment **Chapter 15:** Air, Weather, and Climate **Chapter 19:** Conventional Energy **Chapter 20:** Sustainable Energy
Issue 7: Is Global Warming a Catastrophe That Warrants Immediate Action?	**Chapter 16:** Air Quality Issues **Chapter 19:** Environmental Policy and Decision Making	**Chapter 1:** Understanding Our Environment **Chapter 2:** Environmental Systems: Connections, Cycles, Flows, and Feedback Loops **Chapter 9:** Air: Climate and Pollution **Chapter 15:** Environmental Policy and Sustainability	**Chapter 1:** Understanding Our Environment **Chapter 2:** Principles of Science and Systems **Chapter 15:** Air, Weather, and Climate **Chapter 24:** Environmental Policy, Law, and Planning
Issue 8: Should We Drill for Offshore Oil?	**Chapter 3:** Environmental Risk: Economics, Assessment, and Management **Chapter 8:** Energy and Civilization **Chapter 9:** Energy Sources	**Chapter 11:** Environmental Geology and Earth Resources **Chapter 12:** Energy **Chapter 14:** Economics and Urbanization	**Chapter 19:** Conventional Energy
Issue 9: Is Shale Gas the Solution to Our Energy Woes?	**Chapter 8:** Energy and Civilization **Chapter 9:** Energy Sources	**Chapter 12:** Energy	**Chapter 20:** Sustainable Energy

(Continued)

Taking Sides: Clashing Views on Environmental Issues, 15/e, Expanded	Environmental Science, 13/e by Enger/Smith	Principles of Environmental Science, 6/e by Cunningham/ Cunningham	Environmental Science: A Global Concern, 12/e by Cunningham/ Cunningham
Issue 10: Is Renewable Energy Really Green?	**Chapter 8:** Energy and Civilization	**Chapter 12:** Energy	**Chapter 20:** Sustainable Energy
Issue 11: Are Biofuels a Reasonable Substitute for Fossil Fuels?	**Chapter 8:** Energy and Civilization **Chapter 9:** Energy Sources	**Chapter 12:** Energy	**Chapter 20:** Sustainable Energy
Issue 12: Is It Time to Revive Nuclear Power?	**Chapter 10:** Nuclear Energy	**Chapter 12:** Energy **Chapter 15:** Environmental Policy and Sustainability	**Chapter 19:** Conventional Energy
Issue 13: Do We Have a Population Problem?	**Chapter 7:** Populations: Characteristics and Issues **Chapter 15:** Water Management	**Chapter 4:** Human Populations **Chapter 7:** Food and Agriculture **Chapter 10:** Water: Resources and Pollution	**Chapter 6:** Population Biology **Chapter 7:** Human Populations **Chapter 9:** Food and Hunger **Chapter 10:** Farming: Conventional and Sustainable Practices **Chapter 17:** Water Use and Management
Issue 14: Does Commercial Fishing Have a Future?		**Chapter 7:** Food and Agriculture	**Chapter 9:** Food and Hunger
Issue 15: Can Organic Farming Feed the World?	**Chapter 14:** Agricultural Methods and Pest Management	**Chapter 4:** Human Populations **Chapter 7:** Food and Agriculture **Chapter 14:** Economics and Urbanization	**Chapter 9:** Food and Hunger **Chapter 10:** Farming: Conventional and Sustainable Practices
Issue 16: Should Society Impose a Moratorium on the Use and Release of "Synthetic Biology" Organisms?	**Chapter 2:** Environmental Ethics	**Chapter 15:** Environmental Policy and Sustainability	**Chapter 24:** Environmental Policy, Law, and Planning
Issue 17: Do Environmental Hormone Mimics Pose a Potentially Serious Health Threat?	**Chapter 18:** Environmental Regulations: Hazardous Substances and Wastes	**Chapter 8:** Environmental Health and Toxicology **Chapter 13:** Solid and Hazardous Waste	**Chapter 8:** Environmental Health and Toxicology **Chapter 24:** Environmental Policy, Law, and Planning

Taking Sides: Clashing Views on Environmental Issues, 15/e, Expanded	Environmental Science, 13/e by Enger/Smith	Principles of Environmental Science, 6/e by Cunningham/ Cunningham	Environmental Science: A Global Concern, 12/e by Cunningham/ Cunningham
Issue 18: Should the Superfund Tax Be Reinstated?	**Chapter 18:** Environmental Regulations: Hazardous Substances and Wastes **Chapter 19:** Environmental Policy and Decision Making	**Chapter 8:** Environmental Health and Toxicology **Chapter 13:** Solid and Hazardous Waste **Chapter 15:** Environmental Policy and Sustainability	**Chapter 8:** Environmental Health and Toxicology **Chapter 21:** Solid, Toxic, and Hazardous Waste **Chapter 24:** Environmental Policy, Law, and Planning
Issue 19: Should the United States Reprocess Spent Nuclear Fuel?	**Chapter 10:** Nuclear Energy	**Chapter 12:** Energy **Chapter 15:** Environmental Policy and Sustainability	**Chapter 19:** Conventional Energy
Issue 20: Can "Green" Marketing Claims Be Believed?	**Chapter 19:** Environmental Policy and Decision Making	**Chapter 15:** Environmental Policy and Sustainability	**Chapter 24:** Environmental Policy, Law, and Planning
Issue 21: Does Designating "Wild Lands" Harm Rural Economies?	**Chapter 19:** Environmental Policy and Decision Making	**Chapter 15:** Environmental Policy and Sustainability	**Chapter 24:** Environmental Policy, Law, and Planning

Topic Guide

This topic guide suggests how the selections in this book relate to the subjects covered in your course. You may want to use the topics listed on these pages to search the Web more easily. On the following pages a number of websites have been gathered specifically for this book. They are arranged to reflect the issues of this *Taking Sides* reader. You can link to these sites by going to http://www.mhhe .com/cls.

All the articles that relate to each topic are listed below the bold-faced term.

Business

6. Will Restricting Carbon Emissions Damage the Economy?
18. Should the Superfund Tax Be Reinstated?
20. Can "Green" Marketing Claims Be Believed?

Economics

3. Do Ecosystem Services Have Economic Value?
6. Will Restricting Carbon Emissions Damage the Economy?

Energy

7. Is Global Warming a Catastrophe That Warrants Immediate Action?
8. Should We Drill for Offshore Oil?
9. Is Shale Gas the Solution to Our Energy Woes?
10. Is Renewable Energy Really Green?
11. Are Biofuels a Reasonable Substitute for Fossil Fuels?
12. Is It Time to Revive Nuclear Power?
19. Should the United States Reprocess Spent Nuclear Fuel?

Environmental Law

1. Should the Precautionary Principle Become Part of National and International Law?
5. Should the Military Be Exempt from Environmental Regulations?

Environmental Philosophy

1. Should the Precautionary Principle Become Part of National and International Law?

2. Is Sustainable Development Compatible with Human Welfare?
3. Do Ecosystem Services Have Economic Value?
10. Is Renewable Energy Really Green?
13. Do We Have a Population Problem?

Environmental Policy

5. Should the Military Be Exempt from Environmental Regulations?
18. Should the Superfund Tax Be Reinstated?

Food

14. Does Commercial Fishing Have a Future?
15. Can Organic Farming Feed the World?

Global Warming

7. Is Global Warming a Catastrophe That Warrants Immediate Action?

Population

13. Do We Have a Population Problem?

Politics

5. Should the Military Be Exempt from Environmental Regulations?
12. Is It Time to Revive Nuclear Power?
17. Do Environmental Hormone Mimics Pose a Potentially Serious Health Threat?
18. Should the Superfund Tax Be Reinstated?
19. Should the United States Reprocess Spent Nuclear Fuel?

Precautionary Principle

1. Should the Precautionary Principle Become Part of National and International Law?

Restoration Ecology

4. Should North America's Landscape Be Restored to Its Prehuman State?
21. Does Designating "Wild Lands" Harm Rural Economies?

Superfund

18. Should the Superfund Tax Be Reinstated?

Sustainability

2. Is Sustainable Development Compatible with Human Welfare?
13. Do We Have a Population Problem?

14. Does Commercial Fishing Have a Future?
15. Can Organic Farming Feed the World?

Technology

16. Should Society Impose a Moratorium on the Use and Release of "Synthetic Biology" Organisms?
19. Should the United States Reprocess Spent Nuclear Fuel?

Toxic Chemicals

16. Should Society Impose a Moratorium on the Use and Release of "Synthetic Biology" Organisms?
17. Do Environmental Hormone Mimics Pose a Potentially Serious Health Threat?
18. Should the Superfund Tax Be Reinstated?
19. Should the United States Reprocess Spent Nuclear Fuel?

Introduction

Environmental Issues: The Never-Ending Debate

Thomas A. Easton

When I teach environmental science, I often begin by explaining the roots of the word "ecology," from the Greek *oikos* (house or household), and assigning the students to write a brief paper about their own household. How much, I ask them, do you need to know about the place where you live, and why?

The answers vary. Some of the resulting papers focus on people—roommates if the "household" is a dorm room, spouses and children if the students are older, parents and siblings if they live at home—and the needs to cooperate and get along, and perhaps the need not to overcrowd. Some pay attention to houseplants and pets, and occasionally even bugs and mice. Some focus on economics—possessions, services, and their costs, where the checkbook is kept, where the bills accumulate, the importance of paying those bills, and of course the importance of earning money to pay those bills. Some focus on maintenance—cleaning, cleaning supplies, repairs, whom to call if something major breaks. For some the emphasis is operation—garbage disposal, grocery shopping, how to work the lights, stove, fridge, and so on. A very few recognize the presence of toxic chemicals under the sink and in the medicine cabinet and the need for precautions in their handling. Sadly, a few seem to be oblivious to anything that does not have something to do with entertainment.

Not surprisingly, some students object initially that the exercise seems trivial. "What does this have to do with environmentalism?" they ask. Yet the course is rarely very old before most are saying, "Ah! I get it!" That nice, homey microcosm has a great many of the features of the macrocosmic environment, and the multiple ways people can look at the microcosm mirror the ways people look at the macrocosm. It's all there, as is the question of priorities: What is important? People or fellow creatures or economics or maintenance or operation or waste disposal or food supply or toxics control or entertainment, or all of these?

And how do you decide? I try to illuminate this question by describing a parent trying to teach a teenager not to sit on a woodstove. In July, the kid answers, "Why?" and continues to perch. In August, likewise. And still in September. But in October or November, the kid yells "Ouch!" and jumps off in a hurry.

That is, people seem to learn best when they get burned.

This is surely true in our homely *oikos*, where we may not realize our fellow creatures deserve attention until houseplants die of neglect or cockroaches invade the cupboards. Economics comes to the fore when the phone gets cut off, a pipe ruptures, the air conditioner breaks, strange fumes rise from the basement, trash bags pile up and begin to stink, or the toilet backs up. Toxics control suddenly matters when a child or pet gets into the rat poison.

In the larger *oikos* of environmentalism, such events are paralleled by the loss of a species, or an infestation by another, by floods and droughts, by lakes turned into cesspits by raw sewage, by air turned foul by industrial smokestacks, by groundwater contaminated by toxic chemicals, by the death of industries and the loss of jobs, by famine and plague and even war.

If nothing is going wrong, we are not very likely to realize there is something we should be paying attention to. And this too has its parallel in the larger world. Indeed, the history of environmentalism is, in part, a history of people carrying on with business as usual until something goes obviously awry. Then, if they can agree on the nature of the problem (Did the floor cave in because the joists were rotten or because there were too many people at the party?), they may learn something about how to prevent recurrences.

The Question of Priorities

There is of course a crucial "if" in that last sentence: *If people can agree. . . .* It is a truism to say that agreement is difficult. In environmental matters, people argue endlessly over whether anything is actually wrong, what its eventual impact will be, what if anything can or should be done to repair the damage, and how to prevent recurrence. Not to mention who's to blame and who should take responsibility for fixing the problem! Part of the reason is simple: Different things matter most to different people. Individual citizens may want clean air or water or cheap food or a convenient commute. Politicians may favor sovereignty over international cooperation. Economists and industrialists may think a few coughs (or cases of lung cancer, or shortened life spans) are a cheap price to pay for wealth or jobs.

No one now seems to think that protecting the environment is not important. But different groups—even different environmentalists—have different ideas of what "environmental responsibility" means. To a paper company cutting trees for pulp, it may mean leaving a screen of trees (a "beauty strip") beside the road and minimizing erosion. To hikers following trails through or within view of the same tract of land, that is not enough; they want the trees left alone. The hikers may also object to seeing the users of trail bikes and all-terrain-vehicles on the trails. They may even object to hunters and anglers, whose activities they see as diminishing the wilderness experience. They may therefore push for protecting the land as limited-access wilderness. The hunters and anglers object to that, of course, for they want to be able to use their vehicles to bring their game home, or to bring their boats to their favorite rivers and lakes. They also argue, with some justification, that their license fees support a great deal of environmental protection work.

To a corporation, dumping industrial waste into a river may make perfect sense, for alternative ways of disposing of waste are likely to cost more and diminish profits. Of course, the waste renders the water less useful to wildlife or downstream humans, who may well object. Yet telling the corporation it cannot dump may be seen as depriving it of property. A similar problem arises when regulations prevent people and corporations from using land—and making money—as they had planned. Conservatives have claimed that environmental regulations thus violate the Fifth Amendment to the U.S. Constitution, which says "No person shall . . . be deprived of . . . property, without due process of law; nor shall private property be taken for public use, without just compensation."

One might think the dangers of such things as dumping industrial waste in rivers are obvious. But scientists can and do disagree, even given the same evidence. For instance, a chemical in waste may clearly cause cancer in laboratory animals. Is it therefore a danger to humans? A scientist working for the company dumping that chemical in a river may insist that no such danger has been proven. Yet a scientist working for an environmental group such as Greenpeace may insist that the danger is obvious since carcinogens do generally affect more than one species.

Scientists are human. They have not only employers but also values, often rooted in political ideology and religion. They may feel that the individual matters more than corporations or society, or vice versa. They may favor short-term benefits over long-term benefits, or vice versa.

And scientists, citizens, corporations, and government all reflect prevailing social attitudes. When America was expanding westward, the focus was on building industries, farms, and towns. If problems arose, there was vacant land waiting to be moved to. But when the expansion was done, problems became more visible and less avoidable. People could see that there were "trade-offs" involved in human activity: more industry meant more jobs and more wealth, but there was a price in air and water pollution and human health (among other things).

Nowhere, perhaps, are these trade-offs more obvious than in Eastern Europe. The former Soviet Union was infamous for refusing to admit that industrial activity was anything but desirable. Anyone who spoke up about environmental problems risked jail. The result, which became visible to Western nations after the fall of the Iron Curtain in 1990, was industrial zones where rivers had no fish, children were sickly, and life expectancies were reduced. The fate of the Aral Sea, a vast inland body of water once home to a thriving fishery and a major regional transportation route, is emblematic: Because the Soviet Union wanted to increase its cotton production, it diverted for irrigation the rivers that delivered most of the Aral Sea's fresh water supply. The Aral Sea then began to lose more water to evaporation than it gained, and it rapidly shrank, exposing sea-bottom so contaminated by industrial wastes and pesticides that wind-borne dust is now responsible for a great deal of human illness. The fisheries are dead, and freighters lie rusting on bare ground where once waves lapped.

The Environmental Movement

The twentieth century saw immense changes in the conditions of human life and in the environment that surrounds and supports human life. According to historian J. R. McNeill, in *Something New Under the Sun: An Environmental*

History of the Twentieth-Century World (W.W. Norton, 2000), the environmental impacts that resulted from the interactions of burgeoning population, technological development, shifts in energy use, politics, and economics in that period are unprecedented in both degree and kind. Yet a worse impact may be that we have come to accept as "normal" a very temporary situation that "is an extreme deviation from any of the durable, more 'normal,' states of the world over the span of human history, indeed over the span of earth history." We are thus not prepared for the inevitable and perhaps drastic changes ahead.

Environmental factors cannot be denied their role in human affairs, nor can human affairs be denied their place in any effort to understand environmental change. As McNeill says, "Both history and ecology are, as fields of knowledge go, supremely integrative. They merely need to integrate with each other."

The environmental movement, which grew during the twentieth century in response to increasing awareness of human impacts, is a step in that direction. Yet environmental awareness reaches back long before the modern environmental movement. When John James Audubon (1785–1851), famous for his bird paintings, was young, he was an enthusiastic slaughterer of birds (a few of which he used as models for the paintings). Later in life, he came to appreciate that birds were diminishing in numbers, as were the American bison, and he called for conservation measures. His was a minority voice, however. It was not till later in the century that George Perkins Marsh warned in *Man and Nature* (1864), "We are, even now, breaking up the floor and wainscoting and doors and window frames of our dwelling, for fuel to warm our bodies and seethe our pottage, and the world cannot afford to wait till the slow and sure progress of exact science has taught it a better economy." The Earth, he said, was given to man for "usufruct" (to use the fruit of), not for consumption or waste. Resources should remain to benefit future generations. Stewardship was the point, and damage to soil and forest should be prevented and repaired. He was not concerned with wilderness as such; John Muir (1838–1914; founder of the Sierra Club) was the first to call for the preservation of natural wilderness, untouched by human activities. Marsh's ideas influenced others more strongly. In 1890, Gifford Pinchot (1865–1946) found "the nation . . . obsessed by a fury of development. The American Colossus was fiercely intent on appropriating and exploiting the riches of the richest of all continents." Under President Theodore Roosevelt, he became the first head of the U.S. Forest Service and a strong voice for conservation (not to be confused with preservation; Gifford's conservation meant using nature but in such a way that it was not destroyed; his aim was "the greatest good of the greatest number in the long run"). By the 1930s, Aldo Leopold (1887–1948), best known for his concept of the "land ethic" and his book, *A Sand County Almanac* (1949), could argue that we had a responsibility not only to maintain the environment but also to repair damage done in the past.

The modern environmental movement was kick-started by Rachel Carson's *Silent Spring* (Houghton Mifflin,1962). In the 1950s, Carson realized that the use of pesticides was having unintended consequences—the death of non-pest insects, food-chain accumulation of poisons and the consequent loss of birds, and even human illness—and meticulously documented the case. When her book was published, she and her book were immediately vilified by pesticide proponents in government, academia, and industry (most notably, the pesticides industry). There was no problem, the critics said; the negative effects if any were worth it, and she—a *woman* and a nonscientist—could not possibly know what she was talking about. But the facts won out. A decade later, DDT was banned and other pesticides were regulated in ways unheard of before Carson spoke out.

Other issues have followed or are following a similar course.

The situation before Rachel Carson and *Silent Spring* is nicely captured by Judge Richard Cudahy, who in "Coming of Age in the Environment," *Environmental Law* (Winter 2000), writes, "It doesn't seem possible that before 1960 there was no 'environment'—or at least no environmentalism. I can even remember the Thirties, when we all heedlessly threw our trash out of car windows, burned coal in the home furnace (if we could afford to buy any), and used a lot of lead for everything from fishing sinkers and paint to no-knock gasoline. Those were the days when belching black smoke meant a welcome end to the Depression and little else."

Historically, humans have felt that their own well-being mattered more than anything else. The environment existed to be used. Unused, it was only wilderness or wasteland, awaiting the human hand to "improve" it and make it valuable. This is not surprising at all, for the natural tendency of the human mind is to appraise all things in relation to the self, the family, and the tribe. An important aspect of human progress has lain in enlarging our sense of "tribe" to encompass nations and groups of nations. Some now take it as far as the human species. Some include other animals. Some embrace plants as well, and bacteria, and even landscapes.

The more limited standard of value remains common. Add to that a sense that wealth is not just desirable but a sign of virtue (the Puritans brought an explicit version of this with them when they colonized North America; see Lynn White, Jr., "The Historical Roots of Our Ecological Crisis," *Science*, March 10, 1967), and it is hardly surprising that humans have used and still use the environment intensely. People also tend to resist any suggestion that they should restrain their use out of regard for other living things. Human needs, many insist, must come first.

The unfortunate consequences include the loss of other species. Lions vanished from Europe about 2000 years ago. The dodo of Mauritius was extinguished in the 1600s (see the American Museum of Natural History's account at http://www.amnh.org/exhibitions/expeditions/treasure_fossil/Treasures/Dodo/dodo .html?acts). The last of North America's passenger pigeons died in a Cincinnati zoo in 1914 (see www.amnh.org/exhibitions/expeditions/treasure_fossil/Treasures/ Passenger_Pigeons/pigeons.html?acts). Concern for such species was at first limited to those of obvious value to humans. In 1871, the U.S. Commission on Fish and Fisheries was created and charged with finding solutions to the decline in food fishes and promoting aquaculture. The first federal legislation designed to protect game animals was the Lacey Act of 1900. It was not until 1973 that the U.S. Endangered Species Act was adopted to shield all species from the worst human impacts.

Other unfortunate consequences of human activities include dramatic erosion, air and water pollution, oil spills, accumulations of hazardous (including nuclear) waste, famine, and disease. Among the many "hot stove" incidents that have caught public attention are the following:

- The Dust Bowl—in 1934 wind blew soil from drought-stricken farms in Oklahoma all the way to Washington, DC;
- Cleveland's Cuyahoga River caught fire in the 1960s;
- The Donora, Pennsylvania, smog crisis—in one week of October 1948, 20 died and over 7000 were sickened;
- The London smog crisis in December 1952—4000 dead;
- The Torrey Canyon, Exxon Valdez, and—most recently—the 2010 BP Macondo oil spills, which fouled shores and killed seabirds, seals, and fish;
- Love Canal, where industrial wastes seeped from their burial site into homes and contaminated ground water;
- Union Carbide's toxics release at Bhopal, India—3800 dead and up to 100,000 ill, according to Union Carbide; others claim a higher toll;
- The Three Mile Island, Chernobyl, and Fukushima nuclear accidents;
- The decimation of elephants and rhinoceroses to satisfy a market for tusks and horns;
- The loss of forests—in 1997, fires set to clear Southeast Asian forest lands produced so much smoke that regional airports had to close;
- Ebola, a virus which kills nine-tenths of those it infects, apparently first struck humans because growing populations reached into its native habitat;
- West Nile Fever, caused by transmission of a mosquito-borne virus with a much less deadly record, was brought to North America by travelers or immigrants from Egypt;
- Acid rain, global climate change, and ozone depletion, all caused by substances released into the air by human activities.

The alarms have been raised by many people in addition to Rachel Carson. For instance, in 1968 (when world population was only a little over half of what it is today), Paul Ehrlich's *The Population Bomb* (Ballantine Books) described the ecological threats of a rapidly growing population, and Garrett Hardin's influential essay, "The Tragedy of the Commons," *Science* (December 13, 1968) described the consequences of using self-interest alone to guide the exploitation of publicly owned resources (such as air and water). (In 1974, Hardin introduced the unpleasant concept of "lifeboat ethics," which says that if there are not enough resources to go around, some people must do without.) In 1972, a group of economists, scientists, and business leaders calling themselves "The Club of Rome" published *The Limits to Growth* (Universe Books), an analysis of population, resource use, and pollution trends that predicted difficult times within a century; the study was redone as *Beyond the Limits to Growth: Confronting Global Collapse, Envisioning a Sustainable Future* (Chelsea Green, 1992) and again as *Limits to Growth: The 30-Year Update* (Chelsea Green, 2004), using more powerful computer models, and came to very similar conclusions;

Graham Turner, "A Comparison of *The Limits to Growth* with Thirty Years of Reality," *Global Environmental Change* (August 2008), notes that the *Limits to Growth* projections have been very much on track with actual events.

Among the most recent books is Jared Diamond's *Collapse: How Societies Choose to Fail or Succeed* (Viking, 2005), which uses historical cases to illuminate the roles of human biases and choices in dealing with environmental problems. Among Diamond's important points is the idea that in order to cope successfully with such problems, a society may have to surrender cherished traditions.

The following list of selected U.S. and U.N. laws, treaties, conferences, and reports illustrates the national and international responses to the various cries of alarm:

1967   The U.S. Air Quality Act set standards for air pollution.

1968   The U.N. Biosphere Conference discussed global environmental problems.

1969   The U.S. Congress passed the National Environmental Policy Act, which (among other things) required federal agencies to prepare environmental impact statements for their projects.

1970   The first Earth Day demonstrated so much public concern that the Environmental Protection Agency (EPA) was created; the Endangered Species Act, Clean Air Act, and Safe Drinking Water Act soon followed.

1971   The U.S. Environmental Pesticide Control Act gave the EPA authority to regulate pesticides.

1972   The U.N. Conference on the Human Environment, held in Stockholm, Sweden, recommended government action and led to the U.N. Environment Programme.

1973   The Convention on International Trade in Endangered Species of Wild Fauna and Flora (CITES) restricted trade in threatened species; because enforcement was weak, however, a black market flourished.

1976   The U.S. Resource Conservation and Recovery Act and the Toxic Substances Control Act established control over hazardous wastes and other toxic substances.

1979   The Convention on Long-Range Transboundary Air Pollution addressed problems such as acid rain (recognized as crossing national borders in 1972).

1982 The Law of the Sea addressed marine pollution and conservation.

The second U.N. Conference on the Human Environment (the Stockholm + 10 Conference) renewed concerns and set up a commission to prepare a "global agenda for change," leading to the 1987 Brundtland Report (*Our Common Future*).

1983   The U.S. Environmental Protection Agency and the U.S. National Academy of Science issued reports calling attention to the prospect of global warming as a consequence of the release of greenhouse gases such as carbon dioxide.

1987   The Montreal Protocol (strengthened in 1992) required nations to phase out use of chlorofluorocarbons (CFCs), the chemicals responsible for stratospheric ozone depletion (the "ozone hole").

The Basel Convention controlled cross-border movement of hazardous wastes.

1988 The U.N. assembled the Intergovernmental Panel on Climate Change, which would report in 1995, 1998, and 2001 that the dangers of global warming were real, large, and increasingly ominous.

1992 The U.N. Convention on Biological Diversity required nations to act to protect species diversity.

The U.N. Conference on Environment and Development (also known as the Earth Summit), held in Rio de Janeiro, Brazil, issued a broad call for environmental protections.

The U.N. Convention on Climate Change urged restrictions on carbon dioxide release to avoid climate change.

1994 The U.N. Conference on Population and Development, held in Cairo, Egypt, called for stabilization and reduction of global population growth, largely by improving women's access to education and health care.

1997 The Kyoto Protocol attempted to strengthen the 1992 Convention on Climate Change by requiring reductions in carbon dioxide emissions, but U.S. resistance limited success.

2001 The U.N. Stockholm Convention on Persistent Organic Pollutants required nations to phase out use of many pesticides and other chemicals. It took effect May 17, 2004, after ratification by over 50 nations (not including the United States and the European Union).

2002 The U.N. World Summit on Sustainable Development, held in Johannesburg, South Africa, brought together representatives of governments, nongovernmental organizations, businesses, and other groups to examine "difficult challenges, including improving people's lives and conserving our natural resources in a world that is growing in population, with ever-increasing demands for food, water, shelter, sanitation, energy, health services and economic security."

2003 The World Climate Change Conference held in Moscow, Russia, concluded that global climate is changing, very possibly because of human activities, and the overall issue must be viewed as one of intergenerational justice. "Mitigating global climate change will be possible only with the coordinated actions of all sectors of society."

2005 The U.N. Millennium Project Task Force on Environmental Sustainability released its report, *Environment and Human Well-Being: A Practical Strategy*.

The U.N. Millennium Ecosystem Assessment released its report, *Ecosystems and Human Well-Being: Synthesis* (http://www.millenniumassessment.org/en/index.aspx) (Island Press).

The U.N. Climate Change Conference held in Montreal, Canada, marked the taking effect of the Kyoto Protocol, ratified in 2004 by 141 nations (not including the United States and Australia, which finally ratified the Protocol in 2007).

2007 The Intergovernmental Panel on Climate Change (IPCC) released its Fourth Assessment Report, asserting that global warming is definitely due to human releases of carbon dioxide, the effects on both nature and humanity will be profound, and mitigation, although possible, will be expensive.

2009 In December, the U.N. Climate Change Conference sought increased commitments to reducing carbon dioxide emissions. A follow-up meeting was held in 2010. Both failed to draw binding commitments from the developed world. The United States has not yet ratified the Kyoto Protocol. Another meeting in June 2011 sought even stronger commitments, but without success. Rather than agreeing to international cooperation, governments appear to be choosing to act alone.

Rachel Carson would surely have been pleased by many of these responses, for they suggest both concern over the problems identified and determination to solve those problems. But she would just as surely have been frustrated, for a simple listing of laws, treaties, and reports does nothing to reveal the endless wrangling and the way political and business forces try to block progress whenever it is seen as interfering with their interests. Agreement on banning CFCs was relatively easy to achieve because CFCs were not seen as essential to civilization and there were substitutes available. Restraining greenhouse gas emissions is harder because we see fossil fuels as essential and although substitutes may exist, they are so far more expensive.

The Globalization of the Environment

Years ago, it was possible to see environmental problems as local. A smoke-stack belched smoke and made the air foul. A city sulked beneath a layer of smog. Bison or passenger pigeons declined in numbers and even vanished. Rats flourished in a dump where burning garbage produced clouds of smoke and runoff contaminated streams and groundwater and made wells unusable. Sewage, chemical wastes, and oil killed the fish in streams, lakes, rivers, and harbors. Toxic chemicals such as lead and mercury entered the food chain and affected the health of both wildlife and people.

By the 1960s, it was becoming clear that environmental problems did not respect borders. Smoke blows with the wind, carrying one locality's contamination to others. Water flows to the sea, carrying sewage and other wastes with it. Birds migrate, carrying with them whatever toxins they have absorbed with their food. In 1972, researchers were able to report that most of the acid rain falling on Sweden came from other countries. Other researchers have shown that the rise and fall of the Roman Empire can be tracked in Greenland, where glaciers preserve lead-containing dust deposited over the millennia—the amount rises as Rome flourished, falls with the Dark Ages, and rises again with the Renaissance and Industrial Revolution. Today we know that pesticides and other chemicals can show up in places (such as the Arctic) where they have never been used, even years after their use has been discontinued. The 1979 Convention on Long-Range Transboundary Air Pollution has been strengthened several times with amendments to address persistent organic pollutants, heavy metals, and other pollutants.

We are also aware of new environmental problems that exist only in a global sense. Ozone depletion, first identified in the stratosphere over Antarctica, threatens to increase the amount of ultraviolet light reaching the ground, and thereby increase the incidence of skin cancer and cataracts, among other things. The cause is the use by the industrialized world of CFCs in refrigeration, air conditioning, aerosol cans, and electronics (for cleaning grease off circuit boards). The effect is global. Worse yet, the cause is rooted in northern lands such as the United States and Europe, but the worst effects may be felt where the sun shines brightest—in the tropics, which are dominated by developing nations. Fortunately, the world was able to ban CFC and in time the damage to the ozone layer will heal.

A similar problem arises with global warming, which is also rooted in the industrialized world and its use of fossil fuels. The expected climate effects will hurt worst the poorer nations of the tropics, and perhaps worst of all those on low-lying South Pacific islands, which are expecting to be wholly inundated by rising seas. People who depend on the summertime melting of winter snows and mountain glaciers (including the citizens of California) will also suffer, for already the snows are less and the glaciers are vanishing. According to the Global Humanitarian Forum's "Human Impact Report: Climate Change—The Anatomy of a Silent Crisis" (May 29, 2009; see www.ghf-ge.org/human-impact-report.pdf), global warming is already affecting over 300 million people and is responsible for 300,000 deaths per year. A serious issue of justice or equity is therefore involved.

Both the developed and the developing world are aware of the difficulties posed by environmental issues. In Europe, "green" political parties play a major and growing part in government. In Japan, some environmental regulations are more demanding than those of the United States. Developing nations understandably place dealing with their growing populations high on their list of priorities, but they play an important role in UN conferences on environmental issues, often demanding more responsible behavior from developed nations such as the United States (which often resists these demands; it has refused to ratify international agreements such as the Kyoto Protocol, for example).

Western scholars have been known to suggest that developing nations should forgo industrial development because if their huge populations ever attain the same per-capita environmental impact as the populations of wealthier lands, the world will be laid waste. It is not hard to understand why the developing nations object to such suggestions; they too want a better standard of living. Nor do they think it fair that they suffer for the environmental sins of others.

Are global environmental problems so threatening that nations must surrender their sovereignty to international bodies? Should the United States or Europe have to change energy supplies to protect South Pacific nations from inundation by rising seas? Should developing nations be obliged to reduce birth rates or forgo development because their population growth and industrialization are seen as exacerbating pollution or threatening biodiversity?

Questions such as these play an important part in global debates today. They are not easy to answer, but their very existence says something important about the general field of environmental studies. This field is based in the science of ecology. Ecology focuses on living things and their interactions with each other and their surroundings. It deals with resources and limits and coexistence. It can see problems, their causes, and even potential solutions. And it can turn its attention to human beings as easily as it can to deer mice.

Yet human beings are not mice. We have economies and political systems, vested interests, and conflicting priorities and values. Ecology is only one part of environmental studies. Other sciences—chemistry, physics, climatology, epidemiology, geology, and more—are involved. So are economics, history, law, and politics. Even religion can play a part.

Unfortunately, no one field sees enough of the whole to predict problems (the chemists who developed CFCs could hardly have been expected to realize what would happen if these chemicals reached the stratosphere). Environmental studies is a field for teams. That is, it is a holistic, multidisciplinary field.

This gives us an important basic principle to use when evaluating arguments on either side of any environmental issue: Arguments that fail to recognize the complexity of the issue are necessarily suspect. On the other hand, arguments that endeavor to convey the full complexity of an issue may be impossible to understand. A middle ground is essential for clarity, but any reader or student must realize that something important maybe being left out.

Current Environmental Issues

In 2001, the National Research Council's Committee on Grand Challenges in Environmental Sciences published *Grand Challenges in Environmental Sciences* (National Academy Press, 2001) in an effort to reach "a judgment regarding the most important environmental research challenges of the next generation— the areas most likely to yield results of major scientific and practical importance if pursued vigorously now." These areas include the following:

- Biogeochemical cycles (the cycling of plant nutrients, the ways human activities affect them, and the consequences for ecosystem functioning, atmospheric chemistry, and human activities)
- Biological diversity
- Climate variability
- Hydrologic forecasting (groundwater, droughts, floods, etc.)
- Infectious diseases
- Resource use
- Land use
- Reinventing the use of materials (e.g., recycling)

Similar themes appeared when *Issues in Science and Technology* celebrated its twentieth anniversary with its Summer 2003 issue. The editors noted that over the life of the magazine to date, some problems have hardly changed, nor has our sense of what must be done to solve them. Others have been affected, sometimes drastically, by changes in scientific knowledge, technological capability, and political trends. In the environmental area, the magazine paid special attention to the following:

- Biodiversity
- Overfishing
- Climate change

- The Superfund program
- The potential revival of nuclear power
- Sustainability

Many of the same basic themes were reiterated when *Science* magazine (published weekly by the American Association for the Advancement of Science) published in November and December 2003 a four-week series on the "State of the Planet," followed by a special issue on "The Tragedy of the Commons." In the introduction to the series, H. Jesse Smith began with these words: "Once in a while, in our headlong rush toward greater prosperity, it is wise to ask ourselves whether or not we can get there from here. As global population increases, and the demands we make on our natural resources grow even faster, it becomes ever more clear that the well-being we seek is imperiled by what we do."

Among the topics covered in the series were the following:

- Human population
- Biodiversity
- Tropical soils and food security
- The future of fisheries
- Freshwater resources
- Energy resources
- Air quality and pollution
- Climate change
- Sustainability
- The burden of chronic disease

Many of the topics on these lists are covered in this book. There are of course a great many other environmental issues—many more than can be covered in any one book such as this one. I have not tried to deal here with invasive species, the removal of dams to restore populations of anadromous fishes such as salmon, the depletion of aquifers, floodplain development, urban planning, or many others. My sample of the variety available begins with the more philosophical issues. For instance, there is considerable debate over the "precautionary principle," which says in essence that even if we are not sure that our actions will have unfortunate consequences, we should take precautions just in case (see Issue 1). This principle plays an important part in many environmental debates, from those over the value of restoring damaged ecosystems (Issue 4) or the wisdom of offshore oil drilling (Issue 8) to the folly (or wisdom) of reprocessing nuclear waste (Issue 19).

I said above that many people believed (and still believe) that nature has value only when turned to human benefit. One consequence of this belief is that it may be easier to convince people that nature is worth protecting if one can somehow calculate a cash value for nature "in the raw." Some environmentalists object to even trying to do this, on the grounds that economic value is not the only value, or even the value that should matter (see Issue 3). Related to this question of value is that whether certain human activities should take precedence over environmental protection. Should the military be able to ignore environmental regulations (see Issue 5)? Should profits and jobs come before combating global warming (see Issue 6)?

Should we be concerned about the environmental impacts of specific human actions or products? Here too we can consider offshore oil drilling (see Issue 8), as well as "synthetic biology," so far of concern only in a theoretical way (see Issue 16), the hormone-like effects of some pesticides and other chemicals on both wildlife and humans (see Issue 17). World hunger is a major problem, with people arguing over whether organic farming (see Issue 15) can help. There is also great concern over the future of ocean fisheries (see Issue 14).

Waste disposal is a problem area all its own. It encompasses carbon dioxide from fossil fuel combustion (see Issue 7), hazardous waste (see Issue 18), and nuclear waste (see Issue 19). A new angle on hazardous waste comes from the popularity of the personal computer—or more specifically, from the huge numbers of PCs that are discarded each year.

What solutions are available? Some are specific to particular issues, as shale gas (see Issue 9), renewable energy (see Issue 10), biofuels (see Issue 11), or a revival of nuclear power (see Issue 12) may be to the problems associated with fossil fuels. Some are more general, as we might expect as soon as we hear someone speak of population growth as a primary cause of environmental problems (there is some truth to this, for if the human population were small enough, its environmental impact—no matter how sloppy people were— would also be small) (see Issue 13).

Some analysts argue that whatever solutions we need, government need not impose them all. Private industry may be able to do the job if government can find a way to motivate industry, as with the idea of tradable pollution rights.

The overall aim, of course, is to avoid disaster and enable human life and civilization to continue prosperously into the future. The term for this is "sustainable development" (see Issue 2), and it was the chief concern of the U.N. World Summit on Sustainable Development, held in Johannesburg, South Africa, in August 2002. Exactly how to avoid disaster and continue prosperously into the future are the themes of the UN Millennium Ecosystem Assessment report, *Ecosystems and Human Well-Being: Synthesis* (www .millenniumassessment.org/en/) (Island Press, 2005). The main findings of this report are that over the past half century meeting human needs for food, fresh water, fuel, and other resources has had major negative effects on the world's ecosystems; those effects are likely to grow worse over the next half century and will pose serious obstacles to reducing global hunger, poverty, and disease; and although "significant changes in policies, institutions, and practices can mitigate many of the negative consequences of growing pressures on ecosystems, . . . the changes required are large and not currently under way." Also essential will be improvements in knowledge about the environment, the ways humans affect it, and the ways humans depend upon it, as well as improvements in technology, both for assessing environmental damage and for repairing and preventing damage, as emphasized by Bruce Sterling in "Can Technology Save the Planet?" *Sierra* (July/August 2005). Sterling concludes, perhaps optimistically, that "When we see our historical predicament in its full, majestic scope, we will stir ourselves to great and direly necessary actions. It's not beyond us to think and act in a better way. Yesterday's short-sighted habits are leaving us, the way gloom lifts with the dawn." George Musser, introducing the special September 2005 "Crossroads for Planet Earth" issue of *Scientific American* in "The Climax of Humanity," notes that the next few decades will determine our future. Sterling's optimism may be fulfilled, but if we do not make the right choices, the future may be very bleak indeed.

Internet References . . .

The Ecosystems Services Project

The Ecosystems Services Project studies the services people obtain from their environments, the economic and social values inherent in these services, and the benefits of considering these services more fully in shaping land management policies.

www.ecosystemservicesproject.org/

The Earth Day Network

The Earth Day Network (EDN) promotes environmental citizenship and helps activists in their efforts to change local, national, and global environmental policies. It also makes the Ecological Footprint quiz available.

www.earthday.net/

The International Institute for Sustainable Development

The International Institute for Sustainable Development advances sustainable development policy and research by providing information and engaging in partnerships worldwide. It says it "promotes the transition toward a sustainable future. We seek to demonstrate how human ingenuity can be applied to improve the well-being of the environment, economy and society."

www.iisd.org/

The Natural Resources Defense Council

The Natural Resources Defense Council is one of the most active environmental research and advocacy organizations. Its home page lists its concerns as clean air and water, energy, global warming, toxic chemicals, ocean health, and much more.

www.nrdc.org/

The United Nations Environment Programme

The United Nations Environment Programme "works to encourage sustainable development through sound environmental practices everywhere. Its activities cover a wide range of issues, from atmosphere and terrestrial ecosystems, the promotion of environmental science and information, to an early warning and emergency response capacity to deal with environmental disasters and emergencies."

www.unep.org/

Environmental Philosophy

*E*nvironmental debates are rooted in questions of values—what is right? What is just?—and are inevitably political in nature. It is worth stressing that people who consider themselves to be environmentalists can be found on both sides of most of the issues in this book. They differ in what they see as their own self-interest and even in what they see as humanity's long-term interest.

Understanding the general issues raised in this section is useful preparation for examining the more specific controversies that follow in later sections.

- Should the Precautionary Principle Become Part of National and International Law?
- Is Sustainable Development Compatible with Human Welfare?
- Do Ecosystem Services Have Economic Value?

ISSUE 1

Should the Precautionary Principle Become Part of National and International Law?

YES: Agne Sirinskiene, from "The Status of Precautionary Principle: Moving Towards a Rule of Customary Law," *Jurisprudence* (October 2009)

NO: Ken Cussen, from "Handle with Care: Assessing the Risks of the Precautionary Principle," *Australasian Journal of Environmental Management* (June 2009)

Learning Outcomes

After reading this issue, you should be able to:

- Define the precautionary principle.
- Explain why accepted definitions of the precautionary principle may not be adequate.
- Describe the values that affect the positions people take on the precautionary principle.
- Discuss the difference between the "hazard factor" and the "outrage factor" in people's perceptions of risk.

ISSUE SUMMARY

YES: Agne Sirinskiene argues that the evidence from treaties, legislation, and court cases clearly indicates that the precautionary principle is becoming or has already become a rule of customary national and international law, and international applications of the principle are developing rapidly.

NO: Ken Cussen argues that the precautionary principle is so vague, ill-defined, and value-ridden that it is either vacuous or dangerous. Its underlying assumptions must be clarified before it can be used to guide public policy.

The traditional approach to environmental problems has been reactive. That is, first the problem becomes apparent—wildlife or people sicken and die, drinking water or air tastes foul. Then researchers seek the cause for the problem and regulators seek to eliminate or reduce that cause. The burden is on society to demonstrate that harm is being done and a particular cause is to blame.

An alternative approach is to presume that all human activities—construction projects, new chemicals, new technologies, etc.—have the potential to cause environmental harm. Therefore, those responsible for these activities should prove in advance that they will not do harm and should take suitable steps to prevent any harm from happening. A middle ground is occupied by the "precautionary principle," which has played an increasingly important part in environmental law ever since it first appeared in Germany in the mid-1960s. On the international scene, it has been applied to climate change, hazardous waste management, ozone depletion, biodiversity, and fisheries management. In 1992, the Rio Declaration on Environment and Development, listing it as Principle 15, codified it thus:

> In order to protect the environment, the precautionary approach shall be widely applied by States according to their capabilities. When there are threats of serious or irreversible damage, lack of full scientific certainty shall not be used as a reason for postponing cost-effective measures to prevent environmental degradation.

Other versions of the principle also exist, but all agree that when there is reason to think—but not necessarily absolute proof—that some human activity is or might be harming the environment, precautions should be taken. Furthermore, the burden of proof should be on those responsible for the activity, not on those who may be harmed. This has come to be broadly accepted as a basic tenet of ecologically or environmentally sustainable development. See Marco Martuzzi and Roberto Bertollini, "The Precautionary Principle, Science and Human Health Protection," *International Journal of Occupational Medicine and Environmental Health* (January 2004).

The precautionary principle also contributes to thinking in the areas of risk assessment and risk management in general. Human activities can damage health and the environment. Some people insist that action need not be taken against any particular activity until and unless there is solid, scientific proof that it is doing harm, and even then risks must be weighed against each other. Others insist that mere suspicion should be grounds enough for action. Sainath Suryanarayanan and Daniel Lee Kleinman, "Disappearing Bees and Reluctant Regulators," *Issues in Science and Technology* (Summer 2011), argue that existing risk assessment, particularly in the Environmental Protection Agency (EPA), is biased toward avoiding type I or false positive errors (which incorrectly identify a safe substance as dangerous) rather than type II or false negative errors (which incorrectly identify a dangerous substance as safe). Their particular concern is colony collapse disorder in honeybees, which may be due to exposure to certain insecticides; however, the EPA refuses to call those insecticides dangerous and regulate them accordingly because of a lack of rigorous, experiment-based scientific evidence. They prefer a more precautionary approach, which would shift the balance toward avoiding type II errors. That is, they would rather err by calling safe substances dangerous.

Since solid, scientific proof can be very difficult to obtain, the question of just how much proof is needed to justify action is vital. Not surprisingly, if action threatens an industry, that industry's advocates will argue against taking precautions, generally saying that more proof is needed. Those who feel threatened by an industry or a new technology are more likely to favor the precautionary principle; see John Dryzek, Robert E. Goodin, Aviezer Tucker, and Bernard Reber, "Promethean Elites Encounter Precautionary Publics: The Case of GM Foods," *Science, Technology & Human Values* (May 2009). The "Promethean Elites" are those who—like the Prometheus of myth—favor progress over the status quo and may argue that the precautionary principle holds back progress; see Helene Guldberg, "Challenging the Precautionary Principle," *Spiked-Online* (July 1, 2003) (http://www.spiked-online.com/Articles/00000006DE2F .htm). Yet, says Charles Weiss in "Defining Precaution," a review of The *Precautionary Principle*, UNESCO's World Commission on the Ethics of Scientific Knowledge and Technology Report, *Environment* (October 2007), the principle "is an important corrective to the pressure from enthusiasts and vested interests to push technology in unnecessarily risky directions."

Not everyone agrees. Ronald Bailey, "Precautionary Tale" (*Reason,* April 1999), defines the precautionary principle as "precaution in the face of any actions that may affect people or the environment, no matter what science is able—or unable—to say about that action." "No matter what science says" is not quite the same thing as "lack of full scientific certainty." Indeed, Bailey turns the precautionary principle into a straw man and thereby endangers whatever points he makes that are worth considering. One of those points is that widespread use of the precautionary principle would hamstring the development of the Third World. Roger Scruton, in "The Cult of Precaution," *National Interest* (Summer 2004), calls the precautionary principle "a meaningless nostrum" that is used to avoid risk and says it "clearly presents an obstacle to innovation and experiment," which are essential. Bernard D. Goldstein and Russellyn S. Carruth remind us in "Implications of the Precautionary Principle: Is It a Threat to Science?" *International Journal of Occupational Medicine and Environmental Health* (January 2004), that there is no substitute for proper assessment of risk. Jonathan Adler, "The Precautionary Principle's Challenge to Progress," in Ronald Bailey, ed., *Global Warming and Other Eco-Myths*, (Prima, 2002), argues that because the precautionary principle does not adequately balance risks and benefits, "The world would be safer without it." A. Benedictus, H. Hogeveen, and B. R. Berends, "The Price of the Precautionary Principle: Cost-Effectiveness of BSE Intervention Strategies in the Netherlands," *Preventive Veterinary Medicine* (June 2009), found that measures taken to control the spread of BSE or Mad Cow Disease are a very expensive way to protect human life. Peter M. Wiedemann and Holger Schutz, "The Precautionary Principle and Risk Perception: Experimental Studies in the EMF Area," *Environmental Health Perspectives* (April 2005), report that "precautionary measures may trigger concerns, amplify . . . risk perceptions, and lower trust in public health protection." Cass R. Sunstein, *Laws of Fear: Beyond the Precautionary Principle* (Cambridge, 2005), criticizes the precautionary principle in part because, he says, people overreact to tiny risks. John D. Graham, the dean of the Frederick S. Pardee RAND Graduate School, argues in "The Perils of the Precautionary Principle: Lessons from the American and European Experience," Heritage Lecture #818 (delivered January 15, 2004, at the Heritage Foundation, Washington, DC), that the precautionary principle is so subjective that it permits "precaution without principle" and threatens innovation and public and environmental health. It must therefore be used cautiously.

The 1992 Rio Declaration emphasized that the precautionary principle should be "applied by States according to their capabilities" and that it should be applied in a cost-effective way. These provisions would seem to preclude the draconian interpretations that most alarm the critics. Yet, say David Kriebel et al., "The Precautionary Principle in Environmental Science," *Environmental Health Perspectives* (September 2001): "environmental scientists should be aware of the policy uses of their work and of their social responsibility to do science that protects human health and the environment." Businesses are also conflicted, writes Arnold Brown in "Suitable Precautions," *Across the Board* (January/February 2002), because the precautionary principle tends to slow decision-making, but he maintains that "we will all have to learn and practice anticipation."

Does the precautionary principle make us safer? The January 23, 2009, issue of *CQ Researcher* presents a debate, under that title, between Gary Marchant, who believes that the principle "fails to provide coherent or useful answers on how to deal with uncertain risks," and Wendy E. Wagner, who contends that the existing chemical regulatory system shows the consequences of not taking a precautionary approach. Many people agree with Wagner, and indeed in many parts of the world the precautionary principle is well accepted. Should it become a recognized part of national and international law? In the following selections, Agne Sirinskiene, associate professor in the Faculty of Law, Department of Biolaw, Mykolas Romeris University, Vilnius, Lithuania, argues that the evidence from treaties, legislation, and court cases clearly indicates that the precautionary principle is becoming or has already become a rule of customary national and international law, and international applications of the principle are developing rapidly. Ken Cussen of the Graduate School of the Environment, Macquarie University, Sydney, Australia, argues that the precautionary principle is so vague, ill-defined, and value-ridden that it is either vacuous or dangerous. Its underlying assumptions must be clarified before it can be used to guide public policy.

YES

Agne Sirinskiene

The Status of Precautionary Principle: Moving Towards a Rule of Customary Law

Introduction

Scientific uncertainty regarding the evidence of a link between human activity (as a cause) and its impact on the environment (as a consequence) has been an enormous obstacle for lawmaking in the area of environmental protection. This scientific uncertainty has further increased in the recent decades, as society began using advanced technologies, including biotechnologies. Their long-term impact on the environment and human health is mostly unknown since they have not been studied in longitudinal research. Therefore, some countries have taken "a precautionary" approach in their domestic law, which allows for decision making in the area of environmental protection in case of scientific uncertainty regarding the evidence of cause and consequence. The Federal Republic of Germany has been a pioneer in the area of the "precautionary approach" towards the environment: they formulated the principle of precaution (*Vorsorgeprinzip*) in their domestic law in 1974. A decade later, in 1984, for the first time in history, an indirect reference to the precautionary principle was made in a non-binding international document—the Bremen Ministerial Declaration of the International Conference on the Protection of the North Sea. Consequently, the 1987 London Ministerial Declaration of the International Conference on the Protection of the North Sea already used the term "precautionary approach" explicitly. Although the precautionary principle has received wide international recognition, the status of this principle in law is still under debate. This suspends further application of the principle and allows for a discussion about the principle in the process of its development.

The main goal of this article is to analyse the current status of the precautionary principle in international law and its development into a rule of international customary law. This entails methods of comparative and systematic analysis.

From *Jurisprudence,* vol. 4 issue 118, 2009, pp. 349–364. Copyright © 2009 by Mykolas Romeris University. Reprinted by permission of Mykolas Romeris University School of Law.

The Influence of Problems in the Definition of the Precautionary Principle on the Interpretation of Its Status

The precautionary principle or "precautionary approach" is widely invoked in *soft law* (legally non-binding) documents and *hard law* instruments. According to D. Vanderzwaag, about 14 different definitions of the precautionary principle exist in international law. Such a variety of definitions has even prompted some researchers to assume that the lack of one unanimous definition is one of the properties of the precautionary principle. On the other hand, the variety of formulations is used by critics because it helps uncover problems in the application of the principle.

The most widely known definition of the precautionary principle can be ascribed to the 1992 Rio Declaration. Principle 15 of the Declaration states that "in order to protect the environment, the precautionary approach shall be widely applied by States according to their capabilities. Where there are threats of serious or irreversible damage, lack of full scientific certainty shall [not be] used as a reason for postponing cost-effective measures to prevent environmental degradation." Similarly, the 1992 Framework Convention on Climate Change obliges participating parties "to take precautionary measures to anticipate, prevent or minimize the causes of climate change and mitigate its adverse effects. Where there are threats of serious or irreversible damage, lack of scientific certainty should not be used as a reason for postponing cost-effective measures." The term "approach" instead of "principle" is used in the preamble of the Convention on Biological Diversity (1992): "where there is a threat of significant reduction or loss of biological diversity, lack of full scientific certainty should not be used as a reason for postponing measures to avoid or minimize such a threat." The UN Program for Further Implementation of Agenda 21 speaks of the progress made "in incorporating *principles* contained in the Rio Declaration . . . including . . . the precautionary principle."

The use of different terms and definitions is particularly problematic when interpreting the status of the precautionary principle. For example, in the case of *EC Biotech*, the United States (US) noted that it strongly disagrees that "precaution" has become a rule of international law and that the "precautionary principle" cannot be considered a general principle or norm of international law because it does not have a single, agreed formulation. According to the US, "quite the opposite is true: the concept of precaution has many permutations across a number of different factors. Thus, the United States considers precaution to be an 'approach,' rather than a 'principle' of international law." In another statement at the World Trade Organization (WTO), the US stressed that, even if the precautionary principle were considered a relevant rule of international law under Article 31(3) of the Vienna Convention, it would be useful only for the interpretation of particular treaty terms, and could not override any part of the SPS (Sanitary and Phytosanitary) Agreement. This position is consistent in other cases as well where the US questioned whether "precaution" is a "principle." Consequently, the [US] does not consider the "precautionary principle" to represent customary international law. Such an interpretation of the status of the precautionary principle has been used by the US as a counterargument to the European Communities (EC) position that the precautionary principle is, or has become, "a general customary rule of international law" or at least "a general principle of law." Canada, in the *EC Hormones* case, took a middle position between the EC and the US. On the one hand, Canada declared that the "precautionary approach" is "an *emerging* principle of law" which may crystallize in the future into one of the "general principles of law recognized by civilized nations" within the meaning of Article 38(1)(c) of the Statute of the International Court of Justice. On the other hand, Canada agreed that the precautionary principle has not yet been incorporated into the corpus of public international law.

These arguments from two classical cases reveal that "approach" is generally seen as a softer version of "principle" in international law. This conclusion may also be supported by a case of the International Tribunal for the Law of the Sea (ITLOS), where Judge Laing expressed a dissenting opinion. Judge Laing stated that "adopting an approach, rather than a principle imports a certain degree of flexibility and tends, though not dispositevely, to underscore reticence about making premature pronouncements about desirable normative structures." Another ITLOS [j]udge also associates the term "principle" with legally binding, customary status. Nevertheless, these separate opinions are only representative of the personal views of the judges participating in the said case. ITLOS as a tribunal has never made any statement explaining its position on the status of the precautionary principle. In a dispute between states, the World Trade Organization (WTO) Appellate Body also tried to avoid direct interpretation of the status of the precautionary principle and indicated that "it is unnecessary, and probably imprudent, for the Appellate Body in this appeal to take a position on this important, but abstract, question." The WTO Appellate Body limited itself by saying that "at least outside the field of international environmental law, the precautionary principle still awaits authoritative formulation." Thus, it was never acknowledged on the official WTO level that the precautionary principle is "a general customary rule of international law" or even "a general principle of law." However, the statement that "at least outside the field of international environmental law, the precautionary principle still awaits authoritative formulation" may be interpreted as recognition that the precautionary principle may have the status of the principle in international environmental law.

It may also be acknowledged that certain formulations used in definitions of the precautionary principle also add to the discussions on the status of the principle. In some cases, definitions raise questions of whether they create obligatory rules. For example, some authors hesitate whether principles in the Convention on Climate Change, including the precautionary principle, create an obligation to the member states of the convention, because it is not clear what is meant by "the Parties shall be guided, *inter alia*." In—the text of the Convention uses "should" instead of "must": "the Parties should take precautionary measures to anticipate" (art. 3). The modal verb "must" expresses an obligation and implies that a verb used together with "must" definitely happens, while "should" implies that something may not happen. Another international instrument—the Convention on Biological Diversity—has a more abstract definition which may cause trouble for the implementation and interpretation of the principle. Article 6, General measures, in this legally binding document uses such formulas as "in accordance with its particular conditions and capabilities" and "as far as possible and as appropriate." These and other similar expressions determine that separate norms of legally binding documents have a limited normative character. For them to become effective, corresponding domestic laws or new international agreements must come into effect. This characteristic of the precautionary principle leads some authors to conclude that the precautionary principle is a long way from having legally binding force and stands at the beginning of the so-called "procedural" principles, which may help states to meet their obligations.

Nevertheless, it seems that doubts about the precautionary principle being a "principle" are more common to Anglo-Saxon tradition. The European Union (EU) law does not draw a clear difference between "principle," "approach," and "measures"; these terms are used in parallel to define the same principle and there is nothing to suggest that these three terms cannot be used interchangeably. The European Commission in Communication on the Precautionary Principle also does not differentiate among these terms and recognizes the precautionary principle as a full-fledged and general principle of international law or, as already discussed, even as a general customary rule of international law.

The status of the precautionary principle as a rule of customary law is significant because a rule of customary law creates obligations for all states, except those that have persistently objected to the practice and its legal consequences. Therefore, in cases where the precautionary principle is recognized as a rule of customary law, the application of the principle would acquire a broader scope on the international level. This possible change would be in accordance with EU policy, clearly defined in the articles 6 and 174 of the *Treaty Establishing the European Community*. However, the EC has never explained its statements in the WTO and what reasoning lies behind them. The Communication on the Precautionary Principle and the jurisprudence of European Court of Justice also provide no answers. Bearing in mind this lack of legal certainty, the article will focus [on] further analysis on the criteria for the development of a rule of customary law and how they may be applied to the precautionary principle.

Prerequisites for the Status of the Precautionary Principle as a Rule of Customary Law

The Statute of the International Court of Justice (art. 38, para. 1b) defines customary international law as "evidence of general practice accepted as law." The *Nicaragua* case and the *North Sea Continental Shelf* case complement this article of the Statute and clarify two requirements of customary international law. According to the International Court of Justice (ICJ), customary international law arises when nations follow a practice in an extensive and virtually uniform manner and this practice is followed with the conviction that it is obligatory to do so under international law (*opinio iuris*). Virtually uniform manner is not interpreted in such a way that absolutely all states are supposed to have the same practice during a clearly defined period of time. Consequently, the opposition of some states does not interfere with the development of a customary rule. State practice is usually assessed with the help of defined criteria that indicate how states articulate their recognition of a rule of customary law. These non-exhaustive criteria that serve as evidence of customary international law are: treaties, declarations, decisions of international and national courts, domestic legislation, opinions and statements of states during the preparation of treaties, correspondence between states, and even opinions of lawyers.

However, the best indicators of state practice remain the instruments of international law and state domestic law. As already discussed, there are about 14 different definitions of the precautionary principle in various legally binding and non-binding instruments of international law. The precautionary principle is widely used in agreements and declarations addressing such global problems as climate change, atmospheric and marine pollution, environmental protection and biodiversity and even in legal documents devoted to very specific regional problems such as tourism in Antarctica. After the Maastricht Treaty in 1992, the precautionary principle has become a part of EU environmental law. Currently, the precautionary principle is used in more than 90 international declarations and agreements. In this context, the number of ratifications (majority of the treaties are multilateral) and the number of states signing declarations also reflect broad acceptance of the rule by states. The abundance of treaties and declarations incorporating the precautionary principle provides at least an estimate of state practice and acceptance, which implies that the precautionary principle is crystallizing into a rule of customary environmental law.

Another primary indicator of state practice is domestic law. The precautionary principle is widely used in the domestic environmental law of Germany, Belgium, and the Nordic countries (Denmark, Norway, Sweden, Finland and Island). In 2005, the principle was incorporated into the Preamble of the Constitution of France and is now part of the "Environmental charter" of the Constitution (another part of this preamble is the 1789 Declaration of the Rights of Man and the Citizen). Therefore, in French domestic law the precautionary principle is treated as a constitutional principle, which claims to be on the same level as the principles of the Declaration of the Rights of Man and the Citizen. A systematic analysis of the French Constitution reveals that the relationship between articles 1 and 5 may be interpreted as giving broader application for the precautionary principle and that the principle may also be applied in certain areas of public health.

The precautionary principle is found not only in the domestic laws of European countries. For example, in 1992 the principle became part of the National Strategy for Ecologically Sustainable Development in Australia. In 1993, the principle was incorporated into Australia's Environmental Protection Act. In 1996, the precautionary principle was defined in the Oceans Act of Canada. In 1999, the Environmental Protection Act of Canada, which also regulates the activities of public administration institutions, was also supplemented with the precautionary principle. Even US law makes some indirect allusions to the precautionary principle (as measures) when dealing with questions of food safety and air pollution. Furthermore, as part of environmental impact assessment, the precautionary principle may be found in the local laws of about fifty countries. These examples illustrate the wide implementation of the procedural aspect of the precautionary principle.

In most of these legal acts, the application of the precautionary principle is directly related to the environment and indirectly to human health as under certain circumstances the environment creates risks to human health. The observation of this trend supports the theory that the precautionary principle is crossing the threshold of environmental law and entering new areas of regulations, such as public health.

State practice is further reflected in the applications and decisions of national and international courts where legal parties defend their legal interests based on the precautionary principle. Classical examples of such application of the precautionary principle are the *Gabčikovo-Nagymaros Project (The Danube Dams)* and *French underground nuclear tests* cases in the ICJ. . . .

In the second—*French Underground Nuclear Test*—case, New Zealand as the applicant emphasized that before France can carry out underground nuclear tests near a marine environment, it must provide enough evidence that the tests will not result in the introduction of radioactive material into that environment. According to New Zealand, a risk assessment must be carried out on account of the precautionary principle. In order to prove that the precautionary principle is a rule of customary law, the applicant used a broad range of arguments, such as citing international documents and legal doctrine published in the Sands book on environmental principles: "The legal status of the precautionary principle is evolving. . . . [H]owever, there is sufficient evidence of state practice to justify the conclusion that the principle, as elaborated in the Rio Declaration and the Climate Change and Biodiversity Conventions, has now received sufficiently broad support to allow a good argument to be made that it reflects a principle of customary law." However, in this case the ICJ again did not see a necessity to evaluate the status of the precautionary principle in international law. This position of the court was criticized in a dissenting opinion by Judge Weeramantry. He regretted that the Court had not availed itself of the opportunity to consider the principles of environmental law (including the precautionary principle) and pointed out that "these principles of environmental law do not depend for their validity on treaty provisions. They are part of customary international law" and "they are part of the *sine qua non* for human survival." The latter statement raises the precautionary principle onto the level of newly developed principles which guarantee the survival of the human race; for example, the principle of human dignity has an analogous function in the area of biomedicine. The importance of the principle was also emphasized by other ICJ judges. However, they did not see the precautionary principle as part of customary law and spoke of a different type of status: "the precautionary principle is not an abstraction or an academic component of desirable soft law, but a rule of law within general international law as it stands today."

This tendency of the ICJ to rely on the precautionary principle as a rule of customary law remains still. In 2008, the ICJ received an application from the state of Ecuador. Ecuador complained about the aerial spraying of coca and poppy crops with chemical herbicides carried out by Colombia at locations near, at and across its border with Ecuador. Ecuador claimed that toxic herbicides have caused damage to human health, property and the environment, and therefore Colombia has violated Ecuador's rights under customary and conventional international law: "The harm that has occurred, and is further threatened, includes some with irreversible consequences, indicating that Colombia has failed to meet its obligations of prevention and precaution." Such a reference to the precautionary principle as a rule of customary law and current developments in international law could indicate that in this case the ICJ might express its attitude towards the status of this principle.

In an analysis of state practice, it is worth noting that states have repeatedly invoked the principle as a norm of general international law in international judicial proceedings before the International Tribunal for the Law of the Sea (ITLOS). In the *MOX Plant* case, Ireland submitted that the precautionary principle is now recognized as a rule of customary international law. In the *Southern Bluefin Tuna* case, the International Tribunal recognized the need for the parties to "act with prudence and caution" to ensure that effective conservation measures are taken to prevent serious harm to stocks of southern bluefin tuna; however, the ITLOS abstained from evaluating the status of the principle.

The principle is also used in the domestic courts of European countries, in the European Communities Court of Justice and outside Europe, even in countries with Islamic law systems, such as Pakistan, or in new democratic states in Latin America. For example, in the *Caribbean Conservation Corporation and Others v. Costa Rica* (*Green Turtles*) case, Costa Rica's Constitutional Court upheld the constitutional right to an ecologically balanced environment and the precautionary principle was for the first time invoked by the Court as one of the means to grant such rights. Other examples of application of the precautionary principle at the national level include rulings by the Supreme Courts of India and Canada. Academic discussions are also paying close attention to various aspects of these changes of the principle in international law and interpret the state practice and/or *opinio iuris*. Some academics, like Sands, have even had a direct influence on the practice of international courts and tribunals. Academic support together with recent state practice appears to conclusively endorse the status of the principle as a norm of customary international law.

Based on the facts discussed in this section, we may conclude that the application of the principle by states has truly become widespread and consistent. This means that the discussion on the crystallization of the precautionary principle into a rule of international customary law in general, or at least a rule of international environmental customary law, is appropriate at this time. There is sufficient evidence that the precautionary principle is becoming or perhaps has already become a rule of customary law.

Conclusions

Although the precautionary principle was formulated just three and a half decades ago, state practice has since developed very rapidly. Therefore, there is sufficient state practice and *opinio iuris* to support the position of the EC that the precautionary principle has already crystallized into a general customary rule. Evidence may be found in international documents, national legislation, and judicial decisions of international courts, tribunals, and national courts. All of these legal documents (except a few statements by the US and Canada) indicate strong approval of the obligation to apply the precautionary principle in environmental decision making. Furthermore, international practice in the area of application of the precautionary principle is developing rapidly.

Ken Cussen  **NO**

Handle with Care: Assessing the Risks of the Precautionary Principle

The precautionary principle is often invoked as a way of handling risk in situations of scientific uncertainty. This paper questions the usefulness of the principle, arguing that, depending on how it is interpreted, it is either vacuous or dangerous. Further, it is argued that the principle contains some unwarranted value assumptions.

The Scope of the Precautionary Principle

The precautionary principle is presented by its proponents as a guide for decision-makers on handling risk in the face of scientific uncertainty, and on facilitating democratic decision-making by involving all stakeholders. I argue that there are problems with this characterisation, but initially I describe how widespread is its use and application.

The idea of 'proceeding with caution' has often—no doubt not often enough—been a minor guiding principle of technological and other kinds of innovation. However, its application as a guiding principle for environmental and public health policy-makers and practitioners did not become formalised as a guide for legislative and policy directions until the early 1970s. At that time, the 'Vorsorgeprinzip' ('Foresight,' or more literally 'Pre-care' principle) was developed as a principle of German environmental law. Since then, a principle of precautionary action has emerged in many places, particularly in international policy statements and agreements. It was, for example, recognised in the UN General Assembly's World Charter for Nature in 1982.

The precautionary principle has been linked also with the idea of sustainable development; for example, by its inclusion in the report of the UN Conference on Environment and Development in 1992. Principle 15 of the conference report states that:

> In order to protect the environment, the precautionary approach shall
> be widely applied by states according to their capabilities.

Further, the precautionary principle was formally adopted by the Maastricht Treaty of the European Union in 1992, and has been accepted as the basis of the Cartegena Biosafety Protocol agreed in Montreal in 2000.

From *Australasian Journal of Environmental Management*, June 2009, pp. 66–69. Copyright © 2009 by Environment Institute of Australia and New Zealand. Reprinted by permission of Taylor & Francis Group via Rightslink.

Accordingly, the principle has grown in the last 20 years or so from a vague injunction to 'proceed with caution' or 'look before you leap' into a formal guide for the conduct of states and international bodies in their environmental and health policy and decision-making mechanisms. Because it has the potential to be enormously influential, it would be prudent to be clear about its meaning and the implications of its adoption.

Caution, Risk, Novelty and Uncertainty

Although its proponents claim that the main purpose of the precautionary principle is to offer decision-makers guidance on handling risk in situations of uncertainty, particularly so-called 'scientific uncertainty', I want to dispute that it achieves this. First, life is inherently risky and uncertain—in our daily lives, caution is necessary. That is why the notions of 'risk', 'uncertainty' and 'caution' are clustered conceptually; each element contributes to the meaning of the others. Indeed, caution and risk are a conceptual pair. By that I mean that, if human activities contained no risk, we would have no need of caution and we would probably not need the concept. For example, if gravity worked such that falling from a cliff meant that we slowly descended to the ground unharmed, then we would not need to take care while rock-climbing. If the world was like that, the adage to 'look before you leap' would be pointless advice. However, caution is necessary because all human activities do carry risks, so risk can never be eliminated from our lives. Even breathing puts us at risk of inhaling something undesirable, while eating puts us at risk of choking.

Similarly, risk is meaningless without uncertainty. If we could be certain that we will fall to our death from a cliff-top, then the concept of risk is inappropriate. There is no risk if falling is a certainty. The fact is, though, that we cannot eliminate risk. Every new product or activity inevitably carries risk.

Nor can we eliminate uncertainty, even from our most cherished claims to knowledge. As David Hume argued in the 18th Century, we cannot even be certain that the 'laws of nature' will continue to operate in the future as they have in the past. Consequently, since uncertainty about risk, cause-and-effect relationships, and actual risk itself are ineliminable elements of the human condition, caution is an inherent part of our orientation to the world.

Further, when we embark on novel activities, we seldom know what new risks we may encounter. Nevertheless, since experience shows us that it is reasonable to assume that a new activity may carry new risks, it is sensible to take things cautiously. I don't think anyone will disagree with that presumption, but if that is all that the precautionary principle advises us to do, as some proponents suggest, there would be no problem. As I hope to show, the precautionary principle, if it is not simply vacuous, goes well beyond the idea of 'looking before you leap.'

The Wingspread Statement

Unfortunately, any analysis of the precautionary principle is complicated by the lack of agreement about the precise definition of the term. The debate surrounding the concept is hampered further by its polemical nature: it has

largely fallen into disputing polemical camps which display little attempt at
objective analysis, and with each side denying its own polemical character and
intent. The clearest sign of this is that each side employs a definition of the
precautionary principle that is designed to lend plausibility to their side of the
debate and point of view, and to discredit their opponent's view.

Given this situation, I will examine the most widely accepted and publi-
cised definition. This definition was produced in 1998 by a group of scientists,
philosophers and activists in the United States, and is known as the Wingspread Statement. It represents a mainstream interpretation of the principle.

> When an activity raises threats of harm to the environment or human
> health, precautionary measures should be taken, even if some cause-
> and-effect relationships are not fully established.

This sounds on the face of it like simple common sense. But this perspective is deceptive, as we can see if we examine the three main components of
the statement, and also ask what is missing from it:

1. threat of harm
2. uncertainty (about risks and causal relationships)
3. precautionary response.

These three components are like blank cheques. It is when we try to cash
them, so to speak, that the problems emerge.

What exactly is meant by a 'threat of harm', for instance? Does it mean
that there must be definite evidence of potential or actual dangers, or is it
enough that a stakeholder considers that a novel product or activity *might*
prove harmful, even if there is no evidence to that effect? The problem is that
the phrase, as used here, is ambiguous between two common meanings. To
see this, it is useful to distinguish the way that the idea of 'threat of harm' or
'risk' is used in the public debate as two elements: 'the hazard factor' and 'the
outrage factor.'

By 'the hazard factor,' I mean the actual risk of an activity, as suggested
by evidence. For example, if we found that users of mobile telephones exhibited significantly higher incidence of brain tumours than the population at
large, and research could find no other factor that might explain this correlation, it would be sensible to take various precautionary measures, such as further research, issuing public warnings, or banning such technology.

By 'the outrage factor,' I mean the public perception of hazard. Although
this may sometimes be well-founded and based on evidence (in which case it
is no different from, and is probably based on, the hazard factor), it is often
based on misunderstanding, anecdotal evidence or even prejudice. Additionally, it is often fed by media sensationalism and less-than-responsible reporting of new technologies and scientific discoveries. Every scientist knows how
difficult it is to convey a proper interpretation of uncertain data to the public.

If 'threat of harm' in this formulation of the precautionary principle means 'genuine evidence of hazard,' then the principle can only mean
that we should base our actions and decisions on what we can reasonably
establish of cause-and-effect relationships, even when those relationships are
not fully established. But since cause-and-effect relationships are *never* "fully

established" (i.e. since there is *always* uncertainty in scientific predictions), this says no more than that we should base our decisions on the best available evidence. So, if 'threat of harm' is taken to mean 'some evidence of hazard,' then the principle says too little; it is reduced to a cliché that tells us, 'when there is evidence of hazard, it is prudent to take care'. Since we already knew that, this application of the precautionary principle is useless. Genuine evidence of hazard requires a detailed risk assessment and cost/benefit analysis of the specific case, not vague injunctions to 'take [unspecified] precautionary measures.'

If, on the other hand, 'threat of harm' means 'the outrage factor,' then the precautionary principle says too much; it can be invoked to slow or stop innovative, and possibly very beneficial products or procedures, on the basis of a *lack* of evidence. For example, innovative applications may be prevented merely because someone *believes* (perhaps on no more solid a basis than a 'gut feeling') that some harm *may* occur. This is no trivial matter. By stopping or slowing innovation without good grounds, the principle could cause more harm than it prevents. For example, if the precautionary principle is invoked to stop an activity merely because a lobby group expresses its view that, for example, universal vaccination is dangerous, then people's lives could be endangered.

In any case, what is to count as a 'threat of harm?' Is any threat of harm, ro matter how insignificant, sufficient to cause us to invoke precautionary measures? How serious must the harm be? Vaccination by injection causes temporary pain to individuals. Should we for that reason take precautions to limit or stop vaccinations? Unfortunately, this definition, because it leaves the size and scope of the threat unspecified, leaves this question not only unanswered but unanswerable. This is surely a fatal vagueness.

Again, the principle enjoins us to caution when there is some level of uncertainty about cause-and-effect relationships. But, since all scientific predictions and claims to knowledge contain *some* level of uncertainty, we are left in the dark about what level of uncertainty should trigger precaution.

And what exactly is a precautionary measure? This could be anything from a public announcement about a possible threat (with all the problems that entails) to banning a product or activity outright. Again, this definition is silent, and the silence could be dangerous.

So, because it plays on the ambiguity of the phrase 'threat of harm'—borrowing credibility and plausibility from the 'hazard factor' interpretation and lending it to the 'outrage factor' interpretation—because it is fatally vague about the size and impact of threat, and because it is also fatally vague about what counts as a proper precaution, this most widely quoted formulation of the principle is either empty or potentially dangerous.

It is also important to notice that the principle, as defined in the Wingspread Statement, makes no mention of the possible benefits of the product or activity in question. If we take this omission, together with the vagueness about the scope and level of harm, we find ourselves disbarred from making a balanced judgment of the kind, for example, that allowed penicillin to be made widely available.

The case of penicillin is instructive. It is an unfortunate fact that a small percentage of people have a serious, allergic reaction to penicillin. This was known from the earliest period after its introduction, and so soon afterward alternative antibiotics were trialled and marketed. If the Wingspread version of the precautionary principle had been the determining factor, the introduction of penicillin would have been at least slowed and perhaps even stopped, with obvious, disastrous consequences. Consequently, *a calculation of possible benefits is obviously fundamental to decision-making, even though any such consideration is conspicuously absent from this definition.*

The Wingspread Statement's version of the precautionary principle is therefore both widely influential and deeply flawed. Depending on how we interpret it, it is either vacuous or dangerous.

Saunder's Formulation

Attempts have been made to rescue the precautionary principle from these problems. Peter Saunders, Professor of mathematics at King's College, London, has suggested that the precautionary principle is simply the claim that:

> If there are reasonable scientific grounds for believing that a new process or product may not be safe, it should not be introduced until we have convincing evidence that the risks are small and are outweighed by the benefits

This formulation has several merits and advantages over the Wingspread definition. First, there is no vagueness about what precautions need to be taken; a product or process that fails the scientific criteria should simply not be introduced. Second, it gives due weight to possible benefits of a product or process, and so avoids the problems outlined above, where, absurdly, benefits play no part in decision-making. Finally, with its insistence on 'reasonable scientific grounds' and 'convincing evidence' instead of reliance on the ambiguous 'threat of harm', it removes the problem that someone's gut feeling could stifle innovation. These are large advances.

Saunder's formulation of the precautionary principle is therefore not merely unobjectionable but positively useful. Its only problem is that, since the whole purpose of the precautionary principle is to give guidance on handling risk in situations of scientific uncertainty, Saunder's version—with its reliance on 'convincing evidence' and 'reasonable scientific grounds'—is *not* a precautionary principle at all. It is merely an injunction to carry out a rigorous risk assessment and a cost/benefit analysis before introducing a new product or process. This is sound, sensible advice and is by-and-large what scientists commonly do.

The Values Question

It may be possible to formulate a definition of the precautionary principle that escapes these problems; but I doubt it and have been unable to find one. In any case, there is a problem with the principle that has escaped the notice of even its harshest critics. To make this problem clear, I need to say a couple of things about wealth creation.

Wealth creation is obviously, in itself, a human good. Of course, it is not the only good, and any decision to proceed with a new product or activity on

NO / Ken Cussen **19**

the sole basis that it will create wealth would be irresponsible. Other factors, like the impact of the activity on human or environmental health must always be part of the decision and policy process, as may be questions about fairness, the distribution of the wealth created, and so on. But none of these affect the claim that wealth creation is, in itself, always a human good.

We should notice also that the creation of wealth is a precondition of many other goods. If wealth is not created, we cannot build hospitals, train doctors, educate the poor, or rejuvenate degraded environments.

With all this in mind, let me try to illustrate the problem further by proposing another version of the precautionary principle:

> When a lack of activity raises a threat of harm to wealth precautionary measures should be taken, even if some cause-and-effect relationships are not fully established.

This statement has all the flaws of vagueness and ambiguity that the Wingspread definition has, but is nevertheless its logical equivalent: it is a mirror image. If the Wingspread statement is valid and reasonable, then so is this formulation. Both are precautionary principles—the only difference is in the underlying value assumptions. Where Wingspread has human and environmental health as its fundamental value, this version has wealth creation.

It is important to notice that there is nothing internal to either the Wingspread or the latter formulation to justify their value choices, and cogent and compelling cases could be made for either value choice. This shows two things: first, that both versions are too vague to be useful or practically applicable; and more importantly, that the usual version of the precautionary principle is value-laden in what might be called a 'green' direction, without any justification. An argument is needed to justify this value choice but has not been forthcoming. A principle that is presented by its proponents in the guise of a value neutral guide to policy-making in the face of uncertainty is nothing of the kind. When its underlying value basis is revealed, the principle may be seen as nothing more than a rallying cry for one side of the environment versus development debate. In this guise, it no doubt has its uses—but we should not pretend that the precautionary principle is anything more.

Conclusions

The precautionary principle, as usually formulated, is either vacuous or dangerous. However, even if it is possible to formulate a useful and cogent precautionary principle, its underlying assumptions need to be clarified and made transparent. The fact is that activity and innovation, development and wealth creation, *and their absence* all carry risks and dangers. Any principle that pretends to guide our policy directions and decision-making procedures about such matters must recognise this fact or be discarded.

EXPLORING THE ISSUE

Should the Precautionary Principle Become Part of National and International Law?

Critical Thinking and Reflection

1. To what degree has the precautionary principle already become part of the structure of international law?
2. What should the place of cost-benefit or risk-benefit analysis be in applying the precautionary principle?
3. To what degree should the precautionary principle create legal obligations?
4. How much scientific evidence that harm is likely should be needed before invoking the precautionary principle?

Is There Common Ground?

No one involved in the debates over the value and application of the precautionary principle would argue that it is a bad idea to look before you leap. They are more likely to differ in their visions of where the benefits lie. For instance, proponents of genetically modified crops believe that GMOs benefit the food supply far more than they threaten human or environmental health. Critics see a need for precaution in the potential risk to human or environmental health. Similar polarizations can be seen with many other issues. In this book, look briefly at the following issues and identify opposing visions of benefit or risk:

Issue 5. Should the Military Be Exempt from Environmental Regulations?
Issue 6. Will Restricting Carbon Emissions Damage the Economy?
Issue 8. Should We Drill for Offshore Oil?
Issue 9. Is Shale Gas the Solution to Our Energy Woes?
Issue 16. Should Society Impose a Moratorium on the Use and Release of "Synthetic Biology" Organisms?

ISSUE 2

Is Sustainable Development Compatible with Human Welfare?

YES: Richard Heinberg, from *The End of Growth: Adapting to Our New Economic Reality* (New Society Publishers, 2011)

NO: Ronald Bailey, from "Wilting Greens," *Reason* (December 2002)

Learning Outcomes

After reading this issue, you should be able to:

- Explain what sustainable development is.
- Discuss whether sustainable development must be achieved.
- Describe how unsustainable activities may lead to economic crisis.

ISSUE SUMMARY

YES: Richard Heinberg argues that the era of economic growth as we have known it is over. A major cause of the world's recent (and continuing) economic crisis is depletion of resources such as oil and environmental degradation. We must learn to live sustainably, in "a healthy equilibrium economy."

NO: Ronald Bailey argues that sustainable development results in economic stagnation and threatens both the environment and the world's poor.

O ver the last 50 years, many people have expressed concerns that humanity cannot continue indefinitely to increase population, industrial development, and consumption. The trends and their impacts on the environment are amply described in numerous books, including historian J. R. McNeill's *Something New Under the Sun: An Environmental History of the Twentieth-Century World* (W. W. Norton, 2000).

"Can we keep it up?" is the basic question behind the issue of sustainability. In the 1960s and 1970s, it was expressed as the "Spaceship Earth" metaphor, which said that we have limited supplies of energy, resources, and room, and we must limit population growth and industrial activity, conserve, and recycle if we are not to run into crucial shortages. "Sustainability" entered the global debate in the early 1980s when the secretary general of the United Nations asked Gro Harlem Brundtland, a former prime minister and minister of environment in Norway, to organize and chair a World Commission on Environment and Development and produce a "global agenda for change." The resulting report, *Our Common Future* (Oxford University Press, 1987), defined "sustainable development" as "development that meets the needs of the present without compromising the ability of future generations to meet their own needs." It recognized that limits on population size and resource use cannot be known precisely, that problems may arise not suddenly but rather gradually, marked by rising costs, and that limits may be redefined by changes in technology. But it did also recognize that limits exist and must be taken into account when governments, corporations, and individuals plan for the future.

The Brundtland Report led to the UN Conference on Environment and Development held in Rio de Janeiro in 1992. The Rio conference set sustainability firmly on the global agenda and made it an essential part of efforts to deal with global environmental issues and promote equitable economic development. In brief, sustainability means such things as cutting forests no faster than they can grow back, using ground water no faster than it is recharged by precipitation, stressing renewable energy sources rather than exhaustible fossil fuels, and farming in such a way that soil fertility does not decline. In addition, economics must be revamped to take account of environmental costs as well as capital, labor, raw materials, and energy costs. Many add that the distribution of the Earth's wealth must be made more equitable as well.

The first of the Rio Declaration's 22 principles states, "Human beings are at the center of concerns for sustainable development. They are entitled to a healthy and productive life in harmony with nature." Any solution to the sustainability problem therefore should not infringe human welfare. This makes any solution that involves limiting or reducing human population or blocking improvements in standard of living very difficult to sell. Yet solutions may be possible. David Malin Roodman suggests in *The Natural Wealth of Nations: Harnessing the Market for the Environment* (W. W. Norton, 1998) that taxing polluting activities instead of profit or income would stimulate corporations and individuals to reduce such activities or to discover nonpolluting alternatives. In "Building a Sustainable Society" (Chapter 10 in *State of the World 1999*, W. W. Norton, 1999), he adds recommendations for citizen participation in decision-making, education efforts, and global cooperation, without which we are heading for "a world order [that] almost no one wants." (He is referring to a future of environmental crises, not the "new world order" feared by many conservatives, in which national policies are dictated by international [UN] regulators.) Roodman's recommendations may actually be on the way to reality. Arun Agrawal and Maria Carmen Lemos, "A Greener Revolution in the Making? Environmental Governance in the 21st Century," *Environment* (June 2007), argue that budget cuts and globalization are eroding the power of the state in favor of international "hybrid" arrangements that stress public–private partnerships, markets, and community and local participation.

The World Council of Churches brought to the U.N. World Summit, held in Johannesburg, South Africa, in August 2002, an emphasis on social justice. Martin Robra, in "Justice—The Heart of Sustainability," *Ecumenical Review* (July 2002), writes that the dominant stress on economic growth "has served, first and foremost, the interests of the powerful economic players. It has further marginalized the poor sectors of society, simultaneously undermining their basic security in terms of access to land, water, food, employment, and other basic services and a healthy environment."

Is social justice or equity worth this emphasis? Or is sustainability more a matter of population control, of shielding the natural environment from human impacts, or of economics? A. J. McMichael, C. D. Butler, and Carl Folke, in "New Visions for Addressing Sustainability," *Science* (Dcember 12, 2003), argue that it is wrong to separate—as did the Johannesburg Summit— achieving sustainability from other goals such as reducing fertility and poverty and improving social equity, living conditions, and health. They observe that human population and lifestyle affect ecosystems, ecosystem health affects human health, human health affects population and lifestyle. "A more integrated . . . approach to sustainability is urgently needed," they say, calling for more collaboration among researchers and fields. Yet there remains reason to focus on single threats. In February 2007, Sigma Xi and the United Nations Foundation released the Scientific Expert Group Report on Climate Change and Sustainable Development, *Confronting Climate Change: Avoiding the Unmanageable and Managing the Unavoidable* (http://www.sigmaxi.org/about/news/UNSEGReport.shtml) (Executive Summary, *American Scientist*, February 2007). Among its many points is that climate change from global warming is a huge threat to sustainability. Even in spite of feasible attempts at mitigation and adaptation, there is a serious risk of "intolerable impacts on human well-being."

Lester R. Brown, "Could Food Shortages Bring Down Civilization?" *Scientific American* (May 2009), warns that unsustainable agricultural practices and use of ground water, along with global warming, threaten to diminish food supplies to the point where society may actually break down. Underlining this concern, in October 2008, the European Environment and Sustainable Development Advisory Councils held a conference on "Sustaining Europe for a Long Way Ahead." The EEAC Web site (http://www.eeac-net.org/conferences/sixteen/sixteen_frame.htm) noted, "Addressing the very long term through the lens of sustainable development is now a matter of urgency. The prospect of highly damaging, and extremely costly, effects of global change in climate, in natural hazards caused by human intervention, the loss of biodiversity and disruption of food security, poses serious threats to personal and collective human health and wellbeing. . . . The long term is indeed here already."

Given continuing growth in population and demand for resources, sustainable development is clearly a difficult proposition. Some think it can be done, but others think that for sustainability to work, either population or resource demand must be reduced. Anthony R. Leiserowitz, Robert W. Kates, and Thomas M. Parris, "Do Global Attitudes and Behaviors Support Sustainable Development?" *Environment* (November 2005), find that though the world's people appear to support the component concepts of sustainable development, there is a mismatch between that support and their behavior. In the long term, they say, what is needed is a "shift from materialist to post-materialist values, from anthropocentric to ecological worldview, and a redefinition of the good life." Unfortunately, that shift remains in the future. P. Aarne Vesilind, Lauren Heine, and Jamie Hendry, "The Moral Challenge of Green Technology," *TRAMES: A Journal of the Humanities & Social Sciences* (No.1, 2006), "conclude that the unregulated free market system is incompatible with our search for sustainability. Experience has shown that if green technology threatens profits, green technology loses and profitability wins." See also Bill McKibben, "Breaking the Growth Habit," *Scientific American* (April 2010). Not surprisingly, many people see sustainable development as in conflict with business and industrial activities, private property rights, and such human freedoms as the freedoms to have many children, to accumulate wealth, and to use the environment as one wishes. Economics professor Jacqueline R. Kasun, in "Doomsday Every Day: Sustainable Economics, Sustainable Tyranny," *The Independent Review* (Summer 1999), goes so far as to argue that achieving sustainability will require sacrificing human freedom, dignity, and material welfare on a road to tyranny. Lester R. Brown, "Picking up the Tab," *USA Today Magazine* (May 2007), suggests that because governments currently spend some $700 billion a year subsidizing environmentally destructive activities such as automobile driving, one essential step toward sustainability is to end the subsidies.

The degrowth movement has taken shape to express the environmentalist, anti-consumerist, and anti-capitalist ideas noted above; see Baria Gencer Baykan, "From Limits to Growth to Degrowth within French Green Politics," *Environmental Politics* (June 2007). The First International Degrowth Conference occurred in Paris, France, in 2008. The second was held in Barcelona, Spain, in 2010. A related conference was held in Vancouver, Canada, later in 2010. John Bellamy Foster, "Capitalism and Degrowth—An Impossibility Theorem," *Monthly Review* (January 2011), says that "It is undeniable today that economic growth is the main driver of planetary ecological degradation. But to pin one's whole analysis on overturning an abstract 'growth society' is to lose all historical perspective. . . . [I]t can only take on genuine meaning as part of a critique of capital accumulation and part of the transition to a sustainable, egalitarian, communal order." Unlike most degrowth theorists, Robert Engelman, "Population and Sustainability," *Scientific American Earth 3.0* (Summer 2009), notes that sustainability can only be achieved if population is controlled at the same time as consumption.

In the following selections, author Richard Heinberg argues that the era of economic growth as we have known it is over. A major cause of the world's recent (and continuing) economic crisis and human distress is depletion of resources such as oil and environmental degradation. We must learn to live sustainably, in "a healthy equilibrium economy." Environmental journalist Ronald Bailey argues that sustainable development results in economic stagnation and threatens both the environment and the world's poor.

YES

Richard Heinberg

The End of Growth: Adapting to Our New Economic Reality

Introduction: The New Normal

The central assertion of this book is both simple and startling: *Economic growth as we have known it is over and done with.*

The "growth" we are talking about consists of the expansion of the overall size of the economy (with more people being served and more money changing hands) and of the quantities of energy and material goods flowing through it.

The economic crisis that began in 2007–2008 was both foreseeable and inevitable, and it marks a *permanent, fundamental* break from past decades—a period during which most economists adopted the unrealistic view that perpetual economic growth is necessary and also possible to achieve. There are now fundamental barriers to ongoing economic expansion, and the world is colliding with those barriers.

This is not to say the US or the world as a whole will never see another quarter or year of growth relative to the previous quarter or year. However, when the bumps are averaged out, the general trend-line of the economy (measured in terms of production and consumption of real goods) will be level or downward rather than upward from now on.

Nor will it be impossible for any region, nation, or business to continue growing for a while. Some will. In the final analysis, however, this growth will have been achieved at the expense of other regions, nations, or businesses. From now on, only *relative growth* is possible: the global economy is playing a zero-sum game, with an ever-shrinking pot to be divided among the winners.

Why Is Growth Ending?

Many financial pundits have cited serious troubles in the US economy—including overwhelming, un-repayable levels of public and private debt, and the bursting of the real estate bubble—as immediate threats to economic growth. The assumption generally is that eventually, once these problems are dealt with, growth can and will resume at "normal" rates. But the pundits generally miss factors *external* to the financial system that make a resumption

From *The End of Growth: Adapting to Our New Economic Reality*, by Richard Heinberg (New Society Publishers, 2011), pp. 1–7, 10–12, 14–22. Copyright © 2011 by Richard Heinberg. Reprinted by permission of New Society Publishers. www.newsociety.com

of conventional economic growth a near-impossibility. *This is not a temporary condition; it is essentially permanent.*

Altogether, as we will see in the following chapters, there are three primary factors that stand firmly in the way of further economic growth:

- The *depletion* of important resources including fossil fuels and minerals;
- The proliferation of *negative environmental impacts* arising from both the extraction and use of resources (including the burning of fossil fuels)—leading to snowballing costs from both these impacts themselves and from efforts to avert them; and
- *Financial disruptions* due to the inability of our existing monetary, banking, and investment systems to adjust to both resource scarcity and soaring environmental costs—and their inability (in the context of a shrinking economy) to service the enormous piles of government and private debt that have been generated over the past couple of decades.

Despite the tendency of financial commentators to ignore environmental limits to growth, it is possible to point to literally thousands of events in recent years that illustrate how all three of the above factors are interacting, and are hitting home with ever more force.

Consider just one: the Deepwater Horizon oil catastrophe of 2010 in the US Gulf of Mexico.

The fact that BP was drilling for oil in deep water in the Gulf of Mexico illustrates a global trend: while the world is not in danger of *running out* of oil anytime soon, there is very little new oil to be found in onshore areas where drilling is cheap. Those areas have already been explored and their rich pools of hydrocarbons are being depleted. According to the International Energy Agency, by 2020 almost 40 percent of world oil production will come from offshore. So even though its hard, dangerous, and expensive to operate a drilling rig in a mile or two of ocean water, that's what the oil industry must do if it is to continue supplying its product. That means more expensive oil.

Obviously, the environmental costs of the Deepwater Horizon blowout and spill were ruinous. Neither the US nor the oil industry can afford another accident of that magnitude. So, in 2010 the Obama administration instituted a deepwater drilling moratorium in the Gulf of Mexico while preparing new drilling regulations. Other nations began revising their own deepwater oil exploration guidelines. These will no doubt make future blowout disasters less likely, but they add to the cost of doing business and therefore to the already high cost of oil.

The Deepwater Horizon incident also illustrates to some degree the knock-on effects of depletion and environmental damage upon financial institutions. Insurance companies have been forced to raise premiums on deepwater drilling operations, and impacts to regional fisheries have hit the Gulf Coast economy hard. While economic costs to the Gulf region were partly made up for by payments from BP, those payments forced the company to reorganize and resulted in lower stock values and returns to investors. BP's financial woes in turn impacted British pension funds that were invested in the company.

This is just one event—admittedly a spectacular one. If it were an isolated problem, the economy could recover and move on. But we are, and will be, seeing a cavalcade of environmental and economic disasters, not obviously related to one another, that will stymie economic growth in more and more ways. These will include but are not limited to:

- Climate change leading to regional droughts, floods, and even famines;
- Shortages of energy, water, and minerals; and
- Waves of bank failures, company bankruptcies, and house foreclosures.

Each will be typically treated as a special case, a problem to be solved so that we can get "back to normal." But in the final analysis, they are all related, in that they are consequences of a growing human population striving for higher per-capita consumption of limited resources (including non-renewable, climate-altering fossil fuels), all on a finite and fragile planet.

Meanwhile, the unwinding of decades of buildup in debt has created the conditions for a once-in-a-century financial crash—which is unfolding around us, and which on its own has the potential to generate substantial political unrest and human misery.

The result: we are seeing a perfect storm of converging crises that together represent a watershed moment in the history of our species. We are witnesses to, and participants in, the transition from decades of economic growth to decades of economic contraction.

The End of Growth Should Come as No Surprise

The idea that growth will stall out at some point this century is hardly new. In 1972, a book titled *Limits to Growth* made headlines and went on to become the best-selling environmental book of all time.

That book, which reported on the first attempts to use computers to model the likely interactions between trends in resources, consumption, and population, was also the first major scientific study to question the assumption that economic growth can and will continue more or less uninterrupted into the foreseeable future.

The idea was heretical at the time—and still is. The notion that growth *cannot* and *will not* continue beyond a certain point proved profoundly upsetting in some quarters, and soon *Limits to Growth* was prominently "debunked" by pro-growth business interests. In reality, this "debunking" merely amounted to taking a few numbers in the book completely out of context, citing them as "predictions" (which they explicitly were not), and then claiming that these predictions had failed. The ruse was quickly exposed, but rebuttals often don't gain nearly as much publicity as accusations, and so today millions of people mistakenly believe that the book was long ago discredited. In fact, the original *Limits to Growth* scenarios have held up quite well. (A recent study by Australian Commonwealth Scientific and Industrial Research Organization (CSIRO) concluded, "[Our] analysis shows that 30 years of historical data compares favorably with key features of [the *Limits to Growth*] business-as-usual scenario. . . .")

The authors fed in data for world population growth, consumption trends, and the abundance of various important resources, ran their computer program, and concluded that the end of growth would probably arrive between 2010 and 2050. Industrial output and food production would then fall, leading to a decline in population.

The *Limits to Growth* scenario study has been re-run repeatedly in the years since the original publication, using more sophisticated software and updated input data. The results have been similar each time.

Why Is Growth So Important?

During the last couple of centuries, economic growth became virtually the sole index of national well-being. When an economy grew, jobs appeared and investments yielded high returns. When the economy stopped growing temporarily, as it did during the Great Depression, financial blood-letting ensued.

Throughout this period, world population increased—from fewer than two billion humans on planet Earth in 1900 to over seven billion today; we are adding about 70 million new "consumers" each year. That makes further economic growth even more crucial: if the economy stagnates, there will be fewer goods and services *per capita* to go around.

We have relied on economic growth for the "development" of the world's poorest economies; without growth, we must seriously entertain the possibility that hundreds of millions—perhaps billions—of people will never achieve the consumer lifestyle enjoyed by people in the world's industrialized nations. From now on, efforts to improve quality of life in these nations will have to focus much more on factors such as cultural expression, political freedoms, and civil rights, and much less on an increase in GDP.

Moreover, we have created monetary and financial systems that *require* growth. As long as the economy is growing, that means more money and credit are available, expectations are high, people buy more goods, businesses take out more loans, and interest on existing loans can be repaid. But if the economy is not growing, new money *isn't* entering the system, and the interest on existing loans cannot be paid; as a result, defaults snowball, jobs are lost, incomes fall, and consumer spending contracts—which leads businesses to take out fewer loans, causing still less new money to enter the economy. This is a self-reinforcing destructive feedback loop that is very difficult to stop once it gets going.

In other words, the existing market economy has no "stable" or "neutral" setting: there is only growth or contraction. And "contraction" can be just a nicer name for recession or depression—a long period of cascading job losses, foreclosures, defaults, and bankruptcies.

We have become so accustomed to growth that it's hard to remember that it is actually a fairly recent phenomenon.

Over the past few millennia, as empires rose and fell, local economies advanced and retreated—while world economic activity overall expanded only slowly, and with periodic reversals. However, with the fossil fuel revolution of the past century and a half, we have seen economic growth at a speed and scale unprecedented in all of human history. We harnessed the energies of coal, oil, and natural gas to build and operate cars, trucks, highways, airports, airplanes, and electric grids—all the essential features of modern industrial society. Through the one-time-only process of extracting and burning hundreds of millions of years' worth of chemically stored sunlight, we built what appeared (for a brief, shining moment) to be a perpetual-growth machine. We learned to take what was in fact an extraordinary situation for granted. It became *normal*.

But as the era of cheap, abundant fossil fuels comes to an end, our assumptions about continued expansion are being shaken to their core. The end of growth is a very big deal indeed. It means the end of an era, and of our current ways of organizing economies, politics, and daily life.

It is essential that we *recognize and understand the significance of this historic moment*: if we have in fact reached the end of the era of fossil-fueled economic expansion, then efforts by policy makers to continue pursuing elusive growth really amount to a flight from reality. World leaders, if they are deluded about our actual situation, are likely to delay putting in place the support services that can make life in a non-growing economy tolerable, and they will almost certainly fail to make needed, fundamental changes to monetary, financial, food, and transport systems.

As a result, what could be a painful but endurable process of adaptation could instead become history's greatest tragedy. We can survive the end of growth, and perhaps thrive beyond it, but only if we recognize it for what it is and act accordingly.

But Isn't Growth Normal?

Economies are systems, and as such they follow rules analogous (to a certain extent) to those that govern biological systems. Plants and animals tend to grow quickly when they are young, but then they reach a more or less stable mature size. In organisms, growth rates are largely controlled by genes, but also by availability of food.

In economies, growth seems tied to the availability of resources, chiefly energy ("food" for the industrial system), and credit ("oxygen" for the economy)—as well as to economic planning.

During the past 150 years, expanding access to cheap and abundant fossil fuels enabled rapid economic expansion at an average rate of about three percent per year; economic planners began to take this situation for granted. Financial systems internalized the expectation of growth as a promise of returns on investments.

Most organisms cease growing once they reach adulthood; if curtailment of growth weren't genetically programmed, plants and animals would outgrow a range of practical constraints: imagine, for example, the survival challenges faced by a two-pound hummingbird. If the analogy holds, then economies must eventually stop growing too. Even if planners (society's equivalent of regulatory DNA) dictate more growth, at some point increasing amounts of "food" and "oxygen" will cease to be available. It is also possible for wastes to accumulate to the point that the biological systems that underpin economic activity (such as forests, crops, and human bodies) are smothered and poisoned.

But many economists don't see things this way. That's probably because current economic theories were formulated during the anomalous historical period of sustained growth that is now ending. Economists are merely generalizing from their experience: they can point to decades of steady growth in the recent past, and they simply project that experience into the future. Moreover, they have theories to explain why modern market economies are immune to the kinds of limits that constrain natural systems: the two main ones have to do with *substitution* and *efficiency.*

If a useful resource becomes scarce, its price will rise, and this creates an incentive for users of the resource to find a substitute. For example, if oil gets expensive enough, energy companies might start making liquid fuels from coal. Or they might develop other energy sources undreamed of today. Many economists theorize that this process of substitution can go on forever. It's part of the magic of the free market.

Boosting efficiency means doing more with less. In the US, the number of dollars generated in the economy for every unit of energy consumed has increased steadily over recent decades. Part of this increasing efficiency is a result of outsourcing manufacturing to other nations—which must then burn the coal, oil, or natural gas to make our goods. (If we were making our own running shoes and LCD TVs, we'd be burning that fuel domestically.) Economists also point to another, related form of efficiency that has less to do with energy (in a direct way, at least): the process of identifying the cheapest sources of materials, and the places where workers will be most productive or work for the lowest wages. As we increase efficiency, we use less—of energy, resources, labor, or money—to do more. That enables more economic growth.

Finding substitute resources and upping efficiency are undeniably effective adaptive strategies of market economies. Nevertheless, the question remains as to how long these strategies can continue to work in the real world—which is governed less by economic theories than by the laws of physics. In the real world, some things don't have substitutes, or the substitutes are too expensive, or don't work as well, or can't be produced fast enough. And efficiency follows a law of diminishing returns: the first gains in efficiency are usually cheap, but every further incremental gain tends to cost more, until further gains become prohibitively expensive.

In the end, we can't outsource more than 100 percent of manufacturing, we can't transport goods with zero energy, and we can't enlist the efforts of workers and count on their buying our products while paying them nothing. Unlike most economists, most physical scientists recognize that growth within any functioning, bounded system has to stop sometime.

The Simple Math of Compounded Growth

In principle, the argument for an eventual end to growth is a slam-dunk. If any quantity grows steadily by a certain fixed percentage per year, this implies that it will double in size every so-many years; the higher the percentage growth rate, the quicker the doubling. A rough method of figuring doubling times is known as the rule of 70: dividing the percentage growth rate into 70 gives the approximate time required for the initial quantity to double. If a quantity is growing at 1 percent per year, it will double in 70 years; at 2 percent per year growth, it will double in 35 years; at 5 percent growth, it will double in only 14 years, and so on. If you want to be more precise, you can use the Y^x button on a scientific calculator, but the rule of 70 works fine for most purposes.

Here's a real-world example: Over the past two centuries, human population has grown at rates ranging from less than one percent to more than two percent per year. In 1800, world population stood at about one billion; by 1930 it had doubled to two billion. Only 30 years later (in 1960) it had doubled again to four billion; currently we are on track to achieve a third doubling, to eight billion humans, around 2025. No one seriously expects human population to continue growing for centuries into the future. But imagine if it did—at just 1.3 percent per year (its growth rate in the year 2000). By the year 2780 there would be 148 trillion humans on Earth—one person for each square meter of land on the planet's surface.

It won't happen, of course.

In nature, growth always slams up against non-negotiable constraints sooner or later. If a species finds that its food source has expanded, its numbers will increase to take advantage of those surplus calories—but then its food source will become depleted as more mouths consume it, and its predators will likewise become more numerous (more tasty meals for them!). Population "blooms" (or periods of rapid growth) are nearly always followed by crashes and die-offs. . . .

This discussion has very real implications, because the economy is not just an abstract concept; it is what determines whether we live in luxury or poverty, whether we eat or starve. If economic growth ends, everyone will be impacted, and it will take society years to adapt to this new condition. Therefore it is important to know whether that moment is close at hand or distant in time.

The Peak Oil Scenario

As mentioned, this book will argue that global economic growth is over because of a convergence of three factors—resource depletion, environmental impacts, and systemic financial and monetary failures. However, a single factor may be playing a key role in bringing the age of expansion to a close. That factor is oil.

Petroleum has a pivotal place in the modern world—in transportation, agriculture, and the chemicals and materials industries. The Industrial Revolution was really the Fossil Fuel Revolution, and the entire phenomenon of continuous economic growth—including the development of the financial institutions that facilitate growth, such as fractional reserve banking—is ultimately based on ever-increasing supplies of cheap energy. Growth requires more manufacturing, more trade, and more transport, and those all in turn require more energy. This means that if energy supplies can't expand and energy therefore becomes significantly more expensive, economic growth will falter and financial systems built on expectations of perpetual growth will fail.

As early as 2000, petroleum geologist Colin Campbell discussed a Peak Oil impact scenario that went like this. Sometime around the year 2010, he theorized, stagnant or falling oil supplies would lead to soaring and more volatile oil prices, which would precipitate a global economic crash. This rapid economic contraction would in turn lead to sharply curtailed energy demand, so oil prices would then fall; but as soon as the economy regained strength, demand for petroleum would recover, prices would again soar, and as a result of that the economy would relapse. This cycle would continue, with each recovery phase being shorter and weaker, and each crash deeper and harder, until the economy was in ruins. Financial systems based on the assumption of continued growth would implode, causing more social havoc than the oil price spikes would themselves directly generate.

Meanwhile, volatile oil prices would frustrate investments in energy alternatives: one year, oil would be so expensive that almost any other energy source would look cheap by comparison; the next year, the price of oil would have fallen far enough that energy users would be flocking back to it, with investments in other energy sources looking foolish. But low oil prices would discourage exploration for more petroleum, leading to even worse fuel shortages later on. Investment capital would be in short supply in any case because the banks would be insolvent due to the crash, and governments would be broke due to declining tax revenues. Meanwhile, international competition for dwindling oil supplies might lead to wars between petroleum importing nations, between importers and exporters, and between rival factions within exporting nations.

In the years following the turn of the millennium, many pundits claimed that new technologies for crude oil extraction would increase the amount of oil that can be obtained from each well drilled, and that enormous reserves of alternative hydrocarbon resources (principally tar sands and oil shale) would be developed to seamlessly replace conventional oil, thus delaying the inevitable peak for decades. There were also those who said that Peak Oil wouldn't be much of a problem even if it happened soon, because the market would find other energy sources or transport options as quickly as needed—whether electric cars, hydrogen, or liquid fuel made from coal.

In succeeding years, events appeared to be supporting the Peak Oil thesis and undercutting the views of the oil optimists. Oil prices trended steeply upward—and for entirely foreseeable reasons: discoveries of new oilfields were continuing to dwindle, with most new fields being much more difficult and expensive to develop than ones found in previous years. More oil-producing countries were seeing their extraction rates peaking and beginning to decline despite efforts to maintain production growth using high-tech, expensive extraction methods like injecting water, nitrogen, or carbon dioxide to force more oil out of the ground. Production decline rates in the world's old, super-giant oilfields, which are responsible for the lion's share of the global petroleum supply, were accelerating. Production of liquid fuels from tar sands was expanding only slowly, while the development of oil shale remained a hollow promise for the distant future.

From Scary Theory to Scarier Reality

Then in 2008, the Peak Oil scenario became all too real. Global oil production had been stagnant since 2005 and petroleum prices had been soaring upward. In July 2008, the per-barrel price shot up to nearly $150—half again higher (in inflation-adjusted terms) than the price spikes of the 1970s that had triggered the worst recession since World War II. By summer 2008, the auto industry, the trucking industry, international shipping, agriculture, and the airlines were all reeling.

But what happened next riveted the world's attention to such a degree that the oil price spike was all but forgotten: in September 2008, the global financial system nearly collapsed. The most frequently discussed reasons for this sudden, gripping crisis had to do with housing bubbles, lack of proper regulation of the banking industry, and the over-use of bizarre financial products that almost nobody understood. However, the oil price spike had also played a critical (if largely overlooked) role in initiating the economic meltdown.

In the immediate aftermath of that global financial near-death experience, both the Peak Oil impact scenario proposed a decade earlier and the *Limits to Growth* standard-run scenario of 1972 seemed to be confirmed with uncanny and frightening accuracy. Global trade was falling. The world's largest auto companies were on life support. The US airline industry had shrunk by almost a quarter. Food riots were erupting in poor nations around the world. Lingering wars in Iraq (the nation with the world's second-largest crude oil reserves) and Afghanistan (the site of disputed oil and gas pipeline projects) continued to bleed the coffers of the world's foremost oil-importing nation.

Meanwhile, the dragging debate about what to do to rein in global climate change exemplified the political inertia that had kept the world on track for calamity since the early '70s. It had by now become obvious to a great majority of people familiar with the scientific data that the world has two urgent, incontrovertible reasons to rapidly end its reliance on fossil fuels: the twin threats of climate catastrophe and impending constraints to fuel supplies. Yet at the landmark international Copenhagen climate conference in December 2009, the priorities of the most fuel-dependent nations were clear: carbon emissions should be cut, and fossil fuel dependency reduced, *but only if doing so does not threaten economic growth.*

Bursting Bubbles

. . . [E]xpectations of continuing growth had in previous decades been translated into enormous amounts of consumer and government debt. An ever shrinking portion of America's wealth was being generated by invention of new technologies and manufacture of consumer goods, and an ever greater portion was coming from buying and selling houses, or moving money around from one investment to another.

As a new century dawned, the world economy lurched from one bubble to the next: the emerging-Asian-economies bubble, the dot-com bubble, the real estate bubble. Smart investors knew that these would eventually burst, as bubbles always do, but the smartest ones aimed to get in early and get out quickly enough to profit big and avoid the ensuing mayhem.

If Peak Oil and other limits on resources were closing the spigots on growth in 2007–2008, the pain that ordinary citizens were experiencing seemed to be coming from other directions entirely: loss of jobs and collapsing real estate prices.

In the manic days of 2002 to 2006, millions of Americans came to rely on soaring real estate values as a source of income, turning their houses into ATMs (to use once more the phrase heard so often then). As long as prices kept going up, homeowners felt justified in borrowing to remodel a kitchen or bathroom, and banks felt fine making those loans. Meanwhile, the wizards of Wall Street were finding ways of slicing and dicing sub-prime mortgages into tasty collateralized debt obligations that could be sold at a premium to investors—with little or no risk! After all, real estate values were destined to just keep going up. *God's not making any more land,* went the truism.

Credit and debt expanded in the euphoria of easy money. All this giddy optimism led to a growth of jobs in construction and real estate industries, masking underlying ongoing job losses in manufacturing.

A few dour financial pundits used terms like "house of cards," "tinderbox," and "stick of dynamite" to describe the situation. All that was needed was a metaphoric breeze or rogue spark to produce a catastrophic outcome. Arguably, the oil price spike of mid-2008 was more than enough to do the trick.

But the housing bubble was itself merely a larger fuse: in reality, the entire economic system had come to depend on impossible-to-realize expectations of perpetual growth and was set to detonate. Money was tied to credit, and credit was tied to assumptions about growth. Once growth went sour in 2008, the chain reaction of defaults and bankruptcy began; we were in a slow-motion explosion.

Since then, governments have worked hard to get growth started again. But, to the very limited degree that this effort temporarily succeeded in late 2009 and 2010, it did so by ignoring the underlying contradiction at the heart of our entire economic system—the assumption that we can have unending growth in a finite world.

What Comes After Growth?

The realization that we have reached the point where growth cannot continue is undeniably depressing. But once we have passed that psychological hurdle, there is some moderately good news. The end of economic growth does not necessarily mean we've reached the end of qualitative improvements in human life.

Not all economists have fallen for the notion that growth will go on forever. There are schools of economic thought that recognize nature's limits; and, while these schools have been largely ignored in policy circles, they have developed potentially useful plans that could help society adapt.

The basic factors that will inevitably shape whatever replaces the growth economy are knowable. To survive and thrive for long, societies have to operate within the planet's budget of sustainably extractable resources. This means that even if we don't know in detail what a desirable post-growth economy and lifestyle will look like, we know enough to begin working toward them.

We must discover how life in a non-growing economy can actually be fulfilling, interesting, and secure. The absence of growth does not necessarily imply a lack of change or improvement. Within a non-growing or equilibrium economy there can still be continuous development of practical skills, artistic expression, and certain kinds of technology. In fact, some historians and social scientists argue that life in an equilibrium economy can be superior to life in a fast-growing economy: while 'growth creates opportunities for some, it also typically intensifies competition—there are big winners and big losers, and (as in most boom towns) the quality of relations within the community can suffer as a result. Within a non-growing economy it is possible to maximize benefits and reduce factors leading to decay, but doing so will require pursuing appropriate goals: instead of *more,* we must strive for *better;* rather than promoting increased economic activity for its own sake, we must emphasize that which increases quality of life without stoking consumption. One way to do this is to reinvent and redefine *growth* itself.

The transition to a no-growth economy (or one in which growth is defined in a fundamentally different way) is inevitable, but it will go much better if we plan for it rather than simply watch in dismay as institutions we have come to rely upon fail, and then try to improvise a survival strategy in their absence.

In effect, we have to create a desirable "new normal" that fits the constraints imposed by depleting natural resources. *Maintaining the "old normal" is not an option;* if we do not find new goals for ourselves and plan our transition from a growth-based economy to a healthy equilibrium economy, we will end up with a much less desirable "new normal." Indeed, we are already beginning to see this in the forms of persistent high unemployment, a widening gap between rich and poor, and ever more frequent and worsening environmental crises—all of which translate to profound distress across society.

Ronald Bailey

Wilting Greens

It's clear that we've suffered a number of major defeats," declared Andrew Hewett, executive director of Oxfam Community Aid, at the conclusion of the World Summit on Sustainable Development, held in Johannesburg, South Africa, in September. Greenpeace climate director Steve Sawyer complained, "What we've come up with is absolute zero, absolutely nothing." The head of an alliance of European green groups proclaimed, "We barely kept our heads above water."

It wasn't supposed to be this way. Environmental activists hoped the summit would set the international agenda for sweeping environmental reform over the next 15 years. Indeed, they hoped to do nothing less than revolutionize how the world's economy operates. Such fundamental change was necessary, said the summiteers, because a profligate humanity consumes too much, breeds too much, and pollutes too much, setting the stage for a global ecological catastrophe.

But the greens' disappointment was inevitable because their major goals—preserving the environment, eradicating poverty, and limiting economic growth—are incompatible. Economic growth is a prerequisite for lessening poverty, and it's also the best way to improve the environment. Poor people cannot afford to worry much about improving outdoor air quality, let alone afford to pay for it. Rather than face that reality, environmentalists increasingly invoke "sustainable development." The most common definition of the phrase comes from the 1987 United Nations report *Our Common Future*: development that "meets the needs of the present without compromising the ability of future generations to meet their own needs."

For radical greens, sustainable development means economic stagnation. The Earth Island Institute's Gar Smith told Cybercast News, "I have seen villages in Africa . . . that were disrupted and destroyed by the introduction of electricity." Apparently, the natives no longer sang community songs or sewed together in the evenings. "I don't think a lot of electricity is a good thing," Smith added. "It is the fuel that powers a lot of multinational imagery." He doesn't want poor Africans and Asians "corrupted" by ads for Toyota and McDonald's, or by Jackie Chan movies.

Indian environmentalist Sunita Narain decried the "pernicious introduction of the flush toilet" during a recent PBS/BBC television debate hosted by Bill Moyers. Luckily, most other summiteers disagreed with Narain's curious disdain for sanitation. One of the few firm goals set at the confab was that adequate sanitation should be supplied by 2015 to half of the 2.2 billion people now lacking it.

Sustainable development boils down to the old-fashioned "limits to growth" model popularized in the 1970s. Hence Daniel Mittler of Friends of the Earth International moaned that "the summit failed to set the necessary economic and ecological limits to globalization." The *Jo'burg Memo*, issued by the radical green Heinrich Böll Foundation before the summit, summed it up this way: "Poverty alleviation cannot be separated from wealth alleviation."

The greens are right about one thing: The extent of global poverty is stark. Some 1.1 billion people lack safe drinking water, 2.2 billion are without adequate sanitation, 2.5 billion have no access to modern energy services, 11 million children under the age of 5 die each year in developing countries from preventable diseases, and 800 million people are still malnourished, despite a global abundance of food. Poverty eradication is clearly crucial to preventing environmental degradation, too, since there is nothing more environmentally destructive than a hungry human.

Most summit participants from the developing world understood this. They may be egalitarian, but unlike their Western counterparts they do not aim to make everyone equally poor. Instead, they want the good things that people living in industrialized societies enjoy.

That explains why the largest demonstration during the summit, consisting of more than 10,000 poor and landless people, featured virtually no banners or chants about conventional environmentalist issues such as climate change, population control, renewable resources, or biodiversity. Instead, the issues were land reform, job creation, and privatization.

The anti-globalization stance of rich activists widens this rift. Environmentalists claim trade harms the environment and further impoverishes people in the developing world. They were outraged by the dominance of trade issues at the summit.

"The leaders of the world have proved that they work as employees for the transnational corporations," asserted Friends of the Earth Chairman Ricardo Navarro. Indian eco-feminist Vandana Shiva added, "This summit has become a trade summit, it has become a trade show." Yet the U.N.'s own data underscore how trade helps the developing world. As fact sheets issued by the U.N. put it, "During the 1990s the economies of developing countries that were integrated into the world economy grew more than twice as fast as the rich countries. The 'non-globalizers' grew only half as fast and continue to lag further behind."

By invoking a zero sum version of sustainable development, environmentalists not only put themselves at odds with the developing world; they ignore the way in which economic growth helps protect the environment. The real commons from which we all draw is the growing pool of scientific, technological, and institutional concepts, and the capital they create. Past generations have left us far more than they took, and the result has been an explosion in human well-being, longer life spans, less disease, more and cheaper food, and expanding political freedom.

Such progress is accompanied by environmental improvement. Wealthier is healthier for both people and the environment. As societies become richer and more technologically adept, their air and water become cleaner, they set aside more land for nature, their forests expand, they use less land for agriculture, and more people cherish wild species. All indications suggest that the 21st century will be the century of ecological restoration, as humanity uses physical resources ever more efficiently, disturbing the natural world less and less.

In their quest to impose a reactionary vision of sustainable development, the disappointed global greens will turn next to the World Trade Organization, the body that oversees international trade rules. During the summit, the WTO emerged as the greens' bête noire. As Friends of the Earth International's Daniel Mittler carped, "Instead of using the [summit] to respond to global concerns over deregulation and liberalization, governments are pushing the World Trade Organization's agenda." "See you in Cancun!" promised Greenpeace's Steve Sawyer, referring to the location of the next WTO ministerial meeting in September 2003. That confab will build on the WTO's Doha Trade Round, launched last year, which is aimed at reducing the barriers to trade for the world's least developed countries.

The WTO may achieve worthy goals that eluded the Johannesburg summit, such as eliminating economically and ecologically ruinous farm and energy subsidies and opening developed country markets to the products of developing nations. Free marketeers and greens might even form an alliance on those issues.

But environmentalists want to use the WTO to implement their sustainable development agenda: global renewable energy targets, regulation based on the precautionary principle, a "sustainable consumption and production project," a worldwide eco-labeling scheme. According to Greenpeace's Sawyer, nearly everyone at the Johannesburg summit agreed "there is something wrong with unbridled neoliberal capitalism."

Let's hope the greens fail at the WTO just as they did at the U.N. summit. Their sustainable development agenda, supposedly aimed at improving environmental health, instead will harm the natural world, along with the economic prospects of the world's poorest people. The conflicting goals on display at the summit show that at least some of the world's poor are wise to that fact.

EXPLORING THE ISSUE

Is Sustainable Development Compatible with Human Welfare?

Critical Thinking and Reflection

1. In what sense may economic growth lead to environmental protection?
2. Can a market-based economy exist without continuous growth in production and consumption?
3. How does population relate to the sustainability issue?
4. In what ways is the precautionary principle an essential component of sustainable development?

Is There Common Ground?

It is hard to avoid the conclusion that sustainability—meaning meeting present needs while leaving enough resources unused to permit our children and grandchildren to meet their needs—is important. It is also hard to avoid the conclusion that the environment warrants protection from human impacts. Where people differ is on the best way to achieve these things.

1. In what ways does population control help us achieve both sustainability and environmental protection?
2. An interesting technology currently gaining ground is 3D printing, meaning machines that can make small objects of many kinds. In the future they may be found in community centers or even in homes. Look this up, and consider whether it would help achieve a sustainable world.
3. Would you expect the impact of 3D printing on society to help meet the needs Ronald Bailey says make sustainability incompatible with human welfare?

ISSUE 3

Do Ecosystem Services Have Economic Value?

YES: **Rebecca L. Goldman**, from "Ecosystem Services: How People Benefit from Nature," *Environment* (September/October 2010)

NO: **Marino Gatto and Giulio A. De Leo**, from "Pricing Biodiversity and Ecosystem Services: The Never-Ending Story," *Bioscience* (April 2000)

Learning Outcomes

After reading this issue, you should be able to:

- Explain what ecosystem services are and why they are important to people.
- Discuss whether it makes sense to assess the economic value of ecosystem services.
- Describe how payments for ecosystem services can help in the management of natural ecosystems.

ISSUE SUMMARY

YES: Rebecca L. Goldman argues that ecosystem services are crucial to human well-being, both now and for the sustainable future. They are also affected by human behavior, at both the individual and the national levels. Assessing their economic value is difficult but essential to public decision making.

NO: Professors of applied ecology Marino Gatto and Giulio A. De Leo contend that the pricing approach to valuing nature's services is misleading because it falsely implies that only economic values matter.

Human activities frequently involve trading a swamp or forest or mountainside for a parking lot or housing development or farm (among other things). People generally agree that these developments are worthwhile projects, for they have obvious benefits to people. But are there costs as well? Construction costs, labor costs, and material costs can easily be calculated, but what about the value of the swamp itself, of the forest, and of the species living there?

How much is a species worth? One approach to answering this question is to ask people how much they would be willing to pay to keep a species alive. If the question is asked when there are a million species in existence, few people will likely be willing to pay much. But if the species is the last one remaining, they might be willing to pay a great deal. Most people would agree that both answers fail to get at the true value of a species, for nature is not expressible solely in terms of cash values. Yet some way must be found to weigh the effects of human activities on nature against the benefits gained from those activities. If it is not, we will continue to degrade the world's ecosystems and threaten our own continued well-being. Indeed, human dependence is growing more acute; see Zhongwei Guo, Lin Zhang, and Yiming Li, "Increased Dependence of Humans on Ecosystem Services and Biodiversity," *PLoS One* (www.plosone .org) (October 2010). See also Peter Kareiva, Heather Tallis, Taylor H. Ricketts, Gretchen C. Daily, and Stephen Polasky, eds., *Natural Capital: Theory and Practice of Mapping Ecosystem Services* (Oxford University Press, 2011).

Traditional economics views nature as a "free good." That is, forests generate oxygen and wood, clouds bring rain, and the sun provides warmth, all without charge to the humans who benefit. At the same time, nature has provided ways for people to dispose of wastes—such as dumping raw sewage into rivers or emitting smoke into the air—without paying for the privilege. This "free" waste disposal has turned out to have hidden costs in the form of the health effects of pollution (among other things), but it has been up to individuals and governments to bear the costs associated with those effects. The costs are real, but in general they have not been borne by the businesses and other organizations that caused them. They have thus come to be known as "external" costs.

Environmental economists have recognized the problem of external costs, and government regulators have devised a number of ways to make those who are responsible accept the bill, such as instituting requirements for pollution control and fining those who exceed permitted emissions. Yet some would say that this approach does not help enough.

The *ecosystem services* approach recognizes that undisturbed ecosystems do many things that benefit human beings. A forest, for instance, slows the movement of rain and snowmelt into streams and rivers; if the forest is removed, floods may follow (a connection that a few years ago forced China to de-emphasize forest exploitation). Swamps filter the water that seeps through them. Food chains cycle nutrients necessary for the production of wood and fish and other harvests. Bees pollinate crops and thus make the production of many food crops possible. These services are valuable—even essential—to us, and anything that interferes with them must be seen as imposing costs just as significant as those associated with the illnesses caused by pollution. Yet standard economics encourages the conversion of "free" natural resources into marketable commodities without considering those costs; see Christopher L. Lant, J. B. Ruhl, and Steven E. Kraft, "The Tragedy of Ecosystem Services," *Bioscience* (November 2008).

How can the costs of interfering with ecosystem services be assessed? In 1997, Robert Costanza and his colleagues published an influential paper entitled "The Value of the World's Ecosystem Services and Natural Capital," *Nature* (May 15, 1997). In it, the authors listed a variety of ecosystem services and attempted to estimate what it would cost to replace those services if they were somehow lost. The total bill for the entire biosphere came to $33 trillion (the middle of a $16–54 trillion range), compared to a global gross national product of $25 trillion. Costanza et al. stated that this was surely an underestimate. Janet N. Abramovitz, "Putting a Value on Nature's 'Free' Services," *World-Watch* (January/February 1998), argues that nature's services are responsible for the vast bulk of the value in the world's economy and that attaching economic value to those services may encourage their protection. In "Can We Put a Price on Nature's Services?" *Report from the Institute for Philosophy and Public Policy* (Summer 1997), Mark Sagoff objects that trying to attach a price to ecosystem services is futile because it legitimizes the accepted cost-benefit approach and thereby undermines efforts to protect the environment from exploitation. The March 1998 issue of *Environment* contains environmental economics professor David Pearce's detailed critique of the 1997 Costanza et al. study. Pearce objects chiefly to the methodology, not the overall goal of attaching economic value to ecosystem services. Costanza et al. reply to Pearce's objections in the same issue. Pearce and Edward B. Barbier have published *Blueprint for a Sustainable Economy* (Earthscan, 2000), in which they discuss how governments worldwide are now applying economics to environmental policy.

Despite the controversy over the worth of assigning economic value to various aspects of nature, researchers continue the effort. Gretchen C. Daily et al., in "The Value of Nature and the Nature of Value," *Science* (July 21, 2000), discuss valuation as an essential step in all decision making and argue that efforts "to capture the value of ecosystem assets . . . can lead to profoundly favorable effects." Gretchen C. Daily and Katherine Ellison continue the theme in *The New Economy of Nature: The Quest to Make Conservation Profitable* (Island Press, 2002). In "What Price Biodiversity?" *Ecos* (January 2000), Steve Davidson describes an ambitious program funded by Australia's Commonwealth Scientific and Industrial Research Organization (CSIRO) and the Myer Foundation that is aimed at developing principles and methods for objectively valuing "ecosystem services—the conditions and processes by which natural ecosystems sustain and fulfill human life—and which we too often take for granted. These include such services as flood and erosion control, purification of air and water, pest control, nutrient cycling, climate regulation, pollination, and waste disposal." Jim Morrison, "How Much Is Clean Water Worth?" *National Wildlife* (February/March 2005), argues that such ecosystem services have sufficient economic value to make it profitable to spend millions of dollars to protect natural systems.

Stephen Farber et al., "Linking Ecology and Economics for Ecosystem Management," *Bioscience* (February 2006), find "the valuation of ecosystem services . . . necessary for the accurate assessment of the trade-offs involved in different management options." Ecosystem valuation is currently being used to justify restoration efforts, "linking the science to human welfare," as shown (for example) in Chungfu Tong et al., "Ecosystem Service Values and Restoration in the Urban Sanyang Wetland of Wenzhou, China," *Ecological Engineering*

(March 2007). See also J. R. Rouquette et al., "Valuing Nature-Conservation Interests on Agricultural Floodplains," *Journal of Applied Ecology* (April 2009), and Z. M. Chen et al., "Net Ecosystem Services Value of Wetland: Environmental Economic Account," *Communications in Nonlinear Science & Numerical Simulation* (June 2009). Joshua Farley and Robert Costanza, "Payments for Ecosystem Services: From Local to Global," *Ecological Economics* (September 2010), say that "Payment for Ecosystem Services (PES) is becoming increasingly popular as a way to manage ecosystems using economic incentives." Just as important is the role that assessing the economic value of the environment can play in weaving the environment into "the fabric of economic thinking"; see "Price Fixing: Why It Is Important to Put a Price on Nature," *The Economist* (January 18, 2010) (www.economist.com/node/15321193?story_id=15321193). However, some argue that "narrowing down the complexity of ecosystems to a single service has serious technical difficulties and ethical implications on the way we relate to and perceive nature"; see Nicolas Kosoy and Esteve Corbera, "Payments for Ecosystem Services as Commodity Fetishism," *Ecological Economics* (April 2010).

In 2011, researchers released *The UK National Ecosystem Assessment* (UNEP-WCMC, Cambridge, UK, 2011). Among the key findings were that about 30 percent of ecosystems were declining in their ability to deliver services. They suggest that ecosystem management would benefit if the government "took into account the economic value of a broader array of natural benefits"; see Erik Stokstad, "Appraising U.K. Ecosystems, Report Envisions Greener Horizon," *Science* (June 3, 2011). John W. Day, Jr., et al., "Ecology in Times of Scarcity," *Bioscience* (April 2009), report, "In an energy-scarce future, services from natural ecosystems will assume relatively greater importance in supporting the human economy." As a result, the practice of ecology will become both more expensive and more valuable, and ecological engineering (ecoengineering) and restoration will become more common and necessary. Perhaps the connection between ecosystem services and economic value will become less debatable. Already, the CEO of a major corporation, Andrew Liveris of Dow Chemical, can say "protecting nature can be a profitable corporate priority and a smart global business strategy. . . . The economy and the environment are interdependent"; see Bryan Walso, "Paying for Nature," *Time* (February 21, 2011).

In the following selections, Rebecca L. Goldman of The Nature Conservancy argues that ecosystem services are crucial to human well-being, both now and for the sustainable future. They are also affected by human behavior, at both the individual and the national levels. Assessing their economic value is difficult but essential to public decision making. Professors of applied ecology Marino Gatto and Giulio A. De Leo contend that the pricing approach to valuing nature's services is misleading because it falsely implies that only economic values matter.

YES

Rebecca L. Goldman

Ecosystem Services: How People Benefit from Nature

What do the blue jeans you wear, the hamburger you have for lunch, and the sheet you make your bed with have in common? They all take copious amounts of water to produce. One pair of blue jeans takes 2,900 gallons or about 78 bathtubs of water. Even your morning cup of coffee takes 37 gallons (about one bathtub) of water—not just the one cup you consume. But we don't pay for all the water that goes into our morning cup of coffee. The price of the coffee is based on production and transportation costs (among other costs), but it's much more difficult to value where all the water in one cup of coffee comes from. This difficulty arises from the fact that natural ecosystems are responsible for the retention, release, and regulation of water, but how does a person value a natural ecosystem and the services it provides and put that into the cost of a cup of coffee?

Ecosystem services, or the benefits that nature provides to people, have, in the past decade or two, become a growing focus for the conservation movement, both its science and its policy; see, for example, the Millennium Ecosystem Assessment, the launching of The Ecosystem Marketplace (www .ecosystemmarketplace.com), DIVERSITAS (see http://www.diversitas-international .org), among many others. Uses and definitions of ecosystem services vary, but in general, ecosystem services help demonstrate the link between people and nature and the interdependence of our lives on ecosystem-based processes that create the products we need and use every day. Some examples of ecosystem services are water purification, water retention, soil fertility, carbon sequestration, and coastal protection, among many others.

What are some examples of how ecosystem services are already a part of our lives, how might ecosystem service considerations change daily decisions, and why is this behavior change important? In this article, I answer these questions using three examples—pollination services, flood and natural disaster protection services, and water services—to illustrate the interrelationship between nature and people. With each example I provide a policy or personal decision-making context into which this link could be relevant and/or could lead to a different choice or behavior.

I do this for three reasons. First, I want to introduce the concept of ecosystem services in a more tangible way to demonstrate that ecosystem services are not just relevant to academics and conservation practitioners but to everyone. Second, I want to demonstrate the impact that behaviors and choices have on service provision. Finally, I want to underscore the importance of ecosystem services for a sustainable future and how, by fully considering the tradeoffs associated with daily choices, people can, with small changes, make a potentially large difference. I conclude by illustrating some programs, policies, and reports that have truly enveloped this approach and demonstrate the large impact the "ecosystem services movement" can have on our future. First, however, I begin with some general background about ecosystem services and their importance to sustainable development and to conservation.

Why Are Ecosystem Services Important for Sustainable Development?

The human population is expected to reach 9 billion people by 2050, and with that increase will come a greater demand for many natural resources. Look at freshwater needs, for example. Research has estimated per person per day dietary needs of 2,000–5,000 liters of water, and this does not include water needed for cleaning and other activities. Hand in hand with this growing demand for resources is the conversion of native ecosystems to meet growing needs; this is where a tradeoff assessment in terms of ecosystem services might be useful.

Agricultural and pasture lands represent about 40 percent of global land surface. If people continue to depend on agricultural products as they have in the past, then by 2050, scholars estimate that 10^9 hectares of natural ecosystems will be converted to agriculture. This conversion would include a 2.4–2.7-fold increase in nitrogen- and phosphorus-driven eutrophication of numerous waters with similar increases in pesticide use. Agriculture already accounts for 70 percent of water withdrawals from lakes, rivers, and aquifers.

This dependence and use pattern we have with our land is no different in the oceans. One clear example is oysters. Oysters have been consumed for sustenance for millennia. Reports from the 1800s in England indicate that in one year, 700 million European flat oysters were consumed, a process that employed about 120,000 people. In the Chesapeake Bay on the Eastern Shore of the United States, oyster reefs used to extend for miles. By the 1940s, these reefs had largely disappeared. This is, sadly, true of many ocean creatures, and it underscores our dependence on nature and how a growing demand can affect nature's ability to support itself and us.

We are at a critical juncture. Making choices that can benefit both us and nature may be our best option for securing our livelihoods. Ecosystem services provide a means for people to understand the link between their choices and the natural world. But what exactly are ecosystem services, and how does nature "create" these services? It is to the answers of these questions that I now turn.

What Are Ecosystem Services?

Underlying all the resources we use, the species we see, and the foods we eat are ecosystem processes: the biological, chemical, and physical interactions between components of an ecosystem (e.g., soil, water, species). These processes produce benefits to people in the form of clean water, carbon sequestration, and reductions in erosion, among others. These benefits are ecosystem services. The ability of nature to help filter, regulate the release of, and capture and store water allows us to wear blue jeans, drink coffee, and eat a hamburger, but we rarely think about the true origin of the products we use every day.

The distinction between ecosystem process, services, and goods is one that has received a lot of attention in the literature. The key points emerging from these discussions are that ecosystem processes create our natural world. Ecosystem services are the link between this natural world and people, that is, the specific processes that benefit people. (There can be processes that are not services if there is no person to value that particular process.) Ecosystem goods are created from processes and services and are the tangible, material products we are familiar with, but again the distinction between goods and services is complicated and interrelated.

The appeal of ecosystem services for conservation is the connection to people and people's well-being and how that appeal translates into new and increased interest in conservation across a wide range of resource management issues. Ecosystem services can provide a means to value people's well-being in conservation projects and can help advance a set of on-the-ground actions that are equitable, just, and moral. Ecosystem services can be a basis for sustainable development by providing a means to think through how to retain our natural resources for people and for nature with a growing population and therefore an ever-increasing demand for them.

Ecosystem services, since they are the benefits from *nature*, are often discussed in the context of conservation, but in our daily lives we make choices that depend on and affect flows of services from nature, since all goods and products we use today originate from nature and its services. Each choice we make—drive or ride a bus, buy organic or buy regular vegetables, turn on the heat or put on an extra sweatshirt—has tradeoffs. Conserving nature or converting nature does too, but tradeoffs associated with nature's values are often harder to assess. Not understanding nature's role in the products we use means we won't conserve nature sufficiently; this in turn will compromise our ability to access products we need, or we will have to find sometimes costly alternatives for what nature could otherwise provide to us. Incorporating the full suite of costs and benefits into decision-making means evaluating all costs and benefits associated with nature, too. Economists refer to this full valuation as shadow pricing, but even an informal, "back-of-the envelope" calculation of all values can help to illustrate the importance of ecosystem services in our daily lives.

Ecosystem services help connect people to nature and allow us to make more informed decisions by underscoring all the component pieces of the products we value. How might considering these multiple cost and benefit streams alter people's behaviors? How are people's daily choices linked to eco-system services? I provide concrete examples to answer these questions.

How Might Ecosystem Services Change Which Products We Purchase at the Supermarket?

To provide a more tangible understanding of the behind-the-scene trade-offs affecting choices we make, I will use the example of pollination services from native pollinators supported by natural ecosystems. In 2006, around the world there was a great deal of news and concern about the sudden and extreme disappearance of the honey bee, *Apis mellifera*, due to colony collapse disorder (CCD). The disappearance of the honey bee would have catastrophic financial outcomes, since it is the most economically valuable pollinator worldwide. About 90 percent of commercially grown field crops, citrus and other fruit crops, vegetables, and nut crops currently depend on honey bee pollination services. These crops in the United States are valued at $15–20 billion and include Pennsylvania's apple harvest, which is the fourth largest in the nation with an estimated worth of about $45 million per year and California's almond harvest, which accounts for 80 percent of the world's market share of almonds.

The honey bee most common in the United States is native to Europe, and many of the crops pollinated by honey bees today such as watermelon, almonds, blackberries, and raspberries, among others, could be pollinated by bees native to the landscape (often more efficiently), allowing an ecosystem to generate these important services rather than having to import honey bees from elsewhere. Bees generally require food (flower pollen) and habitat (native vegetation) to survive in a landscape. In addition, bees do not fly great distances, so they require patches of native habitat they can easily fly between. The good news is that studies have shown that retaining even small patches of native habitat provides native bees a home and best promotes pollination eco-system services. If, however, we can ship bees in to provide pollination, why should farmers or the general public care if we eliminate native landscapes and their native pollinators?

One good reason to care is that using nature to provide services can reduce costs for farmers, since wild bee pollinators can be as efficient as man-aged bees. Farmers wouldn't have to pay the costs of transporting and caring for honey bees to successfully produce their crops. In a coffee plantation in Costa Rica, for example, Ricketts et al. found that services from wild polli-nators living in native rainforest patches embedded in the coffee farm were worth $60,000/year, and this is just one service provided by the forest. Other services such as carbon sequestration, soil stabilization, flood mitigation, and water purification could also have added value for the farmer (soil stabiliza-tion), for other people in the region (water purification and flood mitigation), and for people around the world (carbon sequestration). Loss of this rainfor-est, thus, has equivalent costs to society. Yet, right now, despite the plethora of studies that demonstrate which farming practices can greatly impact native bee populations, either encouraging their survival or leading to their demise, many coffee farms depend on the honey bee and not wild pollinators for crop production, making them very vulnerable to extreme consequences of honey bee decline. Why?

The use of native pollinators for crop pollination is not ubiquitous, because this ecosystem service does not come without tradeoffs. Keeping patches of native vegetation on an agricultural landscape means less space for crop production and potentially a reduction in yield. Changing farming practices might mean changing machinery or learning new techniques, which can take time and might have other associated costs. As with any change in practice, there is a risk, and just as consumers have an incentive to buy the least costly vegetables in the market so they can save more money, farmers have an incentive to produce crops as cheaply as possible. Changes in practices sometimes include large, upfront costs that are not offset by benefits or not offset immediately enough, even if in the long run production costs are actually less.

As alluded to, there are tradeoffs associated with destroying native ecosystems and therefore eliminating pollination services. What if more crises occur such as the CCD outbreak leading to further loss of honey bees, and suddenly the cost of importing and keeping these bees becomes unsustainable? Keeping patches of natural ecosystems on the landscape can support a diversity of bee species, making pollinators less vulnerable to being completely annihilated by one disease. What if it would be cheaper to restore some cropland now to ensure some native pollinators are retained, since studies show native bee communities might be an insurance policy in the event of honey bee decline? Who should have to pay for these costs? Should consumers pay $.50 more for a cup of coffee, for example, to help offset potential upfront costs to farmers for using ecosystem pollination services?

According to the soon to be released United Nations TEEB report, conserving nature and ecosystem services might be 10–100 times more valuable than the cost of saving the habitats and species associated with the provision of the services; this demonstrates the potentially major impact of including nature's values. As consumers, it might cost us more at first to buy coffee pollinated by native bees, but in the long run it might be much less costly. This matters for policy, too. For example, should the U.S. farm bill provide financial incentives for farmers to continue growing crops as we do traditionally, or should it subsidize the same crops but provide additional incentives to farmers to restore patches of native habitat to secure native pollination services in appropriate cases? These types of questions not only get at the complexities associated with ecosystem services and how to extract their exact value (how do you know how much extra to charge on the cup of coffee?), but also underscore the importance of broadening our understanding of all the tradeoffs associated with our everyday choices.

How Might Ecosystem Services Save Your Life or Affect the House You Buy?

As with the food we buy in the market, the houses and property we buy may be more or less valuable depending on the impact we have on nature. Nature can provide services that help to mitigate or at least diminish some potentially catastrophic impacts from weather events. Perhaps the most recent and often discussed example of this is the value that mangrove ecosystems have in protecting against coastal flooding and storms. These protection services can enhance or detract from the value of coastal property, and in the case of severe storms like tsunamis, can help save people's lives.

Mangroves are coastal forest systems and make up about 0.4 percent of the world's forests. They are among the most endangered ecosystems on the planet, yet they are frequently cleared so people can make use of the space they occupy for rice paddies, shrimp farms, or other productive activities. Mangroves provide numerous ecosystem services to people, including nurseries for the young of about 30 percent of commercial fishes, coastal protection to prevent erosion and loss of coastal lands, carbon sequestration that helps to reduce the concentration of carbon dioxide in the atmosphere, waste processing, food production, recreation, and protection against large storm surges. For example, a recent study demonstrated that larger mangrove ecosystems led to significantly greater fish catches without having to increase fishing efforts.

Perhaps one of the most significant, yet undervalued, services from mangroves is their ability to help reduce damage caused by tsunamis and tropical storms. Das and Vincent demonstrated that mangrove forests can save lives in the context of large storms like the tsunami of 1999. Even the minimal coverage of mangroves around coastal villages in India reduced the death toll by one-third. Das and Vincent took the study one step further and estimated that the mangroves could have sold for about $67 million (total of 44,000 acres), but their value in terms of life saving services was at least $80 million. Prior to the tsunami, these life saving services may not have been valued, and the consequences could have been grave.

Coastal protection is another important ecosystem service provided by mangroves, which can have important implications for a more basic choice: whether or not to buy a house. Property values are assessed based on a wide range of factors including taxes, location, views, noise pollution, etc., but without factoring in ecosystem services, or the lack thereof, one might pay an inappropriate amount for a house. For example, coastal properties are often the most costly because they have lovely views. If the coastal areas have coral reefs or mangroves, this is no less true; however, through destructive fishing practices and/or overuse, many of these reefs and forests are being destroyed. What once may have been prime real estate right on the water is now a home that is threatened with destruction at any time since it can be easily flooded by a strong storm surge or the ground below it could erode away. The loss of protective ecosystem services from the natural ecosystems may actually mean the coastal home you just bought was vastly overpriced.

Again, the choices are not so black and white. Denying poor coastal communities' access to mangroves for income purposes has costs, just as preserving the mangroves has benefits for human lives. Tradeoffs must be evaluated, and the range of costs and benefits must be included, but without accounting for ecosystem services values, you can't be sure if you should buy the coastal home or if you should cut down the mangroves to be able to sell more fish.

More broadly, governments can't assess what policies might be needed to secure the well-being of their citizens. For example, should the Philippine or Indian governments regulate mangrove cutting more stringently while channeling funds into providing incentives to poor coastal dwellers to compensate them for lost income and further decrease the threat to mangrove systems?

What Do Ecosystem Services Mean for the Water You Drink?

People rely on clean, regular supplies of water for survival, whether for drinking or for the production of other goods and services (agriculture, electricity, etc.). Water users have an incentive to find the lowest cost options for accessing clean water. Interestingly, nature may provide the lowest-cost, longest-term means of providing such water services. Conservationists are increasingly recognizing the value in thinking about these low-cost approaches for financing conservation. Demonstrating the links between nature and people is a way to engage new stakeholders in conservation and potentially find new ways to finance activities.

Increasing numbers of case studies and research studies are emerging about the benefits of payment for ecosystem service approaches as a means to finance conservation. While there is still uncertainty about what factors are likely to contribute to successful ecosystem service projects, these approaches continue to proliferate. Payments for watershed service projects make up a significant portion of implemented ecosystem services schemes (many others relate to carbon). These schemes often involve water users paying "suppliers" for the delivery of clean, consistent water supplies. Using a payment for watershed services approach, called "water funds," developed by The Nature Conservancy (TNC) and several partners, I will provide a tangible example of an ecosystem services approach that has changed the way users are securing access to water.

Water funds are a public/private partnership focused around a long-term, sustainable finance source for conservation. The partnership determines how to fund conservation of the watershed in order to protect valuable biodiversity and to generate vital water services (namely a clean, regular supply of water) upon which large groups of downstream people depend. For context, water funds are proliferating throughout the Andean region of South America, particularly in Colombia and Ecuador though they are not only in this region. The headwaters of important rivers originate in higher altitude natural ecosystems (composed of native grasslands, páramo, and mixed forest) that serve as the hydrologic regulators for the entire water system.

The problems in these Andean systems are three-fold: growing populations of downstream users require increased flows of services, the natural ecosystems that provide the services are not sufficiently protected, and the human communities that threaten the natural ecosystems are poor and depend upon these ecosystems for their livelihoods. In addition, their land use practices can have their own consequences for service provision, as farming and ranching can lead to reduced water retention and increased water pollution. But, water services can't be preserved solely by keeping people out of the natural areas and restoring the working landscapes, as this would compromise many liveli-hoods. Using an ecosystem services framework, TNC and partners used the dependence of downstream people on the services provided by natural ecosys-tems and restored working landscapes to finance conservation and livelihood projects to secure water services sustainably.

In a water fund, water users voluntarily invest money in a trust fund, and the revenue (interest and sometimes part of the principal) from it is used to finance conservation projects in the watershed, which are decided upon by the public/private partnership of users and key stakeholders that oversee the fund. These projects take steps to address the needs of preserving natural ecosystems and maintaining the well-being of watershed communities. Activi-ties and projects can include hiring community-based park guards to maintain the natural areas (to maintain the natural hydrologic regulation of the system, which helps maintain a regular base flow), protecting riparian areas (putting fences up to keep crops and cows away from river banks), revegetating riparian areas (to provide a natural filter for sediments and other pollutants), plant-ing live tree fences to delineate property boundaries, and isolating/fencing off headwaters and steep slopes. These practices can have major impacts on water quality, on the timing and volume of water flows (particularly floods), on fires, and on freshwater biodiversity. One study demonstrated that just maintaining natural vegetation on the landscape can decrease sedimentation tenfold com-pared with converting the area to cropland.

Such management is not without costs, however, and thus the water fund not only finances conservation management projects but also sup-ports community projects to compensate for impacts on livelihoods. Ideally, conservation management activities will enhance farm/ranch productivity through the production of on-farm ecosystem services such as soil stabili-zation and enhanced soil fertility, but these benefits will not be immedi-ate and are not guaranteed. In the shorter term, conservation management agreements include livelihood investments such as environmental educa-tion programs, additional income sources such as guinea pig farms, alterna-tive food sources such as organic vegetable gardens, and expanded capacity for the production of goods such as providing communities with ovens to make the drying of fruit and herbs they sell on the market more efficient and effective.

In this case, taking an ecosystem services approach meant being holistic and recognizing that the downstream users of water have an incentive to find the lowest cost option for continued access to clean water and that nature can provide that service potentially more cheaply and for longer than built infra-structure. The premise of water funds, therefore, is that securing natural eco-systems and improving management of farms and ranches in the watershed will help to ensure users have clean water available to them year round, and in return for these services, users pay for the upstream conservation manage-ment. This approach has had tremendous replication success in this region of South America, with the implementation of seven water funds in the last decade serving cities with combined populations of over 11 million people and helping protect over 1.6 million hectares of land.

How Are Ecosystem Service Approaches Being Leveraged?

The last five years have seen the proliferation of ecosystem services strategies, not just in on-the-ground actions but also in the emergence of new offices, new projects, and new strategies within conservation NGOs, governments, and multilateral donor agencies. This increased attention started with books such as *The New Economy of Nature* and interdisciplinary scholarly investigations such as the *Millennium Ecosystem Assessment* (MA) demonstrating the ecosystem alternatives to resource problems. The MA was called for by the United Nations secretary general in 2000 as a way to assess the impact of ecosystem change on human well-being including a scientific assessment of how to increase the conservation and sustainable use of these ecosystems to secure well-being. Over 1,360 scholars worldwide collaborated on the study publishing the findings in five technical volumes and six synthesis reports (see http://www.millenniumassessment.org/en/index.aspx for more information).

The concept has now been integrated into funding criteria by donor agencies. The World Bank, for example, has an entire strategy dedicated towards developing payment for environmental services (see http://go.worldbank.org/51KUO12O50). In addition, one of the major donor programs of the World Bank, the Global Environment Facility (GEF), has developed a Scientific and Technical Advisory Panel (STAP) to provide guidance on the evaluation of environmental service projects that are seeking funding, since these types of projects are becoming increasingly popular for achieving development and conservation objectives (see http://stapgef.unep.org/resources/sg/PES). This attention from the development community is a clear link between ongoing conservation efforts and efforts to help enhance global development.

Beyond multilaterals, governments are also paying more attention to the benefits of nature to people. This attention may be best illustrated by the relatively recent formation of an entire office in the U.S. government's Department of Agriculture that is dedicated to catalyzing new markets for ecosystem services (see http://www.fs.fed.us/ecosystemservices/OEM/index.shtml). This Office of Environmental Markets was formed in 2008 and aims to help support farm bill programs in the U.S., which can have significant benefits for service provision throughout the country.

There is also a growing and rapidly evolving focus on ecosystem services by conservation nonprofits. Organizations such as Conservation International, TNC, and WWF are changing their missions and/or developing projects that focus on integrating people and nature. New partnerships, notably, the Natural Capital Project—a partnership between Stanford University, TNC, and WWF—are being created to provide tools (see http://www.naturalcapitalproject.org/InVEST.html) to map the flow of various services in geographies around the world and to help more effectively include natural capital in decision-making.

Finally, there is recognition across government, particularly recently by the German Federal Ministry and the European Commission, with other partners, of the importance of making sure we include the true costs of lost biodiversity and associated ecosystem services on our future. In an effort to fully understand the benefits nature has for people's well-being, this group commissioned the TEEB report to sharpen awareness, help facilitate creation of cost-effective policies, and help make better informed decisions.

Over the next decade, measuring the impact of ecosystem service–based approaches both on people's well-being and also on nature is critical. Understanding which activities and actions yield a particular outcome and at what cost will provide us with a more evidence-based suite of options for making informed choices in our daily lives.

**Marino Gatto and
Giulio A. De Leo**

 **NO**

Pricing Biodiversity and Ecosystem Services: The Never-Ending Story

In 1844, the French engineer Jules Juvénal Dupuit introduced cost–benefit analysis to evaluate investment projects. . . . The application of cost–benefit analysis to ecological issues fell out of favor three decades ago, and it was gradually replaced by multicriteria analysis in the decision-making process for projects that have an impact on the environment. Although multicriteria analysis is currently used for environmental impact assessments [EIA] in many nations, [recently] the concept of cost–benefit analysis has again become fashionable, along with the various pricing techniques associated with it, such as contingent valuation methods, hedonic prices, and costs of replacement of ecological services. . . . Economists have generated a wealth of virtuosic variations on the theme of assessing the societal value of biodiversity, but most of these techniques are invariably based on price—that is, on a single scale of values, that of goods currently traded on world markets.

Perhaps the most famous recent study on the issue of pricing biodiversity and ecological services is that by Costanza et al., who argued that if the importance of nature's free benefits could be adequately quantified in economic terms, then policy decisions would better reflect the value of ecosystem services and natural capital. Drawing on earlier studies aimed at estimating the value of a wide variety of ecosystem goods and services, Costanza et al. estimated the current economic value of the entire biosphere at $16–54 trillion per year, with an average value of approximately $33 trillion per year. By contrast, the gross national product of the United States totals approximately $18 trillion per year. The paper, as its authors intended, stimulated much discussion, media attention, and debate. A special issue of *Ecological Economics* (April 1998) was devoted to commentaries on the paper, which, with few exceptions, were laudatory. Some economists have questioned the actual numbers, but many scientists have praised the attempt to value biodiversity and ecosystem functions.

Although Costanza et al. acknowledged that their estimates were crude and imperfect, they also pointed the way to improved assessments. In particular, they noted the need to develop comprehensive ecological economic models that could adequately incorporate the complex interdependencies between ecosystems and economic systems, as well as the complex individual dynamics

From *BioScience*, vol. 50, no. 4, April 2000, by Marino Gatto and Giulio A. De Leo, pp. 347–354. Copyright © 2000 by American Institute of Biological Sciences. Reprinted by permission of University of California Press via Rightslink.

of both types of systems. Despite the authors' caveats and the fact that many economists have been circumspect in applying their own tools to decisions regarding natural systems, the monetary approach is perceived by scientists, policymakers, and the general public as extremely appealing; a number of biologists are also of the opinion that attaching economic values to ecological services is of paramount importance for preserving the biosphere and for effective decision-making in all cases where the environment is concerned.

In this article, we espouse a contrary view, stressing that, for most of the values that humans attach to biodiversity and ecosystem services, the pricing approach is inadequate—if not misleading and obsolete—because it implies erroneously that complex decisions with important environmental impacts can be based on a single scale of values. We contend that the use of cost–benefit analysis as the exclusive tool for decision-making about environmental policy represents a setback relative to the existing legislation of the United States, Canada, the European Union, and Australia on environmental impact assessment, which explicitly incorporates multiple criteria (technical, economic, environmental, and social) in the process of evaluating different alternatives. We show that there are sound methodologies, mainly developed in business and administration schools by regional economists and by urban planners, that can assist decision-makers in evaluating projects and drafting policies while accounting for the nonmarket values of environmental services.

The Limitations of Cost–Benefit Analysis and Contingent Valuation Methods

Historically, the first important implementation of cost–benefit analysis at the political level came in 1936, with passage of the US Flood Control Act. This legislation stated that a public project can be given a green light if the benefits, to whomsoever they accrue, are in excess of estimated costs. This concept implies that all benefits and costs are to be considered, not just actual cash flows from and to government coffers. However, public agencies (e.g., the US Army Corps of Engineers) quickly ran into a problem: They were not able to give a monetary value to many environmental effects, even those that were predictable in quantitative terms. For instance, engineers could calculate the reduction of downstream water flow resulting from construction of a dam, and biologists could predict the river species most likely to become extinct as a consequence of this flow reduction. However, public agencies were not able to calculate the cost of each lost species. Therefore, many ingenious techniques for the monetary valuation of environmental goods and services have been devised since the 1940s. These techniques fall into four basic categories.

- **Conventional market approaches.** These approaches, such as the replacement cost technique, use market prices for the environmental service that is affected. For example, degradation of vegetation in developing countries leads to a decrease in available fuelwood. Consequently, animal dung has to be used as a fuel instead of a fertilizer, and farmers must therefore replace dung with chemical fertilizers. By computing the cost of these chemical fertilizers, a monetary value for the degradation of vegetation can then be calculated.

- **Household production functions.** These approaches, such as the travel cost method, use expenditures on commodities that are substitutes or complements for the environmental service that is affected. The travel cost method was first proposed in 1947 by the economist Harold Hotelling, who, in a letter to the director of the US National Park Service, suggested that the actual traveling costs incurred by visitors could be used to develop a measure of the recreation value of the sites visited.

- **Hedonic pricing.** This form of pricing occurs when a price is imputed for an environmental good by examining the effect that its presence has on a relevant market-priced good. For instance, the cost of air and noise pollution is reflected in the price of plots of land that are characterized by different levels of pollution, because people are willing to pay more to build their houses in places with good air quality and little noise. . . .

- **Experimental methods.** These methods include contingent valuation methods, which were devised by the resource economist Siegfried V. Ciriacy-Wantrup. Contingent valuation methods require that individuals express their preferences for some environmental resources by answering questions about hypothetical choices. In particular, respondents to a contingent valuation methods questionnaire will be asked how much they would be willing to pay to ensure a welfare gain from a change in the provision of a nonmarket environmental commodity, or how much they would be willing to accept in compensation to endure a welfare loss from a reduced provision of the commodity.

Among these pricing techniques, the contingent valuation methods approach is the only one that is capable of providing an estimate of existence values, in which biologists have a special interest. Existence value was first defined by Krutilla as the value that individuals may attach to the mere knowledge that rare and diverse species, unique natural environments, or other "goods" exist, even if these individuals do not contemplate ever making active use of or benefiting in a more direct way from them. The name "contingent valuation" comes from the fact that the procedure is contingent on a constructed or simulated market, in which people are asked to manifest, through questionnaires and interviews, their demand function for a certain environmental good (i.e., the price they would pay for one extra unit of the good versus the availability of the good). . . .

The limits of cost–benefit analysis were discussed in the 1960s, after more than two decades of experimentation. In particular, many authors pointed out that cost–benefit analysis encouraged policymakers to focus on things that can be measured and quantified, especially in cash terms, and to disregard problems that are too large to be assessed easily. Therefore, the associated price might not reflect the "true" value of social equity, environmental services, natural capital, or human health. In particular, economists themselves recognize that the increasingly popular contingent valuation methods are undermined by several conceptual problems, such as free-riding, overbidding, and preference reversal.

When it comes to monetary valuation of the goods and services provided by natural ecosystems and landscapes specifically, a number of additional problems undermine the effectiveness of pricing techniques and cost–benefit analysis. These problems include the very definition of "existence" value, the dependence of pricing techniques on the composition of the reference group, and the significance of the simulated market used in contingent valuation.

The definition of "existence" value A classic example of contingent valuation methods is to ask for the amount of money individuals are willing to pay to ensure the continued existence of a species such as the blue whale. However, the existence value of whales does not take into account potential indirect services and benefits provided by these mammals. It is just the value of the existence of whales for humans, that is, the satisfaction that the existence of blue whales provides to people who want them to continue to exist. Therefore, there is a real risk that species with very low or no aesthetic appeal or whose biological role has not been properly advertised will be given a low value, even if they play a fundamental ecological function. Without adequate information, most people do not understand the extent, importance, and gravity of most environmental problems. As a consequence, people may react emotionally and either underestimate or overestimate risks and effects.

Therefore, it is not surprising that five of the seven guidelines issued by the National Oceanic and Atmospheric Administration [NOAA] about how to conduct contingent valuation discuss how to properly inform and question respondents to produce reliable estimates (e.g., in-person interviews are preferred to telephone surveys to elicit values). Of course, acquisition of reliable and complete information is always possible in theory, but in practice strict adherence to NOAA guidelines makes contingent valuation methods expensive and time consuming.

Difficulties with the reference group for pricing Pricing techniques such as contingent valuation methods provide information about individual willingness to pay or willingness to accept, which must be summed up in the final balance of cost–benefit analysis. Therefore, the outcome of cost–benefit analysis depends strongly on the group of people that is taken as a reference for valuation—particularly on their income. Van der Straaten noted that the Exxon *Valdez* oil spill in 1989 provides a good example of this dependence. The population of the United States was used as a reference group to calculate the damage to the existence value of the affected species and ecosystems using contingent valuation methods. Exxon was ultimately ordered to pay $5 billion to compensate the people of Alaska for their losses. This huge figure was a consequence of the high income of the US population. If the same accident had occurred in Siberia, where salaries are lower, the outcome would certainly have been different.

This example shows that contingent valuation methods simply provide information about the preferences of a particular group of people but do not necessarily reflect the ecological importance of ecosystem goods and services. Moreover, the outcome of cost–benefit analysis depends on which individual willingness to pay or willingness to accept are included in the cost–benefit analysis. If the quality of the Mississippi River is at issue, should the analysis be restricted to US citizens living close to the river, or should the willingness to pay of Californians and New Yorkers be included too? According to Krutilla's definition of existence value, for many environmental goods and ecological services that may ultimately affect ecosystem integrity at the global level, the preferences of the entire human population should potentially be considered in the analysis. Because practical reasons obviously preclude doing so, contingent valuation methods will inevitably only provide information about the preferences of specific groups of people. For many of the ecological services that may be considered the heritage of humanity, contingent valuation methods analyses performed locally in a particular economic situation should be extrapolated only with great caution to other areas. The process of placing a monetary value on biodiversity and ecosystem functioning through nonuser willingness to pay is performed in the same way as for user willingness to pay, but the identification of people who do not use an environmental good directly and still have a legitimate interest in its preservation is problematic.

Significance of the simulated market   Contingent valuation methods are contingent on a market that is constructed or simulated, not real. It is difficult to believe in the efficiency of what Adam Smith called the "invisible hand" of the market for a process that is the artificial production of economic advisors and does not possess the dynamic feedback that characterizes real competitive markets. Is it even possible to simulate a market where units of biodiversity are bought and sold? As Friend stated, "these contingency evaluation methods (CVM) tend to create an illusion of choice based on psychology (willingness) and ideology (the need to pay) which is supposed, somewhat mysteriously, to reflect an equilibrium between the consumer demand for and producer supply of environmental goods and services."

Many additional criticisms of pricing ecological services are more familiar to biologists. For many ecological services, there is simply no possibility of technological substitution. Moreover, the precise contribution of many species is not known, and it may not be known until the species is close to extinction. . . . In addition, specific ecosystem services, as evaluated by Costanza et al., should not be separated from one another and valued individually because the importance of any piece of biodiversity cannot be determined without considering the value of biodiversity in the aggregate. And finally, the use of marginal value theory may be invalidated by the erratic and catastrophic behavior of many ecological systems, resulting in potentially detrimental effects on the health of humans, the productivity of renewable resources, and the vitality and stability of societies themselves.

Despite the efforts of many economists, we believe that some goods and services, especially those related to ecosystems, cannot reasonably be given a monetary value, although they are of great value to humans. Economists coined the term "intangibles" to define these goods. Cost–benefit analysis cannot easily deal with intangibles. As Nijkamp wrote, more than 20 years ago, "the only reasonable way to take account of intangibles in the traditional cost–benefit analysis seems to be the use of a balance with a debit and a credit side in which all intangible project effects (both positive and negative) are represented in their own (qualitative or quantitative) dimensions" as secondary information. In other words, the result of cost–benefit analysis is primarily a single number, the net monetary benefit that comprises all the effects that can be sensibly converted into monetary returns and costs.

Commensurability of Different Objectives and Multicriteria Analysis

Cost–benefit analysis includes intangibles in the decision-making process only as ancillary information, with the main focus being on those effects that can be converted to monetary value. This approach is not a balanced solution to the problem of making political decisions that are acceptable to a wide number of social groups with a range of legitimate interests. . . .

However, even if the attempt to put a price on everything is abandoned, it is not necessary to give up the attempt to reconcile economic issues with social and environmental ones. Social scientists long ago developed multicriteria techniques to reach a decision in the face of multiple different and structurally incommensurable goals. The most important concept in multicriteria analysis was actually conceived by an Italian economist, Vilfredo Pareto, at the end of the nineteenth century. It is best explained by a simple example. Suppose that a natural area hosting several rare species is a target for the development of a mining activity. Alternative mining projects can have different effects in terms of profits from mining (measured in dollars) and in terms of sustained biodiversity (measured in suitable units, for instance, through the Shannon index). Profit from mining can be corrected using welfare economics to include those environmental and social effects that can be priced (e.g., the benefit of providing jobs to otherwise unemployed people, the cost of treating lung disease of miners, and the cost of the loss of the tourists who used to visit the natural area). . . .

The methods of multicriteria analysis are intended to assist the decision-maker in choosing among . . . alternatives . . . (a task that is particularly difficult when there are several incommensurable objectives, not just two). Nevertheless, the initial step of determining [these] alternatives is of enormous importance, for three reasons. First, [doing so] makes perfect sense even if there is no way of pricing a certain environmental good because each objective can be expressed in its own proper units without reduction to a common scale. Second, the determination of all the feasible alternatives . . . requires the joint effort of a multidisciplinary team that includes, for example, economists, engineers, and biologists and that must predict the effects of alternative decisions on all of the different environmental and social components to which humans are sensitive and which, therefore, deserve consideration. Third, the determination of [feasible alternatives] allows the objective elimination of inadequate alternatives because [they are] independent of the subjective perception of welfare . . . [and] in essence describe the tradeoff between the various incommensurable objectives when every effort is made to achieve the best results in all respects; the attention of the authority that must make the final decision is thus directed toward genuine potential solutions because nonoptimal decisions have already been discarded.

It should be noted that a cost–benefit analysis does not elicit trade-offs between incommensurable goods because it also gives a green light to projects . . . , provided that the benefits that can be converted into a monetary scale exceed the costs. . . . Cost–benefit analysis, however, is not useful for elic-iting the tradeoffs between two incommensurable goods, neither of which is monetary. For instance, there might be a conflict between the goals of preserv-ing wildlife within a populated area and minimizing the risk that wild animals are vectors of dangerous diseases. A multicriteria analysis can describe this tradeoff, whereas a cost–benefit analysis cannot.

Another philosophical point concerning the issue of commensurability is the question of implicit pricing. Economists often argue that to make a deci-sion is to put an implicit price on such intangibles as human life or aesthetics and, therefore, to reduce their value to a common scale (as pointed out also by Costanza et al.). . . .

Environmental Impact Assessment and Multiattribute Decision-Making

Because of the flaws of cost–benefit analysis, many countries have taken a dif-ferent approach to decision-making through the use of environmental impact assessment legislation (e.g., the United States in 1970, with the signing of the National Environmental Policy Act, NEPA; France in 1976, with the act 76/629; the European Union in 1985, with the directive 85/337). Environ-mental impact assessment procedures, if properly carried out, represent a wiser approach than setting an a priori value of biodiversity and ecosystem services because these procedures explicitly recognize that each situation, and every regulatory decision, responds to different ethical, economic, political, histori-cal, and other conditions and that the final decision must be reached by giving appropriate consideration to several different objectives. As Canter noted, all projects, plans, and policies that are expected to have a significant environ-mental impact would ideally be subject to environmental impact assessment.

The breadth of goals embraced by environmental impact assessment is much wider than that of cost–benefit analysis. Environmental impact assess-ment provides a conceptual framework and formal procedures for comparing different alternatives to a proposed project (including the possibilities of not developing a site, employing different management rules, or using mitiga-tion measures); for fostering interdisciplinary team formation to investigate all possible environmental, social, and economic consequences of a proposed activity; for enhancing administrative review procedures and coordination among the agencies involved in the process; for producing the necessary documentation to enhance transparency in the decision-making process and the possibility of reviewing all the objective and subjective steps that resulted in a given conclusion; for encouraging broad public participation and the input of different interest groups; and for including monitoring and feedback procedures. Classical multiattribute analysis can be used to rank different alternatives. . . . Ranking usually requires the use of value functions to transform environmental and other indicators (e.g., biological oxygen demand or animal density) to levels of satisfaction on a normalized scale, and the weighting of factors to combine value functions and to rank the alternatives. These weights explicitly reflect the relative importance of the different environmental, social, and economic compartments and indicators.

A wide range of software packages for decision support can assist experts in organizing the collected information; in documenting the various phases of EIA; in guiding the assignment of importance weights; in scaling, rating, and ranking alternatives; and in conducting sensitivity analysis for the overall decision-making process. This last step, of testing the robustness and consistency of multiattribute analysis results, is especially important because it shows how sensitive the final ranking is to small or large changes in the set of weights and value functions, which often reflect different and subjective perspectives. It is important to stress that, although the majority of environmental impact assessments have been conducted on specific projects, such as road construction or the location of chemical plants, there is no conceptual barrier to extending the procedure to evaluation of plans, programs, policies, and regulations. In fact, according to NEPA, the procedure is mandatory for any federal action with an important impact on the environment. The extension of environmental impact assessment to a level higher than a single project is termed "strategic environmental assessment" and has received considerable attention.

Conclusions

An impressive literature is available on environmental impact assessment and multiattribute analysis that documents the experience gained through 30 years of study and application. Nevertheless, these studies seem to be confined to the area of urban planning and are almost completely ignored by present-day economists as well as by many ecologists. Somewhere between the assignment of a zero value to biodiversity (the old-fashioned but still used practice, in which environmental impacts are viewed as externalities to be discarded from the balance sheet) and the assignment of an infinite value (as advocated by some radical environmentalists), lie more sensible methods to assign value to biodiversity than the price tag techniques suggested by the new wave of environmental economists. Rather than collapsing every measure of social and environmental value onto a monetary axis, environmental impact assessment and multiattribute analysis allow for explicit consideration of intangible nonmonetary values along with classical economic assessment, which, of course, remains important. It is, in fact, possible to assess ecosystem values and the ecological impact of human activity without using prices. Concepts such as Odum's eMergy [the available energy of one kind previously required to be used up directly and indirectly to make the product or service] and Rees' ecological footprint [the area of land and water required to support a defined economy or population at a specified standard of living], although perceived by some as naive, may aid both ecologists and economists in addressing this important need.

To summarize our viewpoint, economists should recognize that cost–benefit analysis is only part of the decision-making process and that it lies at the same level as other considerations. Ecologists should accept that monetary valuation of biodiversity and ecosystem services is possible (and even helpful) for part of its value, typically its use value. We contend that the realistic substitute for markets, when they fail, is a transparent decision-making process, not old-style cost–benefit analysis. The idea that, if one could get the price right, the best and most effective decisions at both the individual and public levels would automatically follow is, for many scientists, a sort of Panglossian obsession. In reality, there is no simple solution to complex problems. We fear that putting an a priori monetary value on biodiversity and ecosystem services will prevent humans from valuing the environment other than as a commodity to be exploited, thus reinvigoraing the old economic paradigm that assumes a perfect substitution between natural and human-made capital. As Rees wrote, "for all its theoretical attractiveness, ascribing money values to nature's services is only a partial solution to the present dilemma and, if relied on exclusively, may actually be counterproductive."

EXPLORING THE ISSUE

Do Ecosystem Services Have Economic Value?

Critical Thinking and Reflection

1. In what ways are natural ecosystems valuable?
2. Is it right to destroy natural ecosystems in order to generate money?
3. What is the best way to convince people to protect nature?
4. What is the best way to protect nature?

Is There Common Ground?

No one disagrees that nature provides benefits to human life. The disagreement comes in when discussing whether we have a right to use nature in any way we like, even to the point of destroying it. Indeed the idea that we have such a right has a long history, for nature was long considered worthless wilderness or wasteland, gaining value only when it had been "improved" (built on, farmed, mined, logged, etc.) by human effort. In recent decades, the idea that nature has intrinsic value has been gaining ground, but most people agree that "intrinsic value" is not easily used to convince people to protect nature. Value in terms of benefits to humans works better.

1. Research "carbon offsets." In what way do they provide an example of payment for ecosystem services?
2. Do regulations that require mitigation of undesirable environmental impacts (such as building artificial wetlands to replace natural wetlands destroyed by construction projects) amount to another form of payment for ecosystem services? Can they motivate protection of the original, natural environment?
3. Are there situations where the ecosystem services approach to protection seems unlikely to work?

Internet References . . .

Arc Ecology Project: Military and the Environment

Arc Ecology is a nonprofit, public interest organization concerned with the ecology of humanity and its place in the global ecology. Among the issues it addresses is the environmental threat posed by military activities.

www.arcecology.org/Military.shtml

ECOLEX: A Gateway to Environmental Law

This site, sponsored by the United Nations and the World Conservation Union, is a comprehensive resource for environmental treaties, national legislation, and court decisions.

www.ecolex.org/index.php

Environmental Defense Fund

The Environmental Defense Fund is a leading national nonprofit organization that links science, economics, and law to create innovative, equitable, and cost-effective solutions to society's most urgent environmental problems.

www.edf.org/home.cfm

The Rewilding Institute

The Rewilding Institute's mission is to develop and promote the ideas and strategies to advance continental-scale conservation in North America, particularly the need for large carnivores and a permeable landscape for their movement, and to offer a bold, scientifically credible, practically achievable, and hopeful vision for the future of wild nature and human civilization in North America.

http://rewilding.org/rewildit/

SourceWatch

SourceWatch is a collaborative project of the Center for Media and Democracy. Its primary focus is on documenting the interconnections and agendas of public relations firms, think tanks, industry-funded organizations, and industry-friendly experts that work to influence public opinion and public policy on behalf of corporations, governments, and special interests.

www.sourcewatch.org

Principles versus Politics

In many environmental issues, it is easy to tell what basic principles apply and therefore determine what is the right thing to do. Ecology is clear on the value of species to ecosystem health and the harmful effects of removing or replacing species. Medicine makes no bones about the ill effects of pollution. But are the environmental problems so bad that we must act immediately? Must damaged ecosystems be restored, and to their status of how long ago? Should government agencies or businesses receive exemptions from environmental regulations? Should we go slow on environmental regulations for fear of damaging the economy? Such questions arise in connection with most environmental issues, not just the three listed below, but these three will serve to introduce the theme of principles versus politics.

- Should North America's Landscape Be Restored to Its Prehuman State?

- Should the Military Be Exempt from Environmental Regulations?

- Will Restricting Carbon Emissions Damage the Economy?

ISSUE 4

Should North America's Landscape Be Restored to Its Prehuman State?

YES: C. Josh Donlan, from "Restoring America's Big, Wild Animals," *Scientific American* (June 2007)

NO: Dustin R. Rubenstein, Daniel I. Rubenstein, Paul W. Sherman, and Thomas A. Gavin, from "Pleistocene Park: Does Re-Wilding North America Represent Sound Conservation for the 21st Century?" *Biological Conservation* (vol. 132, 2006)

Learning Outcomes

After reading this issue, you should be able to:

- Explain what rewilding is and how it relates to conservation.
- Explain why African elephants, lions, and other wildlife may be suitable replacements for North American mammoths, lions, and other extinct megafauna.
- Describe several obstacles to rewilding.
- Discuss whether Pleistocene rewilding is likely to work.

ISSUE SUMMARY

YES: C. Josh Donlan proposes that because the arrival of humans in the Americas some 13,000 years ago led to the extinction of numerous large animals (including camels, lions, and mammoths) with major effects on local ecosystems, restoring these animals (or their near-relatives from elsewhere in the world) holds the potential to restore health to these ecosystems. There would also be economic and cultural benefits.

NO: Dustin R. Rubenstein, Daniel I. Rubenstein, Paul W. Sherman, and Thomas A. Gavin argue that bringing African and Asian megafauna to North America is unlikely to restore prehuman ecosystem function and may threaten present species and ecosystems. It would be better to focus resources on restoring species where they were only recently extinguished.

$\mathbf{A}$s far as we can see into the mists of the past, human actions have affected the environment. Desertification, deforestation, erosion, and soil salinization were happening as soon as agriculture began and the first cities appeared. Paul S. Martin, Emeritus Professor of Geosciences at the Desert Laboratory of the University of Arizona, has long argued that even before that time—as far as 50,000 years ago—humans have been the chief cause of extinctions of large animals (see, e.g., "Prehistoric Overkill," in *Quaternary Extinctions: A Prehistoric Revolution,* by Martin and Klein, eds., University of Arizona Press, Tucson, 1984). The basic argument is that soon after human beings arrived in areas such as North America or Australia, the large game animals disappeared, either because humans killed (and presumably ate) them or because humans killed their prey. Critics, apparently unwilling to grant "primitive" people armed with stone-tipped spears and arrows enough potency to wipe out whole species, have argued that the disappearance of large animals was just coincidence, or due to diseases brought by humans and their domestic animals, or due to changes in climate. Since Martin first broached the "prehistoric overkill" hypothesis in the 1960s, the evidence in its favor has accumulated, but it has remained controversial. Less controversial is the idea that when the animals went extinct, the ecosystems in which they lived were affected. Large herbivores, for instance, were responsible for distributing the seeds of many trees. When the herbivores vanished, the fruit piled up under the trees and few seeds had the chance to sprout far enough from the parent tree to thrive. Yet such interactions of plants and herbivores can be restored, as has been demonstrated on Ile aux Aigrettes, an island near Mauritius, where living tortoises have been introduced to replace extinct ones; see Christine J. Griffiths, Dennis M. Hansen, Carl G. Jones, Nicolas Zuel, and Stephen Harris, "Resurrecting Extinct Interactions with Extant Substitutes," *Current Biology* (May 2011). However, whether such substitution can be successful depends on many factors, including climate; see Orien M. W. Richmond, Jay P. McEntee, Robert J. Hijmans, and Justin S. Brashares, "Is the Climate Right for Pleistocene Rewilding? Using Species Distribution Models to Extrapolate Climatic Suitability for Mammals across Continents," *PLoS One* (www.plosone.org) (September 2010).

The history of environmentalism has been marked by three major schools of thought. Conservationists want to see nature's resources protected for future human use. Preservationists want to see nature left alone. The third approach, epitomized by Aldo Leopold (*A Sand County Almanac,* Oxford University Press, 1949, 2001), might be called the "reparationist" school; its theme is that where humans have done damage, they have a responsibility to repair it. Its representatives—restoration ecologists—plant dune grasses to fight desertification, plant trees to restore forests, and advocate for restricting grazing animals to help overgrazed land recover. They also argue for removing dams or increasing water releases to restore rivers to health. Their motives may ultimately be either preservationist (repair and leave alone) or conservationist (repair for future human benefit).

Species that vanished thousands of years ago can hardly be preserved or conserved. Can they be "repaired"? They can't be brought back in their original form, of course, but many such species have relatives elsewhere in the world that might fill similar roles in ecosystems (such as seed distribution) if they were transplanted. Should they be transplanted? How far back does the human responsibility to repair human-caused damage to ecosystems extend? Some ecologists hold that even ancient damage should be repaired. C. Josh Donlan, Joel Berger, Carl E. Bock, Jane H. Bock, David A. Burney, James A. Estes, Dave Foreman, Paul S. Martin, Gary W. Roemer, Felisa A. Smith, Michael E. Soulé, and Harry W. Greene, "Pleistocene Rewilding: An Optimistic Agenda for Twenty-First Century Conservation," *American Naturalist* (November 2006), call for restoring North America's ecology to its prehuman condition (at least in some areas) by introducing equivalents to ancient herbivores and predators. It does not seem likely to happen, but the basic idea has been around awhile. One major presentation of the idea is Dave Foreman's *Rewilding North America* (Island Press, 2004), which calls for the construction of networks of connected protected areas, rather than isolated protected areas as is more common today; see also Caroline Fraser, *Rewilding the World: Dispatches from the Conservation Revolution* (Metropolitan Books, 2009). In essence, networks give threatened species more room, an important consideration especially for predators that may need large home ranges in order to be able to find sufficient food or suitable den sites. The basic idea was developed while Foreman was with the Wildlands Project, of which he was a cofounder in 1991. Today the Wildlands Project works to develop such networks using both public and private lands. Foreman left the project in 2003 to found the Rewilding Institute, which "looks at North America in shades of landscape permeability: the degree to which the land is open and safe for the movement of large carnivores and other wide-ranging and sensitive species between large core habitats." To enhance permeability, it works to create "megalinkages," large networks of protected areas that provide "habitat corridors through a hostile sea" of cities, suburbs, and farms. Foreman's status as a coauthor of the Donlan paper mentioned above suggests that his thinking is evolving from building networks and megalinkages to protect contemporary threatened wildlife to restoring wildlife that vanished long, long ago.

On a more local level, centered on New England, there is RESTORE: The North Woods (www.restore.org), whose aim is "America's first restored landscape: a place of vast recovered wilderness, of forests where the wolf and caribou roam free, of clear waters alive with salmon and trout, of people once again living in harmony with nature." RESTORE's efforts are directed toward creating a large national park in northern Maine and restoring "extirpated and imperiled wildlife, including the eastern timber wolf, Canada lynx, and Atlantic salmon." In the American southwest, the Turner Endangered Species Fund has initiated a project to restore the endangered Bolson tortoise to a portion of its late Pleistocene range; see Joe Truett and Mike Phillips, "Beyond Historic Baselines: Restoring Bolson Tortoises to Pleistocene Range," *Ecological Restoration* (June 2009). In a similar spirit, beavers are being reintroduced in Scotland; see Christopher Werth, "Unleash the Critters," *Newsweek* (April 20, 2009).

Peter Taylor, *Beyond Conservation: A Wildland Strategy* (Earthscan, 2005), calls rewilding "putting a new soul in the landscape," although his focus, like

RESTORE's, is on restoring only what has been lost in the last few centuries. See also J. C. Hallman, "Pleistocene Dreams: A Radical Conservation," *Science & Spirit* (May/June 2008). L. Martin, *Twilight of the Mammoths: Ice Age Extinctions and the Rewilding of America* (University of California Press, 2005), notes that if the Pleistocene overkill hypothesis is true, humans bear a moral responsibility to repair the damage they have caused. This puts him on Donlan's side of the debate, but that debate is hardly over. Indeed, whether it qualifies as conservation is not yet settled; see Howard M. Huynh, "Pleistocene Re-wilding Is Unsound Conservation Practice," *BioEssays* (February 2011).

In the following selections, C. Josh Donlan proposes that because the arrival of humans in the Americas some 13,000 years ago led to the extinction of numerous large animals (including camels, lions, and mammoths) with major effects on local ecosystems, restoring these animals (or their near-relatives from elsewhere in the world) holds the potential to restore health to these ecosystems. There would also be economic and cultural benefits. Dustin R. Rubenstein, Daniel I. Rubenstein, Paul W. Sherman, and Thomas A. Gavin argue that the ecosystems of 13,000 years ago have evolved, adjusting to the changes imposed at that time. Bringing African and Asian megafauna to North America is unlikely to restore prehuman ecosystem function and may threaten present species and ecosystems. It would be better to focus resources on preventing new extinctions and on restoring species where they were only recently extinguished.

In Eric Jaffe's "Brave Old World," *Science News* (November 11, 2006), Paul Sherman is quoted as saying that the aim of the paper by Rubenstein et al. is to stimulate further discussion. However, Rubenstein's group (including Sherman) published "Rewilding Rebuttal," *Scientific American* (October 2007), a few months after the Donlan paper, saying that if the debate is to go forward, Donlan et al. must "abandon sensationalism."

Is it "sensationalism" or "idealism" to speak of undoing damage done thousands of years ago? Whatever one calls it, it may not be very realistic. As conservationists must admit, fighting even to protect contemporary wildlife in isolated areas is difficult. Building networks and megalinkages is even more challenging. The reason is simply that modern civilization imposes a great many competing demands for space and resources. In a special section on "The Rise of Restoration Ecology," *Science* (July 31, 2009), Stephen T. Jackson and Richard J. Hobbs, "Ecological Restoration in the Light of Ecological History," note that a historical viewpoint is essential to understanding ecosystems, but "many historical restoration targets will be unsustainable in the coming decades."

YES

C. Josh Donlan

Restoring America's Big, Wild Animals

In the fall of 2004 a dozen conservation biologists gathered on a ranch in New Mexico to ponder a bold plan. The scientists, trained in a variety of disciplines, ranged from the grand old men of the field to those of us earlier in our careers. The idea we were mulling over was the reintroduction of large vertebrates—megafauna—to North America.

Most of these animals, such as mammoths and cheetahs, died out roughly 13,000 years ago, when humans from Eurasia began migrating to the continent. The theory—propounded 40 years ago by Paul Martin of the University of Arizona—is that overhunting by the new arrivals reduced the numbers of large vertebrates so severely that the populations could not recover. Called Pleistocene overkill, the concept was highly controversial at the time, but the general thesis that humans played a significant role is now widely accepted. Martin was present at the meeting in New Mexico, and his ideas on the loss of these animals, the ecological consequences, and what we should do about it formed the foundation of the proposal that emerged, which we dubbed Pleistocene rewilding.

Although the cheetahs, lions and mammoths that once roamed North America are extinct, the same species or close relatives have survived elsewhere, and our discussions focused on introducing these substitutes to North American ecosystems. We believe that these efforts hold the potential to partially restore important ecological processes, such as predation and browsing, to ecosystems where they have been absent for millennia. The substitutes would also bring economic and cultural benefits. Not surprisingly, the published proposal evoked strong reactions. Those reactions are welcome, because debate about the conservation issues that underlie Pleistocene rewilding merit thorough discussion.

Why Big Animals Are Important

Our approach concentrates on large animals because they exercise a disproportionate effect on the environment. For tens of millions of years, megafauna dominated the globe, strongly interacting and co-evolving with other species and influencing entire ecosystems. Horses, camels, lions, elephants and

From *Scientific American*, vol. 296, issue 6, June 2007, pp. 70–77. Copyright © 2007 by Scientific American, Inc. Reproduced with permission. All rights reserved. www.sciam.com

other large creatures were everywhere: megafauna were the norm. But starting roughly 50,000 years ago, the overwhelming majority went extinct. Today megafauna inhabit less than 10 percent of the globe.

Over the past decade, ecologist John Terborgh of Duke University has observed directly how critical large animals are to the health of ecosystems and how their loss adversely affects the natural world. When a hydroelectric dam flooded thousands of acres in Venezuela, Terborgh saw the water create dozens of islands—a fragmentation akin to the virtual islands created around the world as humans cut down trees, build shopping malls, and sprawl from urban centers. The islands in Venezuela were too small to support the creatures at the top of the food chain—predators such as jaguars, pumas and eagles. Their disappearance sparked a chain of reactions. Animals such as monkeys, leaf-cutter ants and other herbivores, whose populations were no longer kept in check by predation, thrived and subsequently destroyed vegetation—the ecosystems collapsed, with biodiversity being the ultimate loser.

Similar ecological disasters have occurred on other continents. Degraded ecosystems are not only bad for biodiversity; they are bad for human economies. In Central America, for instance, researchers have shown that intact tropical ecosystems are worth at least $60,000 a year to a single coffee farm because of the services they provide, such as the pollination of coffee crops.

Where large predators and herbivores still remain, they play pivotal roles. In Alaska, sea otters maintain kelp forest ecosystems by keeping herbivores that eat kelp, such as sea urchins, in check. In Africa, elephants are keystone players; as they move through an area, their knocking down trees and trampling create a habitat in which certain plants and animals can flourish. Lions and other predators control the populations of African herbivores, which in turn influence the distribution of plants and soil nutrients.

In Pleistocene America, large predators and herbivores played similar roles. Today most of that vital influence is absent. For example, the American cheetah (a relative of the African cheetah) dashed across the grasslands in pursuit of pronghorn antelopes for millions of years. These chases shaped the pronghorn's astounding speed and other biological aspects of one of the fastest animals alive. In the absence of the cheetah, the pronghorn appears "overbuilt" for its environment today.

Pleistocene rewilding is not about recreating exactly some past state. Rather it is about restoring the kinds of species interactions that sustain thriving ecosystems. Giant tortoises, horses, camels, cheetahs, elephants and lions: they were all here, and they helped to shape North American ecosystems. Either the same species or closely related species are available for introduction as proxies, and many are already in captivity in the U.S. In essence, Pleistocene rewilding would help change the underlying premise of conservation biology from limiting extinction to actively restoring natural processes.

At first, our proposal may seem outrageous—lions in Montana? But the plan deserves serious debate for several reasons. First, nowhere on Earth is pristine, at least in terms of being substantially free of human influence. Our demographics, chemicals, economics and politics pervade every part of the planet. Even in our largest national parks, species go extinct without active intervention. And human encroachment shows alarming signs of worsening. Bold actions, rather than business as usual, will be needed to reverse such negative influences. Second, since conservation biology emerged as a discipline more than three decades ago, it has been mainly a business of doom and gloom, a struggle merely to slow the loss of biodiversity. But conservation need not be only reactive. A proactive approach would include restoring natural processes, starting with ones we know are disproportionately important, such as those influenced by megafauna.

Third, land in North America is available for the reintroduction of megafauna. Although the patterns of human land use are always shifting, in some areas, such as parts of the Great Plains and the Southwest, large private and public lands with low or declining human population densities might be used for the project. Fourth, bringing megafauna back to America would also bring tourist and other dollars into nearby communities and enhance the public's appreciation of the natural world. More than 1.5 million people visit San Diego's Wild Animal Park every year to catch a glimpse of large mammals. Only a handful of U.S. national parks receive that many visitors. Last, the loss of some of the remaining species of megafauna in Africa and Asia within this century seems likely—Pleistocene rewilding could help reverse that.

How It Might Be Done

We are not talking about backing up a van and kicking some cheetahs out into your backyard. Nor are we talking about doing it tomorrow. We conceive of Pleistocene rewilding as a series of staged, carefully managed ecosystem manipulations. What we are offering here is a vision—not a blueprint—of how this might be accomplished. And by no means are we suggesting that rewilding should be a priority over current conservation programs in North America or Africa. Pleistocene rewilding could proceed alongside such conservation efforts, and it would likely generate conservation dollars from new funding sources, rather than competing for funds with existing conservation efforts.

The long-term vision includes a vast, securely fenced ecological history park, encompassing thousands of square miles, where horses, camels, elephants and large carnivores would roam. As happens now in Africa and regions surrounding some U.S. national parks, the ecological history park would not only attract ecotourists but would also provide jobs related both to park management and to tourism.

To get to that distant point, we would need to start modestly, with relatively small-scale experiments that assess the impacts of megafauna on North American landscapes. These controlled experiments, guided by sound science and by the fossil record, which indicates what animals actually lived here, could occur first on donated or purchased private lands and could begin immediately. They will be critical in answering the many questions about the reintroductions and would help lay out the costs and benefits of rewilding.

One of these experiments is already under way. Spurred by our 2004 meeting, biologists recently reintroduced Bolson tortoises to a private ranch in New Mexico. Bolson tortoises, some weighing more than 100 pounds, once grazed parts of the southwestern U.S. before disappearing around 10,000 years ago, victims of human hunting. This endangered tortoise now clings to survival, restricted to a single small area in central Mexico. Thus, the reintroduction not only repatriates the tortoise to the U.S., it increases the species' chance for survival. Similar experiments are also occurring outside North America.

The reintroduction of wild horses and camels would be a logical part of these early experiments. Horses and camels originated on this continent, and many species were present in the late Pleistocene. Today's feral horses and asses that live in some areas throughout the West are plausible substitutes for extinct American species. Because most of the surviving Eurasian and African species are now critically endangered, establishing Asian asses and Przewalski's horse in North America might help prevent the extinction of these animals. Bactrian camels, which are critically endangered in the Gobi Desert, could provide a modern proxy for Camelops, a late Pleistocene camel. Camels, introduced from captive or domesticated populations, might benefit U.S. ecosystems by browsing on woody plants that today are overtaking arid grasslands in the Southwest, an ecosystem that is increasingly endangered.

Another prong of the project would likely be more controversial but could also begin immediately. It would establish small numbers of elephants, cheetahs and lions on private property.

Introducing elephants could prove valuable to nearby human populations by attracting tourists and maintaining grasslands useful to ranchers (elephants could suppress the woody plants that threaten southwestern grasslands). In the late Pleistocene, at least four elephant species lived in North America. Under a scientific framework, captive elephants in the U.S. could be introduced as proxies for these extinct animals. The biggest cost involved would be fencing, which has helped reduce conflict between elephants and humans in Africa.

Many cheetahs are already in captivity in the U.S. The greatest challenge would be to provide them with large, securely fenced areas that have appropriate habitat and prey animals. Offsetting these costs are benefits—restoring what must have been strong interactions with pronghorn, facilitating ecotourism as an economic alternative for ranchers, many of whom are struggling financially, and helping to save the world's fastest carnivore from extinction.

Lions are increasingly threatened, with populations in Asia and some parts of Africa critically endangered. Bringing back lions, which are the same species that once lived in North America, presents daunting challenges as well as many potential benefits. But private reserves in southern Africa where lions and other large animals have been successfully reintroduced offer a model— and these reserves are smaller than some private ranches in the Southwest.

If these early experiments with large herbivores and predators show promising results, more could be undertaken, moving toward the long-term goal of a huge ecological history park. What we need now are panels of experts who, for each species, could assess, advise and cautiously lead efforts in restoring megafauna to North America.

A real-world example of how the reintroduction of a top predator might work comes from the wolves of Yellowstone National Park [see "Lessons from the Wolf," by Jim Robbins; *Scientific American,* June 2004]. The gray wolf became extinct in and around Yellowstone during the 1920s. The loss led to increases in their prey—moose and elk—which in turn reduced the distribution of aspens and other trees they eat. Lack of vegetation destroyed habitat for migratory birds and for beavers. Thus, the disappearance of the wolves propagated a trophic cascade from predators to herbivores to plants to birds and beavers. Scientists have started to document the ecosystem changes as reintroduced wolves regain the ecological role they played in Yellowstone for millennia. An additional insight researchers are learning from putting wolves back into Yellowstone is that they may be helping the park cope with climate change. As winters grow milder, fewer elk die, which means less carrion for scavengers such as coyotes, ravens and bald eagles. Wolves provide carcasses throughout the winter for the scavengers to feed on, bestowing a certain degree of stability.

The Challenges Ahead

As our group on the ranch in New Mexico discussed how Pleistocene rewilding might work, we foresaw many challenges that would have to be addressed and overcome. These include the possibility that introduced animals could bring novel diseases with them or that they might be unusually susceptible to diseases already present in the ecosystem; the fact that habitats have changed over the millennia and that reintroduced animals might not fare well in these altered environments; and the likelihood of unanticipated ecological consequences and unexpected reactions from neighboring human communities. Establishing programs that monitor the interactions among species and their consequences for the well-being of the ecosystem will require patience and expertise. And, of course, it will not be easy to convince the public to accept predation as an important natural process that actually nourishes the land and enables ecosystems to thrive. Other colleagues have raised additional concerns, albeit none that seems fatal.

Many people will claim that the concept of Pleistocene rewilding is simply not feasible in the world we live in today. I urge these people to look to Africa for inspiration. The year after the creation of Kruger National Park was announced, the site was hardly the celebrated mainstay of southern African biodiversity it is today. In 1903 zero elephants, nine lions, eight buffalo and very few cheetahs lived within its boundaries. Thanks to the vision and dedication of African conservationists, 7,300 elephants, 2,300 lions, 28,000 buffalo and 250 cheetahs roamed Kruger 100 years later—as did 700,000 tourists, bringing with them tens of millions of dollars.

In the coming century, humanity will decide, by default or design, the extent to which it will tolerate other species and thus how much biodiversity will endure. Pleistocene rewilding is not about trying to go back to the past; it is about using the past to inform society about how to maintain the functional fabric of nature. The potential scientific, conservation and cultural benefits of restoring mega-fauna are clear, as are the costs. Although sound science can help mitigate the potential costs, these ideas will make many uneasy. Yet given the apparent dysfunction of North American ecosystems and Earth's overall state, inaction carries risks as well. In the face of tremendous uncertainty, science and society must weigh the costs and benefits of bold, aggressive actions like Pleistocene rewilding against those of business as usual, which has risks, uncertainties and costs that are often unacknowledged. We have a tendency to think that if we maintain the status quo, things will be fine. All the available information suggests the opposite. . . .

Dustin R. Rubenstein et al. **NO**

Pleistocene Park: Does Re-Wilding North America Represent Sound Conservation for the 21st Century?

Introduction

Ancestors of elephants and lions once roamed much of North America. Recently, a diverse group of conservation biologists has proposed to create a facsimile of this bygone era by reintroducing charismatic African and Asian megafauna to western North America to replace species that disappeared during the Pleistocene extinctions, some 13,000 years ago. Arguing that their vision is justified on "ecological, evolutionary, economic, aesthetic and ethical grounds," Donlan et al. believe that modern "Pleistocene Parks" would provide refuges for species that are themselves threatened or endangered, and that repopulating the American west with these large mammals would improve local landscapes, restore ecological and evolutionary potential, and make amends for the ecological excesses of our ancestors.

To understand the uniqueness of this proposal, a terminological clarification is necessary. The "re-wilding" of ecosystems is the practice of reintroducing extant species (captive-bred or wild caught) back to places from which they were extirpated in historical times (i.e., in the past several hundred years). Because re-wilding deals with recently extirpated species and short evolutionary time scales, it is reasonable to assume that there have been minimal evolutionary changes in the target species and their native habitats. Re-wilding of ecosystems is not a new conservation practice and, indeed, it has become a standard management tool.

By contrast, "Pleistocene re-wilding" of ecosystems is a revolutionary idea that would involve introducing to present-day habitats either (1) extant species that are descended from species that occurred in those habitats during the Pleistocene, but that went extinct about 13,000 years ago, or (2) modern-day ecological proxies for extinct Pleistocene species. Pleistocene re-wilding is thus a novel plan for ecological restoration on a more grandiose temporal and spatial scale than is re-wilding.

Pleistocene re-wilding has been discussed for many years, and in 1989, it was attempted in Siberia, Russia, when mega-herbivores including wood bison (*Bison bison athabascae*), Yakutian horses (*Equus* sp.), and muskoxen (*Ovibos moschatus*) were introduced in an effort to recreate the grassland ecosystem of the

From *Biological Conservation*, vol. 132, issue 2, October 2006, pp. 232–238. Copyright © 2006 by Elsevier Science Ltd. Reprinted by permission via Rightslink.

Pleistocene. However, the North American Pleistocene re-wilding proposal of Donlan et al. is far more ambitious than this because it aims to reconstruct an ancient ecosystem by translocating a more diverse array of African and Asian megafauna to geographical regions and plant communities that have evolved without such creatures since the Pleistocene. Species targeted for introduction span several trophic levels and include predators such as African cheetahs (*Acinonyx jubatus*) and lions (*Panthera leo*), and large herbivores like African (*Loxodonta africana*) and Asian (*Elephas maximus*) elephants, various equids (*Equus* spp.), and Bactrian camels (*Camelus bactranus*). This plan includes animals that are both descendant species of extinct taxa and ecological proxies for extinct species.

Pleistocene re-wilding of North America has two principal goals: (1) to restore some of the evolutionary and ecological potential that was lost from North America 13,000 years ago; and (2) to help prevent the extinction of some of the world's existing megafauna by creating new, and presumably better protected, populations in North America. Discussion of the proposal is just beginning and, although some initial concerns have been raised, supporters and detractors agree that Pleistocene re-wilding is a bold and innovative idea, deserving of careful consideration. This paper was developed with the intent of extending healthy and fruitful scientific debate about Pleistocene re-wilding of North America.

The Ecology and Evolution of Pleistocene Re-Wilding: Restoring Ecological Potential to North American Ecosystems

Although an ethical desire to redress the excesses of our ancestors might serve as an initial justification for Pleistocene re-wilding, the ecological and evolutionary merits of such a plan must be considered carefully. Pleistocene re-wilding of North America would involve a monumental introduction of large mammals into areas where they have been extinct for millennia, and into habitats that have existed without such creatures for similarly long periods of evolutionary time. The potential negative ecological effects of transplanting exotic species to non-native habitats are well-known. The results of Pleistocene re-wilding in North America are unknown and might well be catastrophic; ecosystem functioning could be disrupted, native flora and fauna, including species of conservation value, could be negatively impacted, and a host of other unanticipated ecological problems could arise.

Pleistocene re-wilders believe that it is possible to enhance ecological potential, that is, to recreate evolutionarily-relevant mammalian species assemblages and restore ecosystem functioning to Pleistocene levels, because they believe that the flora of North American ecosystems is essentially unchanged since the Pleistocene. However, plant communities are dynamic and constantly in flux, genotypically and phenotypically, and there has been over 13,000 years for grassland and shrub-steppe communities to evolve and plant assemblages to change in the absence of the full suite of Pleistocene mega-herbivores. When managers discuss restoring ecological potential, or simply ecosystem restoration, it is important for them to be clear about what they are trying to restore and to what level of restoration they are trying to reach. Whereas Pleistocene re-wilding could potentially increase the ecological potential of some of North America's ecosystems by reintroducing predators on species like pronghorn or bighorn sheep (and thus, indirectly restoring the evolutionary potential of these prey species), or by restoring herbivorous keystone species like elephants to the temperate grasslands, it is questionable whether it would restore ecological potential to Pleistocene levels.

Indeed, rather than restoring our "contemporary" wild ecosystems to the "historic" wild ecosystems of the Pleistocene and their original levels of ecosystem functioning, which are unknown, Pleistocene re-wilding could instead result in "re-wilded" novel, or emerging, ecosystems with unique species compositions and new or altered levels of ecosystem functioning. Biogeographic assemblages and evolutionary lineages would be co-mingled in novel ways; new parasites and diseases could be introduced; and food chains would be disrupted. Moreover, without really knowing how Pleistocene ecosystems functioned, there will be no way to determine whether Pleistocene re-wilding restored ancient ecosystems or disrupted contemporary ones.

While the reintroduction of large grazers can, in some cases, shape and restore grassland ecosystems, this will depend on whether the grazers are indigenous or exotic. Modern introductions of exotic feral horses have dramatically altered vegetation in marsh and grassland ecosystems throughout the New World, and these changes have had direct impacts on a variety of native animal species, some positive, but some negative. Moreover, exotic grazers, such as the one-humped camel (*Camelus dromedarius*), have wreaked havoc upon desert ecosystems in Australia by selectively eating rare plant species. Similarly, the reintroduction of large predators can also have unexpected results on populations of prey species. For instance, wolves reintroduced to Yellowstone National Park, USA preyed upon elk more, and other species of ungulates less, than what was predicted prior to reintroduction.

Of course, it might be argued that these problems would quickly become apparent if Pleistocene re-wilding were first attempted on a small-scale, experimental basis. However, experiments of this nature cannot be done quickly and may take decades and generations to play out. For instance, the Siberian Pleistocene Park experiment began in 1989, and as of yet, few of the results have been published. Moreover, it may not be possible to conduct adequate, meaningful experiments on small spatial scales because many of these species have large home ranges or migrate great distances. For instance, African cheetahs can have home ranges of nearly 200 km^2, and African elephants can migrate distances of up to nearly 150 km or more.

Despite the potential dangers to ecosystem functioning, the reintroductions proposed by Donlan et al. would place many of the animals in temperate grasslands and shrub-steppe habitats, which are among the most threatened, but least protected, ecosystems in the world. If the reintroduction of exotic megafauna could help preserve these ecosystems, conservationists must weigh the possibility of preserving disrupted or novel North American ecosystems against the possibility of losing those ecosystems altogether.

The Ethics and Aesthetics of Pleistocene Re-Wilding: Protecting and Restoring the Evolutionary Potential of Threatened Megafauna

Humans were at least partly responsible for exterminating some species of Pleistocene megafauna and, today, anthropogenic impacts continue to contribute to the extinction of the world's remaining megafauna. Donlan et al. argue that humans bear an ethical responsibility to prevent future megafaunal extinctions and redress past losses. They suggest that introducing large Asian and African vertebrates to North America will not only ensure their long-term survival, but also restore their evolutionary potential (i.e., increase the number of individuals worldwide to allow them greater chances to radiate and generate new phenotypic and genotypic variants). Although this plan is certainly well-intentioned, the underlying reasoning is flawed. In essence, it is an attempt to preserve charismatic African and Asian species that are being driven to extinction by humans in their native habitats by refocusing efforts in places where those species have never occurred and where humans drove their distant ancestors extinct. Although Donlan et al. do not advocate giving up on conserving megafauna in developing nations, diverting attention from some of the world's most economically poor, but most biologically rich, countries to make amends for the ecological excesses of our North American ancestors could cripple, rather than assist, the conservation movement worldwide.

The human population is growing and natural habitats are declining in extent and diversity everywhere. Couple this with the political and economic strife that is occurring in many developing nations and it is not difficult to see why native megafauna, especially large mammals, are declining in numbers worldwide. Despite this dire situation, Pleistocene re-wilding of North America is not the only viable solution to preserve the world's megafauna. In the developing world, new conservation models are being implemented that go hand-in-hand with human development as wildlife must pay for itself by generating economic benefits for local citizenry to help alleviate poverty. Although there are many challenges in developing such programs, there is much to be gained by overcoming them because most of the native megafauna in developing regions inhabit private, often unprotected, lands outside of parks. For instance, across Africa, 84% of African elephant habitat is outside of protected areas, and in Kenya, 70% of the wildlife lives outside of protected areas for at least part of the year.

Conserving African and Asian megafauna does not require relocating them to North America. However, it will require new conservation plans that ensure local citizenry receive economic benefits from wildlife. Available human and financial resources might be better expended on preserving land, promoting ecotourism, building fences in areas of high human–wildlife conflict, and establishing educational and research programs in areas of Africa and Asia where indigenous megafauna are most at risk, rather than on introducing those same large, exotic species to North America.

In addition, the question of how the Pleistocene re-wilding plan would affect existing conservation efforts in North America must be considered. Conservationists often struggle with local opposition to re-wilding with native predators, and even the reintroduction of relatively benign large mammals (e.g., moose) meets resistance. The introduction of modern relatives of extinct predators will be opposed even more strongly by state governments and locally-affected citizens. And, with good reason: escapes are inevitable, resulting in human–wildlife conflict as often occurs near protected areas in Africa and Asia. . . . It is difficult enough for North American conservationists to address the real concerns of local citizens about attacks by mountain lions (native predators) on joggers. One can only imagine the anti-conservation backlash that would be generated by news coverage of farmers coping with crop destruction by herds of elephants, or lions and cheetahs attacking cattle, or even children.

While Pleistocene re-wilding may help maintain the evolutionary potential of modern, extant species, it cannot restore the evolutionary potential of extinct species that no longer exist. And even attempting to restore evolutionary potential of endangered species using modern-day species from foreign continents as proxies for creatures that went extinct in North America is controversial. Donlan et al. highlighted the peregrine falcon (*Falco peregrinus*) to illustrate how using similar, but not genetically identical, (sub-)species can indeed serve as proxies for nearly-extinct taxa. Moreover, for nearly two decades, conservation biologists have proposed introducing closely related proxy species for extinct birds on New Zealand and other Pacific islands. However, species that went extinct some 13,000 years ago are probably more genetically different from their modern-day proxies, who have continued to evolve for millennia, than are two sub-species of modern falcons or modern Pacific island birds. For instance, although recent molecular data suggest that the common horse (*Equus caballus*) is genetically similar to its evolutionary ancestral species, modern elephants, cheetahs, and lions are quite genetically distinct from their extinct Pleistocene relatives.

Rather than use modern-day species from foreign continents as proxies for creatures that went extinct in North America, conservation efforts should focus on re-wilding native species into their historical ranges throughout North America to restore ecosystems and increase the evolutionary potential of indigenous species. For instance, native herbivores like bison (*Bison bison*), pronghorn (*Antilcapra americana*), elk (*Cervus elaphus*), jack rabbits (*Lepus townsendii*), and various ground-dwelling squirrels (*Spermophilus* spp.) and prairie dogs (*Cynomys* spp.), as well as native predators like black-footed ferrets (*Mustela nigripes*), bobcats (*Lynx rufus*), badgers (*Taxidea taxus*), and swift foxes (*Vulpes velox*) are likely candidates for reintroduction to geographic regions from which they were extirpated in the past several hundred years.

Donlan et al. (2005) suggested that another appropriate candidate for reintroduction is the Bolson tortoise (*Gopherus avomarginatus*). Because this animal once lived throughout the southwestern United States and still persists in small areas of similar habitat in Mexico, it may not differ greatly from its ancestral form and, therefore, it might be a reasonable candidate for reintroduction. However, before attempting such a reintroduction, one would also have to consider how much the plant and animal communities in the tortoise's native geographic habitats have changed (evolved) since this reptile went locally extinct.

If reintroducing charismatic megafauna is an important goal of Pleistocene re-wilding because of its possible galvanizing effect on public support for conservation, then one might consider expanding reintroductions of some of North America's own megafauna like wolves (*Canis lupus*) or grizzly bears (*Urus arctos*) to other portions of their known recent (i.e., historical) ranges. And if more predators are deemed necessary, an even better candidate for re-wilding would be the puma (*Puma concolor*), because it is more genetically similar to the long-extinct American cheetah (*Miracinonyx trumani*) than the African cheetah is to the American cheetah. Moreover, the puma is a native mammalian predator that barely survived the Pleistocene extinctions 13,000 years ago, and still remains threatened throughout much of its North American range.

The Economics and Politics of Pleistocene Re-Wilding: Uncertainty and Tradeoffs

The political and economic ramifications of Pleistocene re-wilding of North America are unclear. Certainly, it will be expensive because land acquisition and preparation, translocation, monitoring, protection, and containment require considerable human and financial resources. Moreover, all of these efforts would likely cost proportionally more in North America than they would in Africa or Asia, given the higher prices of salaries and supplies. Because conservation funding is limited, Pleistocene re-wilding may compete for resources that might otherwise have gone to local conservation efforts. Although it is possible that introducing charismatic African and Asian megafauna to North America could ignite public and political support, ultimately leading to an overall increase in funding for conservation projects worldwide, other new ideas might increase the pool of resources with less risk to North American ecosystems and conservation efforts worldwide.

Donlan et al. are careful to point out that the initial steps of Pleistocene re-wilding can occur without large-scale translocations of proxy species because many of these animals are already in captivity in the United States. This would potentially reduce costs, as well as avoid potential political problems between the United States and developing nations, the ultimate "sources" for the animals. However, reintroductions from wild populations have been more successful than those from captive populations, and Pleistocene re-wilding inevitably would involve translocating animals from Asia and Africa to North America, either to increase population numbers or to improve population viability by augmenting genetic diversity. Such translocations of megafauna into areas where they were recently extirpated occur routinely throughout Africa and Asia, and learning from these examples could shed some light on the practicality of Pleistocene re-wilding of North America.

The Practicality of Pleistocene Re-Wilding: The Reality of Reintroductions

One of the goals of Pleistocene re-wilding of North America is to ". . . restore equid species to their evolutionary homeland" and, indeed, some of the best known and most successful reintroductions of endangered species to their historical ranges involve equids. For example, the Tahki, or Przewalski's horse (*Equus ferus przewalskii*), which is endemic to Mongolia and China, was considered extinct in the wild by the end of the 1960s and fewer than 400 Tahki remained in captivity in 1979. By the beginning of the 1990s, however, efforts began to reestablish populations in the wild and two reintroduction sites were chosen in Mongolia. By 2000, the population at one site had declined only slightly, while that at the other had increased by 50%. These encouraging trends led to a third introduction in 2005 and suggest that the re-wilding of the Tahki's historic range is likely to succeed.

In an attempt to repopulate Israel with recently extirpated biblical animals, onagers, a race of Asian asses (*Equus hemionus*), were translocated from Israel's Hai Bar breeding reserve to a nearby erosional crater in the Negev desert. Between 1968 and 1993 multiple reintroductions of 50 individuals took place. It was not until the end of the 1990s, however, before the population started to expand numerically and spatially. Low fertility of translocated adult females relative to that of their wild-born daughters and male-biased sex ratios among the progeny limited recruitment. These unanticipated biological constraints suggest that even reintroductions of native species to their historical habitats are not assured of succeeding.

Repopulating the historic range of the endangered Grevy's zebra (*Equus grevyi*) in east Africa is viewed as critically important to saving the species from extinction. Fewer then 2000 Grevy's zebras remain in small areas of Ethiopia and northern Kenya, whereas only 35 years ago, over 20,000 individuals inhabited areas all the way to the horn of Africa. Efforts to repopulate areas of the Grevy's historic range have involved capturing and moving small groups of appropriate sex ratios and age structures to holding areas before subsequent release. While two such reintroductions in Kenya, one to Tsavo National Park and one to Meru National Park, began successfully, neither has led to expanding populations. In fact, in Meru, differences in the composition and abundance of mammal species in the Grevy's zebras' new range have, at times, led to rapid declines in their numbers. Therefore, even reintroductions within the natural geographic regions of a species are often fraught with surprises due to diseases, unexpected differences in environmental conditions, and naïveté toward predators. . . .

Another Jurassic Park?

We all remember "Jurassic Park", Crichton's fictional account of re-wilding an isolated island with extinct dinosaurs recreated from ancient DNA. Pleistocene re-wilding of North America is only a slightly less sensational proposal. It is a little like proposing that two wrongs somehow will make a right: both the modern-day proxy species are "wrong" (i.e., different genetically from the species that occurred in North America during the Pleistocene), and the ecosystems into which they are to be reintroduced are "wrong" (i.e., different in composition from the Pleistocene ecosystems, as well as from those in which the modern-day proxy species evolved). Pleistocene re-wilding of North America will not restore evolutionary potential of North America's extinct megafauna because the species in question are evolutionarily distinct, nor will it restore ecological potential of North America's modern ecosystems because they have continued to evolve over the past 13,000 years. In addition, there is a third and potentially greater "wrong" proposed: adding these exotic species to current ecological communities could potentially devastate populations of indigenous, native animals and plants.

Although Donlan et al. argued that Pleistocene re-wilding of North America is justified for ecological, evolutionary, economic, aesthetic, and ethical reasons, there are clearly numerous ecological and evolutionary concerns. On the one hand, the plan might help conserve and maintain the evolutionary potential of some endangered African and Asian megafauna, as well as indirectly enhance the evolutionary potential of native North American prey species that have lacked appropriate predators since the Pleistocene. On the other hand, the plan cannot restore the evolutionary potential of extinct species and it is unlikely to restore the ecological potential of western North America's grassland and shrub-steppe communities. Instead, it may irreparably disrupt current ecosystems and species assemblages. Moreover, there are many potential practical limitations to Donlan et al.'s plan. Reintroduced camels did not survive for long in the deserts of the American West. Could African megafauna, especially large carnivores, really populate the same areas? Would the genetically depauperate cheetah succumb to novel diseases? Would elephants survive the harsh prairie winters, lacking the thick coats of their mastodon ancestors?

Answering these questions and accomplishing Pleistocene re-wilding of North America would require a massive effort and infusion of funds and could take more time to experimentally test than some of these critically endangered species have left to survive in their existing native habitats. If financial and physical resources were available on this scale, they would be better spent on developing and field-testing new ways to manage and conserve indigenous populations of African, Asian, and North American wildlife in their historically-populated native habitats, on conducting ecological, behavioral, and demographic studies of these organisms in the environments in which they evolved, and on educating the public on each continent about the wonders of their own dwindling flora and fauna.

EXPLORING THE ISSUE

Should North America's Landscape Be Restored to Its Prehuman State?

Critical Thinking and Reflection

1. Do people have a moral or ethical responsibility to attempt to repair damage done to the environment?
2. Is ancient or more recent damage to the environment more worthy of repair?
3. How could ancient people possibly have wiped out mammoths and other large animals?

Is There Common Ground?

Not everyone is concerned about ancient extinctions, whether caused by humans or not. More recent extinctions provoke more concern, even guilt over the human role in those extinctions. Yet they cannot be undone, and rewilding is not always practical or even possible, depending on how ecosystems have changed and on what other creatures have filled vacant niches.

Assign students to consult regional history books in the library to learn what wildlife were common in your area, say, 200 years ago. The list may include bison, mountain lions, beavers, otters, mink, condors, grizzly bears, wolves, salmon, sturgeon, eagles, and many more. They may then write reports answering the following questions:

1. What animals were common in your area in the past?
2. Which ones are now extinct (at least locally)?
3. What roles did they play in local ecosystems?
4. What animals—if any—fill those roles today?
5. How would you propose to rewild your local environment?
6. How likely is it that local rewilding would succeed?

ISSUE 5

Should the Military Be Exempt from Environmental Regulations?

YES: Benedict S. Cohen, from "Impact of Military Training on the Environment," Testimony before the Senate Committee on Environment and Public Works (April 2, 2003)

NO: Jamie Clark, from "Impact of Military Training on the Environment," Testimony before the Senate Committee on Environment and Public Works (April 2, 2003)

Learning Outcomes

After reading this issue, you should be able to:

- Describe how military activities affect the environment.
- Discuss how similar military training must be to military action.
- Explain why the Department of Defense wants military activities exempted from environmental regulations.
- Explain why some people object to exempting military activities from environmental regulations.

ISSUE SUMMARY

YES: Benedict S. Cohen argues that environmental regulations interfere with military training and other "readiness" activities, and that although the U.S. Department of Defense will continue "to provide exemplary stewardship of the lands and natural resources in our trust," those regulations must be revised to permit the military to do its job without interference.

NO: Jamie Clark argues that reducing the Department of Defense's environmental obligations is dangerous because both people and wildlife would be threatened with serious, irreversible, and unnecessary harm.

Most of us have heard of "scorched earth" wars, in which an army destroys forests and farms in order to deny the enemy their benefit. We have surely seen the images of a Europe laid waste by World War II. More recently, we may recall, the Gulf War saw oil deliberately released to flood desert sands and the waters of the Persian Gulf. Enough smoke poured from burning oil wells to threaten both local climate change and human health. See, e.g., Randy Thomas, "Eco War," *Earth Island Journal* (Spring 1991), B. Ruben, "Gulf Smoke Screens," *Environmental Action* (July/August 1991), and Jeffrey L. Lange, David A. Schwartz, Bradley N. Doebbeling, Jack M. Heller, and Peter S. Thorne, "Exposures to the Kuwait Oil Fires and their Association with Asthma and Bronchitis among Gulf War Veterans," *Environmental Health Perspectives* (November 2002). Weaponry can have environmental effects by destroying dams, by physically destroying plants and animals, by causing erosion, and by disseminating toxic materials; see Henryk Bem and Firyal Bou-Rabee, "Environmental and Health Consequences of Depleted Uranium Use in the 1991 Gulf War," *Environment International* (March 2004).

The environmental impact of war would seem impossible to deny. But even after the Gulf War, the U.S. Department of Defense tried to suppress satellite photos showing the extent of the damage; see Shirley Johnston, "Gagged on Smoke," *Earth Island Journal* (Summer 1991). And for many years, it insisted that nuclear war was survivable, until researchers made it clear that even a small nuclear war would produce a "nuclear winter" that would probably destroy civilization, if not the human species. See T. Rueter and T. Kalil, "Nuclear Strategy and Nuclear Winter" *World Politics* (July 1991), and Carl Sagan and Richard Turco, *A Path Where No Man Thought: Nuclear Winter and Its Implications* (Random House, 1990).

Preparations for war may also have serious environmental impacts. Puerto Rico's island of Vieques was long a bomb depot and bombing range for the U.S. Navy. Local residents protested vigorously and documented heavy-metal contamination of the local ecosystem. After the Navy left the island, it "has continued to deny that it has been anything but an excellent environmental steward in Vieques." See Shane DuBow and Scott S. Warren, "Vieques on the Verge," *Smithsonian* (January 2004).

In 2002, the U.S. Congress, through the Readiness and Range Preservation Initiative, granted the Department of Defense a temporary exemption to the Migratory Bird Treaty Act that allowed the "incidental taking" of endangered birds during bombing and other training on military lands. In 2003, the Department of Defense asked Congress for additional exemptions from environmental regulations, specifically the Clean Air Act, Marine Mammal Protection Act, Endangered Species Act, Migratory Bird Treaty Act, and federal toxic waste laws. Paul Mayberry, Deputy Undersecretary of Defense for Readiness, said the exemptions were justified because many environmental restrictions were putting the nation's military readiness at stake. See "Pentagon Seeks Clarity in Environmental Laws Affecting Ranges," Agency Group 09, FDCH Regulatory Intelligence Database (March 21, 2003). The United States Senate Committee on Environment and Public Works held a hearing on "The Impact of Military Training on the Environment" on April 2, 2003. After this hearing, the issue received some press attention as environmentalists and congressional Democrats prepared to oppose the Bush Administration. See "War on the Environment," *The Ecologist* (May 2003), and John Stanton, "Activists, Democrats Brace for Defense

Environment Showdown," *CongressDaily AM* (May 13, 2003). The Senate voted for the exemptions, and early in 2004, the issue came before the House of Representatives. The Congressional Research Service summarized the issue in a report to Congress (David M. Bearden, "Exemptions from Environmental Law for the Department of Defense: Background and Issues for Congress" [updated May 15, 2007] [www.fas.org/sgp/crs/natsec/RS22149.pdf]). Dan Miller, First Assistant Attorney General, Natural Resources and Environmental Section, Colorado Department of Law, testified against the exemptions before the House Committee on Energy and Commerce Subcommittee on Energy and Air Quality, saying:

> Even read in the narrowest possible fashion, the [proposed exemptions] would hamstring state and EPA cleanup authorities at over 24 million acres of 'operational ranges,' an area the size of Maryland, Massachusetts, New Jersey, Hawaii, Connecticut and Rhode Island combined. As a practical matter, environmental regulators would likely be precluded from using RCRA, CERCLA, and related state authorities to require any investigation or cleanup of groundwater contamination on these ranges, even if the contamination had migrated off-range, polluted drinking or irrigation water supplies, and even if it posed an imminent and substantial endangerment to human health. And it is likely that DOD's amendments would be construed more broadly to exempt even more contamination from state and EPA oversight. . . . If we have learned anything in the past thirty years of environmental regulation, it is that relying on federal agencies to 'voluntarily' address environmental contamination is often fruitless. One need look no further than the approximately 130 DOD facilities on the Superfund National Priorities List, or DOD's poor record of compliance with state and federal environmental laws to see that independent, legally enforceable state oversight of federal agencies is required to achieve effective results.

In May, the House Committees on Armed Services Readiness and Energy and Commerce announced that they were not about to consider the exemption. See "Hefley: No Plans to Exempt Military from Enviro Laws," *Congress-Daily AM* (May 6, 2004).

But the issue is not about to go away. The Government Accounting Office prepared a background paper, "Military Training: DOD Approach to Managing Encroachment on Training Ranges Still Evolving," GAO-03-621T (April 2, 2003), delivered by Barry W. Holman, Director, Defense Infrastructure Issues, as testimony before the Senate Committee on Environment and Public Works. It discussed eight encroachment issues, including urban growth around military bases, air and noise pollution, unexploded ordnance and other munitions, endangered species habitat, and protected marine resources. Since urban growth is not likely to cease and the number of endangered species and protected marine resources is sure to increase, encroachment is not about to diminish. The Department of Defense, says the GAO, must better document the impact of the encroachment on training and costs; it has not yet produced required reports to Congress. So far, "workarounds" have been enough to deal with the problem, but that may not remain sufficient. E. G. Willard, Tom Zimmerman, and Eric Bee, "Environmental Law and National Security: Can

Existing Exemptions in Environmental Laws Preserve DOD Training and Operational Prerogatives without New Legislation?" *Air Force Law Review* (2004), conclude that existing exemptions are not enough to support military readiness and say that more are needed. "The bottom line is that we must be able to train the way we fight, and we must be able to operate to defend the country and its interests." Paul D. Thacker, "Are Environmental Exemptions for the U.S. Military Justified?" *Environmental Science & Technology* (October 15, 2004), noted that "many critics of the administration say that the campaign is more about undermining environmental laws than protecting military readiness."

Recent incidents involving beached whales, apparently due to use of sonar by Navy ships, have returned the issue to the public eye. E. C. M. Parsons et al., "Navy Sonar and Cetaceans: Just How Much Does the Gun Need to Smoke Before We Act?" *Marine Pollution Bulletin* (June 2008), note that such strandings are not uncommon and that they often occur near where sonar is being used. Parsons et al., charge "senior government officials" with obstructing measures to prevent harm to the whales by limiting sonar use. According to Joan Biskupic, "Justices to Debate Whether Navy Sonar Harms Whales," *USA Today* (June 24, 2008), after the Natural Resources Defense Council sued, the courts restricted the use of sonar. When the Bush administration appealed, saying that the restrictions jeopardize "the Navy's ability to train sailors and Marines for wartime deployment [which is] essential to the national security," the Supreme Court agreed to hear the case. The Supreme Court removed the restrictions in November 2008; Joan Biskupic, "High Court OKs Sonar Training off Calif. Coast," *USA Today* (November 13, 2008). See also "Can the U.S. Navy Proceed With Its Plan for Sonar Training Exercises off the Coast of Southern California?" *Supreme Court Debates* (November 2008). Research on the effect of sonar on whales continues; see Chris Hardman, "Marine Mammal Aid," *Americas* (March/April 2011). Recently, beaked whales were shown to flee Navy sonar use; see Peter L. Tyack et al., "Beaked Whales Respond to Simulated and Actual Navy Sonar," *PLoS One* (www.plosone.org) (2011).

In the following selections, both drawn from testimony at the United States Senate Committee on Environment and Public Works hearing on "The Impact of Military Training on the Environment" held on April 2, 2003, Benedict S. Cohen, Deputy General Counsel for Environment and Installations, Department of Defense, argues in his testimony before the Committee that environmental regulations interfere with military training and other "readiness" activities, and that although the Department of Defense will continue "to provide exemplary stewardship of the lands and natural resources in our trust," those regulations must be revised to permit the military to do its job without interference. Jamie Clark, Senior Vice President for Conservation Programs, National Wildlife Federation, argues that reducing the Department of Defense's environmental obligations is dangerous for two reasons. First, both people and wildlife will be threatened with serious, irreversible, and unnecessary harm. Second, other federal agencies and industry sectors with important missions, using the same logic as used here by the Department of Defense, would demand similar exemptions from environmental laws.

YES

Benedict S. Cohen

Impact of Military Training on the Environment

Mr. Chairman and distinguished members of this Committee, I appreciate the opportunity to discuss with you the very important issue of sustaining our test and training capabilities, and the legislative proposal that the Administration has put forward in support of that objective. In these remarks I would like particularly to address some of the comments and criticisms offered concerning these legislative proposals.

Addressing Encroachment

We have only recently begun to realize that a broad array of encroachment pressures at our operational ranges are increasingly constraining our ability to conduct the testing and training that we must do to maintain our technological superiority and combat readiness. Given world events today, we know that our forces and our weaponry must be more diverse and flexible than ever before. Unfortunately, this comes at the same time that our ranges are under escalating demands to sustain the diverse operations required today, and that will be increasingly required in the future.

This current predicament has come about as a cumulative result of a slow but steady process involving many factors. Because external pressures are increasing, the adverse impacts to readiness are growing. Yet future testing and training needs will only further exacerbate these issues, as the speed and range of our weaponry and the number of training scenarios increase in response to real-world situations our forces will face when deployed. We must therefore begin to address these issues in a much more comprehensive and systematic fashion and understand that they will not be resolved overnight, but will require a sustained effort.

Environmental Stewardship

Before I address our comprehensive strategy, let me first emphasize our position concerning environmental stewardship. Congress has set aside 25 million acres of land—some 1.1% of the total land area in the United States. These lands were entrusted to the Department of Defense (DoD) to use efficiently and to care for properly. In executing these responsibilities we are committed

From Testimony before the Senate Committee on Environment and Public Works, April 2, 2003.

to more than just compliance with the applicable laws and regulations. We are committed to protecting, preserving, and, when required, restoring, and enhancing the quality of the environment.

- We are investing in pollution prevention technologies to minimize or reduce pollution in the first place. Cleanup is far more costly than prevention.
- We are managing endangered and threatened species, and all of our natural resources, through integrated natural resource planning.
- We are cleaning up contamination from past practices on our installations and are building a whole new program to address unexploded ordnance on our closed, transferring, and transferred ranges.

Balance

The American people have entrusted these 25 million acres to our care. Yet, in many cases, these lands that were once "in the middle of nowhere" are now surrounded by homes, industrial parks, retail malls, and interstate highways.

On a daily basis our installation and range managers are confronted with a myriad of challenges—urban sprawl, noise, air quality, air space, frequency spectrum, endangered species, marine mammals, and unexploded ordnance. Incompatible development outside our fence-lines is changing military flight paths for approaches and take-offs to patterns that are not militarily realistic—results that lead to negative training and potential harm to our pilots. With over 300 threatened and endangered species on DoD lands, nearly every major military installation and range has one or more endangered species, and for many species, these DoD lands are often the last refuge. Critical habitat designations for an ever increasing number of threatened or endangered species limit our access to and use of thousands of acres at many of our training and test ranges. The long-term prognosis is for this problem to intensify as new species are continually added to the threatened and endangered list.

Much too often these many encroachment challenges bring about unintended consequences to our readiness mission. This issue of encroachment is not going away. Nor is our responsibility to "train as we fight."

2003 Readiness and Range Preservation Initiative (Rrpi)

Overview

DoD's primary mission is maintaining our Nation's military readiness, today and into the future. DoD is also fully committed to high-quality environmental stewardship and the protection of natural resources on its lands. However, expanding restrictions on training and test ranges are limiting realistic preparations for combat and therefore our ability to maintain the readiness of America's military forces.

Last year, the Administration submitted to Congress an eight-provision legislative package, the Readiness and Range Preservation Initiative (RRPI).

Congress enacted three of those provisions as part of the National Defense Authorization Act for Fiscal Year 2003. Two of the enacted provisions allow us to cooperate more effectively with local and State governments, as well as private entities, to plan for growth surrounding our training ranges by allowing us to work toward preserving habitat for imperiled species and assuring development and land uses that are compatible with our training and testing activities on our installations.

Under the third provision, Congress provided the Department a regulatory exemption under the Migratory Bird Treaty Act for the incidental taking of migratory birds during military readiness activities. We are grateful to Congress for these provisions, and especially for addressing the serious readiness concerns raised by recent judicial expansion of the prohibitions under the Migratory Bird Treaty Act. I am pleased to inform this Committee that as a direct result of your legislation, Air Force B-1 and B-52 bombers, forward deployed to Anderson Air Force Base, Guam, are performing dry run training exercises over the Navy's Bombing Range at Farallon de Medinilla in the Commonwealth of the Northern Mariana Islands.

Last year, Congress also began consideration of the other five elements of our Readiness and Range Preservation Initiative. These five proposals remain essential to range sustainment and are as important this year as they were last year—maybe more so. The five provisions submitted this year reaffirm the principle that military lands, marine areas, and airspace exist to ensure military preparedness, while ensuring that the Department of Defense remains fully committed to its stewardship responsibilities. These [. . .] remaining provisions:

- Authorize use of Integrated Natural Resource Management Plans in appropriate circumstances as a substitute for critical habitat designation;
- Reform obsolete and unscientific elements of the Marine Mammal Protection Act, such as the definition of "harassment," and add a national security exemption to that statute;
- Modestly extend the allowable time for military readiness activities like bed-down of new weapons systems to comply with Clean Air Act; and
- Limit regulation of munitions on operational ranges under the Comprehensive Environmental Response, Compensation, and Liability Act (CERCLA) and Resource Conservation and Recovery Act (RCRA), if and only if those munitions and their associated constituents remain there, and only while the range remains operational.

Before discussing the specific elements of our proposal, I would like to address some overarching issues. A consistent theme in criticisms of our proposal is that it would bestow a sweeping or blanket exemption for the Defense Department from the Nation's environmental laws. No element of this allegation is accurate.

First, our initiative would apply only to military readiness activities, not to closed ranges or ranges that close in the future, and not to "the routine operation of installation operating support functions, such as administrative offices, military exchanges, commissaries, water treatment facilities, storage, schools, housing, motor pools . . . nor the operation of industrial activities, or the construction or demolition of such facilities." Our initiative thus is not applicable to the Defense Department activities that have traditionally been of greatest concern to state and federal regulators. It does address only uniquely military activities—what DoD does that is unlike any other governmental or private activity. DoD is, and will remain, subject to precisely the same regulatory requirements as the private sector when we perform the same types of activities as the private sector. We seek alternative forms of regulation only for the things we do that have no private-sector analogue: military readiness activities.

Moreover, our initiative largely affects environmental regulations that don't apply to the private sector or that disproportionately impact DoD:

- Endangered Species Act "critical habitat" designation has limited regulatory consequences on private lands, but can have crippling legal consequences for military bases.
- Under the Marine Mammal Protection Act, the private sector's Incidental Take Reduction Plans give commercial fisheries the flexibility to take significant numbers of marine mammal each year, but are unavailable to DoD—whose critical defense activities are being halted despite far fewer marine mammal deaths or injuries a year.
- The Clean Air Act's "conformity" requirement applies only to federal agencies, not the private sector.

Our proposals therefore are of the same nature as the relief Congress afforded us last year under the Migratory Bird Treaty Act, which environmental groups are unable to enforce against private parties but, as a result of a 2000 circuit court decision were able and willing to enforce, in wartime, against vital military readiness activities of the Department of Defense.

Nor does our initiative "exempt" even our readiness activities from the environmental laws; rather, it clarifies and confirms existing regulatory policies that recognize the unique nature of our activities. It codifies and extends EPA's existing Military Munitions Rule; confirms the prior Administration's policy on Integrated Natural Resource Management Plans and critical habitat; codifies the prior Administration's policy on "harassment" under the Marine Mammal Protection Act; ratifies longstanding state and federal policy concerning regulation under RCRA and CERCLA of our operational ranges; and gives states and DoD temporary flexibility under the Clean Air Act. Our proposals are, again, of the same nature as the relief Congress provided us under the Migratory Bird Treaty Act last year, which codified the prior Administration's position on DoD's obligations under the Migratory Bird Treaty Act.

Ironically, the alternative proposed by many of our critics—invocation of existing statutory emergency authority—would fully exempt DoD from the waived statutory requirements for however long the exemption lasted, a more far-reaching solution than the alternative forms of regulation we propose.

Accordingly, our proposals are neither sweeping nor exemptive; to the contrary, it is our critics who urge us to rely on wholesale, repeated use of emergency exemptions for routine, ongoing readiness activities that could easily be accommodated by minor clarifications and changes to existing law.

Existing Emergency Authorities

As noted above, many of our critics state that existing exemptions in the environmental laws and the consultative process in 10 USC 2014 render the Defense Department's initiative unnecessary.

Although existing exemptions are a valuable hedge against unexpected future emergencies, they cannot provide the legal basis for the Nation's everyday military readiness activities.

- The Marine Mammal Protection Act, like the Migratory Bird Treaty Act the Congress amended last year, has no national security exemption.
- 10 USC 2014, which allows a delay of at most five days in regulatory actions significantly affecting military readiness, is a valuable insurance policy for certain circumstances, but allows insufficient time to resolve disputes of any complexity. The Marine Corps' negotiations with the Fish and Wildlife Service over excluding portions of Camp Pendleton from designation as critical habitat took months. More to the point, Section 2014 merely codifies the inherent ability of cabinet members to consult with each other and appeal to the President. Since it does not address the underlying statutes giving rise to the dispute, it does nothing for readiness in circumstances where the underlying statute itself—not an agency's exercise of discretion—is the source of the readiness problem. This is particularly relevant to our RRPI proposal because none of the five amendments we propose have been occasioned by the actions of state or federal regulators. Four of the five proposed amendments (RCRA, CERCLA, MMPA, and ESA), like the MBTA amendment Congress passed last year, were occasioned by private litigants seeking to overturn federal regulatory policy and compel federal regulators to impose crippling restrictions on our readiness activities. The fifth, our Clean Air Act amendment, was proposed because DoD and EPA concluded that the Act's "general conformity" provision unnecessarily restricted the flexibility of DoD, state, and federal regulators to accommodate military readiness activities into applicable air pollution control schemes. Section 2014, therefore, although useful in some circumstances, would be of no use in addressing the critical readiness issues that our five RRPI initiatives address.
- Most of the environmental statutes with emergency exemptions clearly envisage that they will be used in rare circumstances, as a last resort, and only for brief periods.
- Under these statutes, the decision to grant an exemption is vested in the President, under the highest possible standard: "the paramount interest of the United States," a standard understood to involve exceptionally grave threats to national survival. The exemptions are also usually limited to renewable periods of a year (or in some cases as much as three years for certain requirements).
- The ESA's section 7(j) exemption process, which differs significantly from typical emergency exemptions, allows the Secretary of Defense to direct the Endangered Species Committee to exempt agency actions

in the interest of national security. However, the Endangered Species Committee process has given rise to procedural litigation in the past, potentially limiting its usefulness—especially in exigent circumstances. In addition, because it applies only to agency actions rather than to ranges themselves, any exemption secured by the Department would be of limited duration and benefit: because military testing and training evolve continuously, such an exemption would lose its usefulness over time as the nature of DoD actions on the range evolved.

- The exemption authorities do not work well in addressing those degradations in readiness that result from the cumulative, incremental effects of many different regulatory requirements and actions over time (as opposed to a single major action).
- Moreover, readiness is maintained by thousands of discrete test and training activities at hundreds of locations. Many of these are being adversely affected by environmental provisions. Maintaining military readiness through use of emergency exemptions would therefore involve issuing and renewing scores or even hundreds of Presidential certifications annually.
- And although a discrete activity (e.g., a particular carrier battle group exercise) might only rarely rise to the extraordinary level of a "paramount national interest," it is clearly intolerable to allow all activities that do not individually rise to that level to be compromised or ended by overregulation.
- Finally, to allow continued unchecked degradation of readiness until an external event like Pearl Harbor or September 11 caused the President to invoke the exemption would mean that our military forces would go into battle having received degraded training, with weapons that had received degraded testing and evaluation. Only the testing and training that occurred after the emergency exemption was granted would be fully realistic and effective.

The Defense Department believes that it is unacceptable as a matter of public policy for indispensable readiness activities to require repeated invocation of emergency authority—particularly when narrow clarifications of the underlying regulatory statutes would enable both essential readiness activities and the protection of the environment to continue. Congress would never tolerate a situation in which another activity vital to the Nation, like the practice of medicine, was only permitted to go forward through the repeated use of emergency exemptions.

That having been said, I should make clear that the Department of Defense is in no way philosophically opposed to the use of national security waivers or exemptions where necessary. We believe that every environmental statute should have a well-crafted exemption, as an insurance policy, though we continue to hope that we will seldom be required to have recourse to them. [. . .]

Specific Proposals

This year's proposals do include some clarifications and modifications based on events since last year. Of the five, the Endangered Species Act (ESA) and Clean Air Act provisions are unchanged. Let me address the changed provisions first.

RCRA and CERCLA

The legislation would codify and confirm the longstanding regulatory policy of EPA and every state concerning regulation of munitions use on operational ranges under RCRA and CERCLA. It would confirm that military munitions are subject to EPA's 1997 Military Munitions Rule while on range, and that cleanup of operational ranges is not required so long as material stays on the range. If such material moves off range, it still must be addressed promptly under existing environmental laws. Moreover, if munitions constituents cause an imminent and substantial endangerment on range, EPA will retain its current authority to address it on range under CERCLA section 106. (Our legislation explicitly reaffirms EPA's section 106 authority.) The legislation similarly does not modify the overlapping protections of the Safe Drinking Water Act, NEPA, and the ESA against environmentally harmful activities at operational military bases. The legislation has no effect whatsoever on DoD's cleanup obligations under RCRA or CERCLA at Formerly Used Defense Sites, closed ranges, ranges that close in the future, or waste management practices involving munitions even on operational ranges (such as so-called OB/OD activities).

The core of our concern is to protect against litigation the longstanding, uniform regulatory policy that (1) use of munitions for testing and training on an operational range is not a waste management activity or the trigger for cleanup requirements, and (2) that the appropriate trigger for DoD to address the environmental consequences of such routine test and training uses involving discharge of munitions is (a) when the range closes, (b) when munitions or their elements migrate or threaten to migrate off-range, or (c) when munitions or their elements create an imminent and substantial endangerment on-range. [. . .]

This legislation is needed because of RCRA's broad definition of "solid waste," and because states possess broad authority to adopt more stringent RCRA regulations than EPA (enforceable both by the states and by environmental plaintiffs). EPA therefore has quite limited ability to afford DoD regulatory relief under RCRA. Similarly, the broad statutory definition of "release" under CERCLA may also limit EPA's ability to afford DoD regulatory relief. And the President's site-specific, annually renewable waiver (under a paramount national interest standard in RCRA and a national security standard in CERCLA) is inapt for the reasons discussed above. [. . .]

Marine Mammal Protection Act

Although I realize this Committee is not centrally concerned with the Marine Mammal Protection Act (MMPA), I would like to take a moment to discuss it for purposes of completeness. This year's MMPA proposal includes some new provisions. This year's proposal, like last year's, would amend the term "harassment" in the MMPA, which currently focuses on the mere "potential" to injure or disturb marine mammals.

Our initiative adopts verbatim a reform proposal developed during the prior Administration by the Commerce, Interior, and Defense Departments and applies it to military readiness activities. That proposal espoused a recommendation by the National Research Council (NRC) that the currently overbroad definition of "harassment" of marine mammals—which includes "annoyance" or "potential to disturb"—be focused on biologically significant effects. As recently as 1999, the National Marine Fisheries Service (NMFS) asserted that under the sweeping language of the existing statutory definition harassment "is presumed to occur when marine mammals react to the generated sounds or visual cues"—in other words, whenever a marine mammal notices and reacts to an activity, no matter how transient or benign the reaction. As the NRC study found, "If [this] interpretation of the law for level B harassment (detectable changes in behavior) were applied to shipping as strenuously as it is applied to scientific and naval activities, the result would be crippling regulation of nearly every motorized vessel operating in U.S. waters."

Under the prior Administration, NMFS subsequently began applying the NRC's more scientific, effects-based definition. But environmental groups have challenged this regulatory construction as inconsistent with the statute. As you may know, the Navy and the National Oceanic and Atmospheric Administration suffered an important setback last year involving a vital anti-submarine warfare sensor—SURTASS LFA, a towed array emitting low-frequency sonar that is critical in detecting ultra-quiet diesel-electric submarines while they are still at a safe distance from our vessels. In the SURTASS LFA litigation environmental groups successfully challenged the new policy as inconsistent with the sweeping statutory standard, putting at risk NMFS' regulatory policy, clearly substantiating the need to clarify the existing statutory definition of harassment that we identified in our legislative package last year. [. . .]

The last change we are proposing, a national security exemption process, also derives from feedback the Defense Department received from environmental advocates last year after we submitted our proposal, as I discussed above. Although DoD continues to believe that predicating essential military training, testing, and operations on repeated invocations of emergency authority is unacceptable as a matter of public policy, we do believe that every environmental statute should have such authority as an insurance policy. The comments we received last year highlighted the fact that the MMPA does not currently contain such emergency authority, so this year's submission does include a waiver mechanism. Like the Endangered Species Act, our proposal would allow the Secretary of Defense, after conferring with the Secretaries of Commerce or Interior, as appropriate, to waive MMPA provisions for actions or categories of actions when required by national security. This provision is not a substitute for the other clarifications we have proposed to the MMPA, but rather a failsafe mechanism in the event of emergency.

The only substantive changes are those described above. The reason that the text is so much more extensive than last year's version is that last year's version was drafted as a freestanding part of title 10—the Defense Department title—rather than an amendment to the text of the MMPA itself. This year, because we were making several changes, we concluded that as a drafting matter we should include our changes in the MMPA itself. That necessitated a lot more language, largely just reciting existing MMPA language that we are not otherwise modifying.

The environmental impacts of our proposed reforms would be minimal. Although our initiative would exclude transient, biologically insignificant effects from regulation, the MMPA would remain in full effect for biologically significant effects—not only death or injury but also disruption of significant activities. The Defense Department could neither harm marine mammals nor disrupt their biologically significant activities without obtaining authorization from FWS or NMFS, as appropriate.

Nor does our initiative depart from the precautionary premise of the MMPA. The Precautionary Principle holds that regulators should proceed conservatively in the face of scientific uncertainty over environmental effects. But our initiative embodies a conservative, science-based approach validated by the National Research Council. By defining as "harassment" any readiness activities that "injure or have the significant potential to injure," or "disturb or are likely to disturb," our initiative includes a margin of safety fully consistent with the Precautionary Principle. The alternative is the existing grossly overbroad, unscientific definition of harassment, which sweeps in any activity having the "potential to disturb." As the National Research Council found, such sweeping overbreadth is unscientific and not mandated by the Precautionary Principle. [. . .]

The Defense Department already exercises extraordinary care in its maritime programs: all DoD activities worldwide result in fewer than 10 deaths or injuries annually (as opposed to 4800 deaths annually from commercial fishing activities). And DoD currently funds much of the most significant research on marine mammals, and will continue this research in future.

Although the environmental effects of our MMPA reforms will be negligible, their readiness implications are profound. Application of the current hair-trigger definition of "harassment" has profoundly affected both vital R&D efforts and training. Navy operations are expeditionary in nature, which means world events often require planning exercises on short notice. To date, the Navy has been able to avoid the delay and burden of applying for a take permit only by curtailing and/or dumbing down training and research/testing. For six years, the Navy has been working on research to develop a suite of new sensors and tactics (the Littoral Advanced Warfare Development Program, or LWAD) to reduce the threat to the fleet posed by ultraquiet diesel submarines operating in the littorals and shallow seas like the Persian Gulf, the Straits of Hormuz, the South China Sea, and the Taiwan Strait. These submarines are widely distributed in the world's navies, including "Axis of Evil" countries such as Iran and North Korea and potentially hostile great powers. In the 6 years that the program has operated, over 75% of the tests have been impacted by environmental considerations. In the last 3 years, 9 of 10 tests have been affected. One was cancelled entirely, and 17 different projects have been scaled back.

Endangered Species Act

Our Endangered Species Act provision is unchanged from last year. The legislation would confirm the prior Administration's decision that an Integrated Natural Resources Management Plan (INRMP) may in appropriate circumstances obviate the need to designate critical habitat on military installations. These plans for conserving natural resources on military property, required by the Sikes Act, are developed in cooperation with state wildlife agencies, the U.S. Fish and Wildlife Service, and the public. In most cases they offer comparable or better protection for the species because they consider the base's environment holistically, rather than using a species-by-species analysis. The prior Administration's decision that INRMPs may adequately provide for appropriate endangered species habitat management is being challenged in court by environmental groups, who cite Ninth Circuit caselaw suggesting that other habitat management programs provided an insufficient basis for the Fish and Wildlife Service to avoid designating Critical Habitat. These groups claim that no INRMP, no matter how protective, can ever substitute for critical habitat designation. This legislation would confirm and insulate the Fish and Wildlife Service's policy from such challenges.

Both the prior and current Administrations have affirmed the use of INRMPs as a basis for possible exclusion from critical habitat. Such plans are required to provide for fish and wildlife management, land management, forest management, and fish and wildlife-oriented recreation; fish and wildlife habitat enhancement; wetland protection, enhancement, and restoration; establishment of specific natural resource management goals, objectives, and timeframes; and enforcement of natural resource laws and regulations. And unlike the process for designation of critical habitat, INRMPs assure a role for state regulators. Furthermore, INRMPs must be reviewed by the parties on a regular basis, but not less than every five years, providing a continuing opportunity for FWS input.

By contrast, in 1999, the Fish and Wildlife Service stated in a Notice of Proposed Rulemaking that "we have long believed that, in most circumstances, the designation of 'official' critical habitat is of little additional value for most listed species, yet it consumes large amounts of conservation resources. [W]e have long believed that separate protection of critical habitat is duplicative for most species."

Our provision does not automatically eliminate critical habitat designation, precisely because under the Sikes Act, the statute giving rise to INRMPs, the Fish & Wildlife Service is given approval authority over those elements of the INRMP under its jurisdiction. This authority guarantees the Fish & Wildlife Service the authority to make a case-by-case determination concerning the adequacy of our INRMPs as a substitute for critical habitat designation. And if the Fish & Wildlife Service does not approve the INRMP, our provision will not apply to protect the base from critical habitat designation.

Our legislation explicitly requires that the Defense Department continue to consult with the Fish and Wildlife Service and the National Marine Fisheries Service under Section 7 of the Endangered Species Act (ESA); the other provisions of the ESA, as well as other environmental statutes such as the National Environmental Policy Act, would continue to apply, as well.

The Defense Department's proposal has vital implications for readiness. Absent this policy, courts, based on complaints filed by environmental litigants, compelled the Fish and Wildlife Service to re-evaluate "not prudent" findings for many critical habitat determinations, and as a result FWS proposed to designate over 50% of the 12,000-acre Marine Corps Air Station (MCAS) Miramar and over 56% of the 125,000-acre Marine Corps Base (MCB) Camp Pendleton. Prior to adoption of this policy, 72% of Fort Lewis and 40% of the Chocolate Mountains Aerial Gunnery Range were designated as critical habitat for various species, and analogous habitat restrictions were imposed on 33% of Fort Hood. These are vital installations.

Unlike Sikes Act INRMPs, critical habitat designation can impose rigid limitations on military use of bases, denying commanders the flexibility to manage their lands for the benefit of both readiness and endangered species.

Clean Air Act General Conformity Amendment

Our Clean Air Act amendment is unchanged since last year. The legislation would provide more flexibility for the Defense Department in ensuring that emissions from its military training and testing are consistent with State Implementation Plans under the Clean Air Act by allowing DoD and the states a slightly longer period to accommodate or offset emissions from military readiness activities.

The Clean Air Act's "general conformity" requirement, applicable only to federal agencies, has repeatedly threatened deployment of new weapons systems and base closure/realignment despite the fact that relatively minor levels of emissions were involved. [. . .]

Conclusion

In closing Mr. Chairman, let me emphasize that modern warfare is a "come as you are" affair. There is no time to get ready. We must be prepared to defend our country wherever and whenever necessary. While we want to train as we fight, in reality our soldiers, sailors, airmen and Marines fight as they train. The consequences for them, and therefore for all of us, could not be more momentous.

DoD is committed to sustaining U.S. test and training capabilities in a manner that fully satisfies that military readiness mission while also continuing to provide exemplary stewardship of the lands and natural resources in our trust. [. . .]

 **NO**

Impact of Military Training on the Environment

Prior to arriving at the National Wildlife Federation in 2001, I served for 13 years at the U.S. Fish and Wildlife Service, with the last 4 years as the Director of the agency. Prior to that, I served as Fish and Wildlife Administrator for the Department of the Army, Natural and Cultural Resources Program Manager for the National Guard Bureau, and Research Biologist for the U.S. Army Medical Research Institute. I am the daughter of a U.S. Army Colonel, and lived on or near military bases throughout my entire childhood.

Based on this experience, I am very familiar with the Defense Department's long history of leadership in wildlife conservation. On many occasions during my tenures at FWS and the Defense Department, DOD rolled up its sleeves and worked with wildlife agency experts to find a way to comply with environmental laws and conserve imperiled wildlife while achieving military preparedness objectives.

The Administration now proposes in its Readiness and Range Preservation Initiative that Congress scale back DOD's responsibilities to conserve wildlife and to protect people from the hazardous pollution that DOD generates. This proposal is both unjustified and dangerous. It is unjustified because DOD's longstanding approach of working through compliance issues on an installation-by-installation basis works. As DOD itself has acknowledged, our armed forces are as prepared today as they ever have been in their history, and this has been achieved without broad exemptions from environmental laws.

The DOD proposal is dangerous because, if Congress were to broadly exempt DOD from its environmental protection responsibilities, both people and wildlife would be threatened with serious, irreversible and unnecessary harm. Moreover, other federal agencies and industry sectors with important missions, using the same logic as used here by DOD, would line up for their own exemptions from environmental laws.

My expertise is in the Endangered Species Act (ESA), so I would like to focus my testimony on why exempting the Defense Department from key provisions of the ESA would be a serious mistake. I will rely on my fellow witnesses to explain why the proposed exemptions from other environmental and public health and safety laws is similarly unwise.

From Testimony before the Senate Committee on Environment and Public Works, April 2, 2003.

Concerns with the ESA Exemption

The Defense Department's proposed ESA exemption suffers from three basic flaws: it would severely weaken this nation's efforts to conserve imperiled species and the ecosystems on which all of us depend; it is unnecessary for maintaining military readiness; and it ignores the Defense Department's own record of success in balancing readiness and conservation objectives under existing law.

1. Section 2017 Removes a Key Species Conservation Tool

Section 2017 of the Administration's Readiness and Range Preservation Initiative would preclude designations of critical habitat on any lands owned or controlled by DOD if DOD has prepared an Integrated Natural Resources Management Plan (INRMP) pursuant to the Sikes Act and has provided "special management consideration or protection" of listed species pursuant to Section 3(5)(A) of the ESA.

This proposal would effectively eliminate critical habitat designations on DOD lands, thereby removing an essential tool for protecting and recovering species listed under the ESA. Of the various ESA protections, the critical habitat provision is the only one that specifically calls for protection of habitat needed for recovery of listed species. It is a fundamental tenet of biology that habitat must be protected if we ever hope to achieve the recovery of imperiled fish, wildlife and plant species.

Section 2017 would replace this crucial habitat protection with management plans developed pursuant to the Sikes Act. The Sikes Act does not require the protection of listed species or their habitats; it simply directs DOD to prepare INRMPs that protect wildlife "to the extent appropriate." Moreover, the Sikes Act provides no guaranteed funding for INRMPs and the annual appropriations process is highly uncertain. Even the best-laid management plans can go awry when the anticipated funding fails to come through. Yet, under Section 2017, even poorly designed INRMPs that allow destruction of essential habitat and put fish, wildlife or plant species at serious risk of extinction would be substituted for critical habitat protections.

Section 2017 contains one minor limitation on the substitution of INRMPs for critical habitat designations: such a substitution is allowed only where the INRMP provides "special management consideration or protection" within the meaning of Section 3(5)(A) of the ESA. Unfortunately, this limitation does nothing to ensure that INRMPs truly conserve listed species.

The term "special management consideration or protection" was never intended to provide a biological threshold that land managers must achieve in order to satisfy the ESA. The term is found in Section 3(5) of the ESA, which sets forth a two-part definition of critical habitat. Section 3(5)(A) states that critical habitat includes areas occupied by a listed species that are "essential for the conservation of the species" and "which may require special management consideration or protection." Section 3(5)(B) states that critical habitat also includes areas not currently occupied by a listed species that are simply "essential for the conservation of the species."

As this language makes clear, an ESA §3(5) finding by the U.S. Fish and Wildlife Service or National Marine Fisheries Service (Services) that a parcel of land "may require special management consideration or protection" is not the same as finding that it is already receiving adequate protection. Such a finding simply highlights the importance of a parcel of land to a species, and it should lead to designation of that land as critical habitat. See *Center for Biological Diversity* v. *Norton,* 240 F. Supp. 2d 1090 (D. Ariz. 2003) (rejecting, as contrary to plain meaning of ESA, defendant's interpretation of "special management consideration or protection" as providing a basis for substituting a U.S. Forest Service management plan for critical habitat protection). By allowing DOD to substitute INRMPs for critical habitat designations whenever it unilaterally makes a finding of "special management consideration or protection," Section 2017 significantly weakens the ESA.

Section 2017 is also problematic because it would eliminate many of the ESA Section 7 consultations that have stimulated DOD to "look before it leaps" into a potentially harmful training exercise. As a result of Section 7 consultations, DOD and the Services have routinely developed what is known as "work-arounds," strategies for avoiding or minimizing harm to listed species and their habitats while still providing a rigorous training regimen.

Section 2017 purports to retain Section 7 consultations. However, the duty to consult only arises when a proposed federal action would potentially jeopardize a listed species or adversely modify or destroy its critical habitat. By removing critical habitat designations on lands owned or controlled by DOD, Section 2017 would eliminate one of the two possible justifications for initiating a consultation, reducing the likelihood that consultations will take place. This would mean that DOD and the Services would pay less attention to species concerns and would be less effective in conserving imperiled species and maintaining the sustainability of the land.

The reductions in species protection proposed by DOD would have major implications for our nation's rich natural heritage. DOD manages approximately 25 million acres of land on more than 425 major military installations. These lands are home to at least 300 federally listed species. Without the refuge provided by these bases, many of these species would slide rapidly toward extinction. These installations have played a crucial role in species conservation and must continue to do so.

2. The ESA Exemption Is Not Necessary to Maintain Military Readiness

The ESA already has the flexibility needed for the Defense Department to balance military readiness and species conservation objectives. Three key provisions provide this flexibility. First, under the consultation provision of Section 7(a)(2) of the Act, DOD is provided with the opportunity to develop solutions in tandem with the Services to avoid unnecessary harm to listed species from military activities. Typically, the Services conclude, after informal consultation, that the proposed action will not adversely affect a listed species or its designated critical habitat or, after formal consultation, that it will not likely jeopardize a listed species or destroy or adversely modify its critical habitat. See, e.g., U.S. Army Environmental Center, Installation Summaries from the FY 2001 Survey of Threatened and Endangered Species on Army Lands (August 2002) at 9 (noting successful conclusion of 282 informal consultations and 36 formal consultations, with no "jeopardy" biological opinions). In both informal and formal consultations, the Services either will recommend that the action go forward without changes, or it will work with DOD to design "work arounds" for avoiding and minimizing harm to the species and its habitat. In either case, DOD accomplishes its readiness objectives while achieving ESA compliance.

Second, under Section 4(b)(2) of the ESA, the Services are authorized to exclude any area from critical habitat designation if they determine that the benefits of exclusion outweigh the benefits of specifying the area. (An exception is made for when the Services find that failure to designate an area as critical habitat will result in the extinction of a species—a finding that the Services have never made.) In making this decision, the Services must consider "the economic impact, and any other relevant impact" of the critical habitat designation. DOD has recently availed itself of this provision to convince the U.S. Fish and Wildlife Service to exclude virtually all of the habitat at Camp Pendleton—habitat deemed critical to five listed species in proposed rulemakings—from final critical habitat designations. Thus, for situations where the Section 7(a)(2) consultation procedures place undue burdens on readiness activities, DOD already has a tool for working with the Services on excluding land from critical habitat designation. Attached to my testimony is a factsheet that shows how the Services have worked cooperatively with DOD on these exclusions, and another factsheet showing the importance of maintaining the Services' role in evaluating proposed exclusions.

Third, under Section 7(j) of the ESA an exemption "shall" be granted for an activity if the Secretary of Defense finds the exemption is necessary for reasons of national security. To this date, DOD has never sought an exemption under Section 7(j), highlighting the fact that other provisions of the ESA have provided DOD with all the flexibility it needs to reconcile training needs with species conservation objectives.

Where there are site-specific conflicts between training needs and species conservation needs, the ESA provides these three mechanisms for resolving them in a manner that allows DOD to achieve its readiness objectives. Granting DOD a nationwide ESA exemption, which would apply in many places where no irreconcilable conflicts between training needs and conservation needs have arisen, would be harmful to imperiled species and totally unnecessary to achieve readiness objectives.

a. DOD Has Misstated the Law Regarding Its Ability to Continue with a Cooperative, Case-by-Case Approach to Critical Habitat Designations
DOD has stated that the ESA exemption is necessary because a recent court ruling in Arizona would prevent DOD from taking the cooperative, case-by-case approach to critical habitat designations that was developed when I served as

Director of the Fish and Wildlife Service. This description of the court ruling is inaccurate—the ruling clearly allows DOD to continue the cooperative, case-by-case approach if it wishes.

The court ruling at issue is entitled *Center for Biological Diversity* v. *Norton*, 240 F. Supp. 2d 1090 (D. Ariz. 2003). In this case, FWS excluded San Carlos Apache tribal lands from a critical habitat designation pursuant to ESA §4(b)(2) because the tribal land management plan was adequate and the benefits of exclusion outweighed the benefits of inclusion. The federal district court upheld the exclusion as within FWS's broad authority under ESA §4(b)(2). At the same time, the court held that lands could not legitimately be excluded from a critical habitat designation on the basis of the "special management" language in ESA §3(5).

Under the court's reasoning, FWS continues to have the broad flexibility to exclude DOD lands from a critical habitat designation on the basis of a satisfactory INRMP and the benefits to military training that the exclusion would provide. The ruling simply clarifies that such exclusions must be carried out pursuant to ESA §4(b)(2) rather than ESA §3(5). Thus, DOD's assertion that the Center for Biological Diversity ruling prevents it from working with FWS to secure exclusions of DOD lands from critical habitat designations is inaccurate.

b. DOD's Anecdotes Do Not Demonstrate That the ESA Has Reduced Readiness

The DOD has offered a series of misleading anecdotes describing difficulties it has encountered in balancing military readiness and conservation objectives. Before Congress moves forward with any exemption legislation, the appropriate Congressional committees should get a more complete picture of what is really happening at DOD installations.

Some of DOD's anecdotes are simply unpersuasive on their face, such as DOD's repeated assertion that environmental laws have prevented the armed services from learning how to dig foxholes and that troops abroad have been put at greater risk as a result. There is simply no evidence that environmental laws have ever prevented foxhole digging. Moreover, given its vast and varied landholdings and the many management options available, the Defense Department certainly can find places on which troops can learn to dig foxholes without encountering endangered species or other environmental issues.

Other anecdotes have simply disregarded the truth. For example, DOD and its allies have repeatedly argued that more than 50 percent of Camp Pendleton may not be available for training due to critical habitat designations. In fact, only five species have been proposed for critical habitat designations at Camp Pendleton. In each of these five instances, DOD raised concerns about impacts to military readiness, and in each instance, FWS worked closely with DOD to craft a solution. FWS ultimately excluded virtually all of the habitats for the five listed species on Camp Pendleton from critical habitat designations—even though FWS had earlier found that these habitats were essential to the conservation of the species. As a result of FWS's exclusion decisions, less than one percent of the training land at Camp Pendleton, and less than 4 percent of all of Camp Pendleton, is designated critical habitat. (Most of the critical habitat designated at Camp Pendleton is non-training land leased to San Onofre State Park, agricultural operations, and others. DOD's repeated suggestion that more than 50 percent of Camp Pendleton is at risk of being rendered off-limits to training due to critical habitat is simply inaccurate.

DOD also has argued that training opportunities and expansion plans at Fort Irwin have been thwarted by the desert tortoise. Yet just two weeks ago this official line was contradicted by the reality on the ground. In an article dated March 21, 2003, Fort Irwin spokesman Army Maj. Michael Lawhorn told the Barstow Desert Dispatch that he is unaware of any environmental regulations that interfere with troops' ability to train there. He also said there isn't any environmental law that hinders the expansion. [. . .]

These examples of misleading anecdotes highlight the need for Congress to look behind the reasons that are being put forward by DOD as the basis for weakening environmental laws. DOD uses the anecdotes in an attempt to demonstrate that conflicts between military readiness and species conservation objectives are irreconcilable. However, solutions to these conflicts are within reach if DOD is willing to invest sufficient time and energy into finding them. DOD has vast acres of land on which to train and vast stores of creativity and expertise among its land managers. With careful inventorying and planning, DOD can find a proper balance.

Has DOD made the necessary effort to inventory and plan for its training needs? In June 2002, the General Accounting Office issued a report entitled "Military Training: DOD Lacks a Comprehensive Plan to Manage Encroachment on Training Ranges," suggesting that the answer is no. The GAO found:

- DOD has not fully defined its training range requirements and lacks information on training resources available to the Services to meet those requirements, and that problems at individual installations may therefore be overstated.
- The Armed Services have never assessed the overall impacts of encroachment on training.
- DOD's readiness reports show high levels of training readiness for most units. In those few instances of when units reported lower training readiness, DOD officials rarely cited lack of adequate training ranges, areas or airspace as the cause.
- DOD officials themselves admit that population growth around military installations is responsible for past and present encroachment problems.
- The Armed Services' own readiness data do not show that environmental laws have significantly affected training readiness.

Ten months after the issuance of the GAO report, DOD still has not produced evidence that environmental laws are at fault for any of the minor gaps in readiness that may exist. EPA Administrator Whitman confirmed this much at a recent hearing. At a February 26, 2003, Senate Environment and Public Works Committee hearing on EPA's budget, EPA Administrator Whitman stated that she was "not aware of any particular area where environmental protection regulations are preventing the desired training."

To this date, DOD has not provided Congress with the most basic facts about the impacts of ESA critical habitat requirements on its readiness activities. Out of DOD's 25 million acres of training land, how many acres are designated critical habitat? At which installations? Which species? In what ways have the critical habitat designations limited readiness activities? What efforts did DOD make to alert FWS to these problems and to negotiate resolutions? Without answers to these most basic questions, Congress cannot fairly conclude that the ESA is at fault for any readiness gaps or that a sweeping ESA exemption is warranted.

3. DOD has Worked Successfully with the Services to Balance Readiness and Species Conservation Objectives

The third reason why enacting DOD's proposed ESA changes would be a mistake is because the current approach—developing solutions at the local level, rather than relying on broad, national exemptions—has worked. My experience at both FWS and DOD has shown me that solutions developed at the local level are sometimes difficult to arrive at, but they are almost always more intelligent and long-lasting than one-size-fits-all solutions developed at the national level.

Allow me to provide a few brief examples. At the Marine Corps Base at Camp Lejeune in North Carolina, every colony tree of the endangered red-cockaded woodpecker is marked on a map, and Marines are trained to operate their vehicles as if those mapped locations are land mines. Here is the lesson that Major General David M. Mize, the Commanding General at Camp Lejeune, has drawn from this experience:

> "Returning to the old myth that military training and conservation are mutually exclusive; this notion has been repeatedly and demonstrably debunked. In the overwhelming majority of cases, with a good plan along with common sense and flexibility, military training and the conservation and recovery of endangered species can very successfully coexist."
>
> "Military installations in the southeast are contributing to red-cockaded woodpecker recovery while sustaining our primary mission of national military readiness."
>
> "I can say with confidence that the efforts of our natural resource managers and the training community have produced an environment in which endangered species management and military training are no longer considered mutually exclusive, but are compatible."

These sentiments, which I share, were relayed by Major General Mize just eight weeks ago at a National Defense University symposium sponsored by the U.S. Army Forces Command (FORSCOM) and others. At that symposium, representatives of Camp Lejeune Marine Corps Base, Eglin Air Force Base, Fort Bragg Army Base, Fort Stewart Army Base, Camp Blanding Training Center in Florida, the U.S. Army Environmental Center, and other Defense facilities—some of the most heavily utilized training bases in the country—heralded the success that Defense Department installations have had in furthering endangered species conservation while maintaining military readiness.

On the Mokapu Peninsula of Marine Corps Base Hawaii, the growth of non-native plants, which can decrease the reproductive success of endangered waterbirds, is controlled through annual "mud-ops" maneuvers by Marine Corps Assault Vehicles. Just before the onset of nesting season, these 26 ton vehicles are deployed in plow-like maneuvers that break the thick mats of invasive plants, improving nesting and feeding opportunities while also giving drivers valuable practice in unusual terrain. [. . .]

These success stories highlight a major trend that I believe has been missed by those promoting the DOD exemptions. In recent years, DOD has increasingly recognized the importance of sustainability because it meets several importance objectives at once. Sustainable use of the land helps DOD achieve not only compliance with environmental laws, but also long-term military readiness and cost-effectiveness goals. For example, by operating tanks so that they avoid the threatened desert tortoise, DOD prevents erosion, a problem that is extremely difficult and costly to remedy. If DOD abandons its commitment to environmental compliance, it will incur greater long-term costs for environmental remediation and will sacrifice land health and military readiness.

A November 2002 policy guidance issued by the then-Secretary of the Navy to the Chief of Naval Operations and the Commandant of the Marine Corps suggests that certain members of DOD's leadership are indeed willing to abandon the sustainability goal. The policy guidance on its face seems fairly innocuous—it purports to centralize at the Pentagon all decision making on proposed critical habitat designations and other ESA actions. However, the Navy Secretary's cover memo makes clear that its purpose is also to discourage any negotiation of solutions to species conservation challenges by Marines or Navy personnel in the field, lest these locally-developed "win-win" solutions undercut DOD's arguments on Capitol Hill that the ESA is broken. According to paragraph 2 of the cover memo, "concessions [. . .] could run counter to the legislative relief that we are continuing to pursue with Congress."

Similar sentiments were voiced by Deputy Defense Secretary Paul Wolfowitz in his March 7, 2003, memo to the chiefs of the Army, Navy and Air Force. Deputy Secretary Wolfowitz argued that "it is time for us to give greater consideration to requesting exemptions" from environmental laws and pleaded for specific examples of instances in which environmental regulations hamper training. The implicit message is that efforts at the installation level to resolve conflicts between conservation and training objectives should be suspended, and that such conflicts instead should be reported to the Pentagon, where environmental protections will simply be overridden.

These messages to military personnel in the field mark a very unfortunate abdication of DOD's leadership in wildlife conservation. To maintain its leadership role as steward of this nation's endangered wildlife, DOD must encourage its personnel to continue developing innovative solutions and not thwart those efforts.

Conclusion

With the Iraq war ongoing and terrorism threats always present, no one can dismiss the importance of military readiness. However, there is no justification for the Defense Department to retreat from its environmental stewardship commitments at home. As base commanders have been telling us, protecting endangered species and other important natural resources is compatible with maintaining military readiness.

Surveys show that the American people today want environmental protection from the federal government, including the Defense Department, as much as ever. According to an April 2002 Zogby Poll, 85% of registered voters believe that the Defense Department should be required to follow America's environmental and public health laws and not be exempt. Americans believe that no one, including the Defense Department, should be above the law.

Congress should reject the proposed environmental exemptions in the Administration's defense authorization package. This proposal, along with the parallel proposal in the Administration's FY04 budget request that Congress cut spending on DOD's environmental programs by $400 million, are a step in the wrong direction.

DOD has a long and impressive record of balancing readiness activities with wildlife conservation. The high quality of wildlife habitats at many DOD installations provides tangible evidence of DOD's positive contribution to the nation's conservation goals. At a time when environmental challenges are growing, DOD should be challenged to move forward with this successful model and not to sacrifice any of the progress that has been made. [. . .]

EXPLORING THE ISSUE

Should the Military Be Exempt from Environmental Regulations?

Critical Thinking and Reflection

1. It is one thing to debate exempting military training from environmental regulations. Should we also be debating exempting actual combat from environmental regulations?
2. If military activities warrant exemptions from environmental regulations, are there other activities that warrant similar exemptions? Name three, and attempt to justify their exemptions.
3. If military activities *do not* warrant exemptions from environmental regulations, are there other activities that might? (This amounts to asking what is more important to national survival than military defense.)
4. In what ways is the precautionary principle an essential component of sustainable development?

Is There Common Ground?

Even military spokespersons agree that protecting the environment is important, and they frequently insist that the Department of Defense is an "exemplary" steward of the lands and resources under its care. Yet they are clearly prepared to set that stewardship aside when it interferes with the military mission. Their critics are not, although they may be less vocal when the discussion shifts from training or readiness to actual military action.

It is worth noting that changes in military technology may in time reduce the environmental impacts of military training and action. Ask students to research and report on the use of computer games (see, e.g., http://www.modelbenders.com/papers/RSmith_Escapist_080829.pdf), drones, and robots.

ISSUE 6

Will Restricting Carbon Emissions Damage the Economy?

YES: Paul Cicio, from "Competitiveness and Climate Policy: Avoiding Leakage of Jobs and Emissions," testimony before the House Committee on Energy and Commerce, Subcommittee on Energy and Environment (March 18, 2009)

NO: Aaron Ezroj, from "How Cap and Trade Will Fuel the Global Economy," *Environmental Law Reporter* (July 2010)

Learning Outcomes

After reading this issue, you should be able to:

- Explain how a cap-and-trade system can reduce pollution levels over time.
- Explain how cap-and-trade systems motivate corporate polluters to pollute less.
- Explain why restricting carbon emissions would stimulate the development of new energy-related technology.
- Discuss what factors are most important in making policy decisions such as whether to restrict emissions of greenhouse gases.

ISSUE SUMMARY

YES: Paul Cicio argues that lacking global agreements, capping greenhouse gas emissions of the industrial sector will make domestic production less competitive in the global market, drive investment and jobs offshore, increase exports, and damage the economy. The real greenhouse gas problem lies with other sectors of the economy, and that is where attention should be focused.

NO: Aaron Ezroj argues that although restricting emissions (as in a cap-and-trade program) may increase costs for some businesses, it will create many business opportunities in the financial sector, low-carbon technologies, carbon capture-and-storage projects, advanced-technology vehicles, and legal and nonlegal consulting. The overall effect will be to fuel the global economy.

Following World War II, the United States and other developed nations experienced an explosive period of industrialization accompanied by an enormous increase in the use of fossil fuel energy sources and a rapid growth in the manufacture and use of new synthetic chemicals. In response to growing public concern about pollution and other forms of environmental deterioration resulting from this largely unregulated activity, the U.S. Congress passed the National Environmental Policy Act of 1969. This legislation included a commitment on the part of the government to take an active and aggressive role in protecting the environment. The next year the Environmental Protection Agency (EPA) was established to coordinate and oversee this effort. During the next two decades, an unprecedented series of legislative acts and administrative rules were promulgated, placing numerous restrictions on industrial and commercial activities that might result in the pollution, degradation, or contamination of land, air, water, food, and the workplace.

Such forms of regulatory control have always been opposed by the affected industrial corporations and developers as well as by advocates of a free-market policy, who prefer reliance on voluntary measures. More moderate critics of the government's regulatory program recognize that adequate environmental protection will not result from completely voluntary policies. They suggest that a new set of strategies is needed. Arguing that "top-down, federal, command and control legislation" is not an appropriate or effective means of preventing ecological degradation; they propose a wide range of alternative tactics, many of which are designed to operate through the economic marketplace. The first significant congressional response to these proposals was the incorporation of tradable pollution emission rights into the 1990 Clean Air Act amendments as a means for achieving the set goals for reducing acid-rain-causing sulfur dioxide emissions. More recently, the 1997 international negotiations on controlling global warming in Kyoto, Japan, resulted in a protocol that includes emissions trading as one of the key elements in the plan to limit the atmospheric buildup of greenhouse gases.

Charles W. Schmidt, "The Market for Pollution," *Environmental Health Perspectives* (August 2001), argues that emissions trading schemes represent "the most significant developments" in the use of economic incentives to motivate corporations to reduce pollution. In "A Low-Cost Way to Control Climate Change," *Issues in Science and Technology* (Spring 1998), Byron Swift argues that the "cap-and-trade" feature of the U.S. Acid Rain Program has been so successful that a similar system for implementing the Kyoto Protocol's emissions trading mandate as a cost-effective means of controlling greenhouse gases should work. In March 2001, the U.S. Senate Committee on Agriculture, Nutrition, and Forestry held a "Hearing on Biomass and Environmental Trading Opportunities for Agriculture and Forestry," in which witnesses urged Congress to encourage trading for both its economic and its environmental benefits. Richard L. Sandor, chairman and chief executive officer of Environmental Financial Products LLC, said that "200 million tons of CO_2 could be sequestered through soils and forestry in the United States per year. At the most conservative prices of $20–$30 per ton, this could potentially generate $4–$6 billion in additional agricultural income."

Cap-and-trade systems work by setting a limit (the cap) on how much of a pollutant can be emitted per year. Permits to emit a portion of the pollutant are then made available—either free or as the result of an auction process—to businesses that emit that pollutant. Businesses that do not emit as much as their permits allow can then sell their unused permits. This provides an incentive to reduce emissions by spending money to improve efficiency. But a cap-and-trade system does require spending money for initial permits, for improved technology, and/or for additional permits when a company cannot or does not control emissions. This means added expenses, with an impact on profitability. Many businesses are concerned that this will affect their competitive position and even their ability to stay in business. A crucial question thus becomes whether environmental or economic protection should come first.

Europe is already implementing its own Greenhouse Gas Emissions Trading Scheme, although Marianne Lavelle, "The Carbon Market Has a Dirty Little Secret," *U.S. News and World Report* (May 14, 2007), reports that in Europe the value of tradable emissions allowances fell so low at one point, partly because too many allowances were issued, that it was cheaper to burn more fossil fuel and emit more carbon than to burn and emit less. Future trading schemes will need to be designed to avoid the problem, and the U.S. Congress is actively considering ways to address the issue (see "Support Grows for Capping and Trading Carbon Emissions," *Issues in Science and Technology* (Summer 2007)). David G. Victor and Danny Cullenward, "Making Carbon Markets Work," *Scientific American* (December 2007), argue that what is needed is to combine a trading program with limits on emissions and careful management. Bill McKibben, "The Greenback Effect," *Mother Jones* (May/June 2008), agrees, noting that although markets may not be perfect, when they work, they work fast.

Meanwhile, there is also interest in what is known as "carbon offsets," by which corporations, governments, and even individuals compensate for carbon dioxide emissions by investing in activities that remove carbon dioxide from the air or reduce emissions from a different source. See Anja Kollmuss, "Carbon Offsets 101," *World Watch* (July/August 2007). Unfortunately, present carbon-offset schemes contain loopholes that mean they may do little to reduce overall emissions; see Madhusree Mukerjee, "A Mechanism of Hot Air," *Scientific American* (June 2009).

On May 21, 2009, the House Energy and Commerce Committee approved H.R. 2454, "The American Clean Energy and Security Act." The goal of the Act, said Committee Chair Henry A. Waxman (D-CA), was to "break our dependence on foreign oil, make our nation the world leader in clean energy jobs and technology, and cut global warming pollution. I am grateful to my colleagues who supported this legislation and to President Obama for his outstanding leadership on these critical issues." Among other things, the Act called for Title VII of the Clean Air Act to provide a declining limit on global warming pollution (a "cap" as in "cap-and-trade") and to hold industries accountable for pollution reduction under the limit. The aim was to cut global warming pollution by 17 percent compared with 2005 levels in 2020, by 42 percent in 2030, and by

83 percent in 2050. (See the summary of the Act at http://energycommerce.house .gov/Press_111/20090515/hr2454_summary.pdf). In June 2009, the House of Representatives passed the bill, which also called for utilities to use more renewable energy sources. Unfortunately, the Senate refused to pass a corresponding bill, and "cap and trade" is dead, at least for now; see Jeff Johnson, "Cap and Trade Dies in Senate," *Chemical & Engineering News* (August 2, 2010). Part of the reason was the perception of the bill as imposing additional taxes in a time of economic difficulty; see C. Boyden Gray, "The Problem with Cap and Trade," *American Spectator* (June 2010).

Many people feel that it is about time that the United States took such action. According to the Global Humanitarian Forum's "Human Impact Report: Climate Change—The Anatomy of a Silent Crisis" (May 29, 2009), global warming is already affecting more than 300 million people and is responsible for 300,000 deaths per year. Action now is clearly appropriate, even though it does seem inevitable that this deadly impact of global warming must grow worse for many years before it can be stopped. However, the debate over the proper actions to take is by no means over. Some analysts argue that a carbon tax would be more effective; see Bettina B. F. Wittneben, "Exxon Is Right: Let Us Re-Examine Our Choice for a Cap-and-Trade System over a Carbon Tax," *Energy Policy* (June 2009). William B. Bonvillian, "Time for Climate Plan B," *Issues in Science and Technology* (Winter 2011), calls for a major effort to stimulate the development of new technologies.

In the YES selection, Paul Cicio, president of the Industrial Energy Consumers of America, argues that lacking global agreements, capping greenhouse gas emissions of the industrial sector will make domestic production less competitive in the global market, drive investment and jobs offshore, increase exports, and damage the economy. The real greenhouse gas problem lies with other sectors of the economy, and that is where attention should be focused. In the NO selection, Aaron Ezroj, Law Fellow at Adams Broadwell Joseph & Cardozzo, argues that although restricting emissions (as in a cap-and-trade program) may increase costs for some businesses, it will create many business opportunities in the financial sector, low-carbon technologies, carbon capture-and-storage projects, advanced-technology vehicles, and legal and nonlegal consulting. The overall effect will be to fuel the global economy.

YES

Paul Cicio

Competitiveness and Climate Policy: Avoiding Leakage of Jobs and Emissions

Key Points

Capping the greenhouse gas (GHG) emissions of the industrial sector will drive investment and jobs offshore and increase imports. It will not bring major developing countries to the table but they will benefit through increased exports to the US. Even the third phase of the EU Emissions Trading Scheme (ETS) contains a provision to ensure their trade exposed industries receive compensation in order to prevent job loss and emissions leakage. Regulating the US industrial sector "before" negotiating an international agreement undermines our ability to achieve a fair and effective GHG reduction agreement for US industry.

For the industrial sector, climate policy is also trade, energy, [and] economic and employment policy. They are all intrinsically linked and inseparable. It is for this reason that regulating GHG emissions for the industrial sector be negotiated with both developed and developing countries in the context of a fair trade and productivity.

The US industrial sector is not the problem. In the US, the industrial sector's GHG emissions have risen only 2.6% above 1990 levels while emissions from the residential sector are up 29%, commercial up 39%, transportation up 27% and electricity generation up 29%.

The industrial sector competes globally and requires a global GHG policy solution that is based on productivity, something that the developing countries industrial sector can potentially agree to. A GHG cap is an unacceptable policy alternative for them and for us.

The US cannot grow the economy without using more volume of our products. The only question is whether the product will be supplied from domestic sources or imports. In fact a cap limits economic efficiency because it even limits the ability to maximize production from existing facilities that are not running at installed capacity. Since 2000, US manufacturing has been losing ground. From 2000 to 2008, imports are up 29% and manufacturing unemployment fell 22%, losing 3.8 million jobs, a direct statistical correlation.

The use of energy by the industrial sector is value-added. Our products enable GHG emission reductions. Lifecycle studies show that they save much

U.S. Senate, March 18, 2009.

more energy and GHG emissions than what is used/emitted in their production. Raising energy costs raises the cost of these valuable products.

The industrial sector already has a price signal for GHG emissions, it is called global competition and because we are energy intensive, we either drive down our energy costs or go out of business.

Under cap and trade, the industrial sector pays twice: [t]hrough the additional cost of carbon embedded in energy purchases and through the higher cost of natural gas and electricity. Higher demand for natural gas will result in higher prices for all consumers. Since natural gas power generation sets the marginal price of electricity, higher natural gas prices will mean higher electricity prices for all consumers.

A cap will damage the ability of the US industrial sector to take back market share from imports and increase exports.

Cap and trade does not address our country's fundamental need to significantly increase the availability, affordability and reliability of low carbon sources of supply.

Carbon trading and market manipulation is of great concern. The US government has proven unable to prevent market manipulation for mature energy and food commodities and credit default swaps—carbon markets will be much harder to regulate.

If the US proceeds to cap GHGs, it must provide to industry free allowances equal to the resulting increased direct and indirect costs due to GHG regulation until major competing developing countries have similar cost increases.

Congressional Justification for Not Capping GHG Emissions of the Industrial Sector

Congress has a choice to make and it is a decision it cannot afford to make incorrectly. It must decide whether to maintain and possibly increase US manufacturing jobs by not capping GHG emissions on the industrial sector—or create jobs in foreign countries by importing manufacturing products to supply the needs of our economy.

The Industrial Energy Consumers of America is an association of leading manufacturing companies with $510 billion in annual sales and with more than 850,000 employees nationwide. It is an organization created to promote the interests of manufacturing companies. IECA membership represents a diverse set of industries including plastics, cement, paper, food processing, brick, chemicals, fertilizer, insulation, steel, glass, industrial gases, pharmaceuticals, aluminum and brewing.

The decision should not be hard because there is very sound economic and environmental justification for Congress to not act in the short term to cap GHG reductions on the industrial sector but to forge a different policy path that will provide sustained GHG reductions globally by harnessing real market forces called competition.

The industrial sector needs a globally level playing field that lets the best companies win. Adding costs by unilateral action helps "all" of our competitors in other countries take our business and our jobs. **We need US leadership to forge a global effort to address industrial sector GHG emission reductions that is focused on "fair trade" and "productivity." This is the only way to potentially bring developing nations to the table.**

Productivity is a language that all manufacturers understand and fundamental to competition. We believe that all governments want increased productivity by their industrial sector. We urge you to take action in this more realistic direction.

The world in which the industrial manufacturing company operates is diverse and business is often won or lost on the difference between pennies per unit of product. Competitiveness is everything. Some segments of industry, such as the power producers, may support cap and trade, but that's because they don't compete globally and they simply pass through their increased costs, we don't have that luxury.

Unlike that vision that many Americans have of China building coal-fired power plants using antiquated technology, it is vitally important that the Congress understand that a great number of companies that we compete with from developing countries are top-in-class competitors. They are utilizing the latest, world class technology. Some of these facilities are state owned or supported. Many also have subsidized energy costs. Energy costs most often determine our competitiveness and it can be our largest non-controllable cost.

The Congress can act in the public interest to consider both the cost and benefits of not imposing the cap on the industrial sector. The benefits of not imposing a GHG cap include good paying jobs, exports that reduce our balance of payments and the domestic production of products that are solutions to our climate challenges.

So far, only the environmental costs have been debated. We caution you to consider that your policy decision can lead to a further acceleration of the loss of the industrial sector. Just look at the facts. Due to the loss of competitiveness since 2000, the manufacturing sector has lost 3.8 million jobs thru 2008. During this same time period, imports rose 29%, a direct statistical correlation.

President Obama rightfully points to the disappearing middle class as troubling. We agree. The US began to lose the middle class when the industrial sector began to lose competitiveness along with our high paying jobs that most often pay benefits. The timing is consistent. We encourage the president and Congress to work with us to put new industrial policies in place that will increase competitiveness and grow the industrial sector and greatly restore the middle class.

To their credit, Representative Inslee and Doyle have rightfully recognized the need to protect manufacturing competitiveness. They are well intentioned but their solution is not really a solution for an industry that competes globally. We will still be burdened with costs and uncertainty. Most importantly, it does not do anything to bring the industrial sectors of developing countries into a climate agreement. Instead, a global solution is warranted that puts us on equal footing with our competitors. The international agreement should be negotiated first, not second. Regulating the US industrial sector in advance of negotiations completely removes our negotiation leverage.

The global reality is that developing nations place a significant priority on their manufacturing sector for both domestic economic growth and exports. They have a long history of providing all types of subsidies that include energy and trade credits. If they subsidize energy costs for their manufacturers, why wouldn't they also subsidize the cost of GHG reductions to enable exports to the US? US industry needs a level playing field—and then let us compete.

The justification is obvious and in the best interests of the country. The industrial sector's absolute GHG emissions are only 2.6% above 1990 levels and the rate of change has been flat due to energy efficiency improvements and a declining manufacturing presence. In contrast, according to the EPA, the transportation sector emissions are up 27%, residential up 29%, commercial up 39% and power generation up 29%. The point is that the industrial sector is *not* a contributor to growing GHG emissions and should not be a high priority for GHG reduction mandates.

Secondly, the products we produce are essential for economic growth of the country and a vibrant opportunity to create new high paying jobs. As the economy rebounds, our country will require significant volumes of the products that we produce such as cement, steel, aluminum, chemicals, plastics, paper, glass, and fertilizer which are all energy intensive. You can't produce renewable energy without our products. The question Congress must answer is whether it wants these products to be supplied by production facilities in the US or imported from foreign countries.

If Congress places a declining GHG cap on the industrial sector, you can be pretty confident that US companies will "not" invest their capital nor create jobs in the US. The reason is obvious. There is a lack of confidence that other countries will place a GHG cap on their manufacturers any time soon which would place US industry at a significant competitive disadvantage. Setting a starting date of 2012 for a GHG cap will result in industrial companies making pre-emptive capital decisions on where to locate and increase the production of their products that anticipates these assumptions.

Third, products from the manufacturing sector provide the "enabling solutions" to the challenges of climate change and it is important that GHG regulation does not increase the cost of these products to deter consumer purchases.

It takes energy to save energy. Insulation can be made from glass, plastic or paper, all of which are energy intensive. Double pane windows use twice the amount of glass but save an enormous amount of energy over the life of a building. Reducing the weight of autos, trucks and aircraft is an essential solution but requires greater use of aluminum, composite plastics and different grades of steel. More steel and plastics are needed for wind turbines. The production of solar silicon used to make solar panels is energy intensive. There are literally a thousand examples of how manufacturing products contribute to the climate solution and it is important to keep the cost of these products low.

The industrial sector is the "green sector." Manufacturing has a remarkable track record of reducing energy while continuing to increase the output of product. They predominantly use natural gas as a fuel versus coal. They are the largest consumer of biomass that is used for making paper and as a fuel for producing energy efficient steam and power. They utilize combined heat and power extensively and substantial quantities of recycled steel, aluminum, glass and paper which is extraordinarily energy efficient.

Fourth, placing a GHG cap on manufacturing makes it much more difficult for our sector to reclaim domestic market share and increase exports. The US has a significant trade deficit in part due to declining manufacturing product exports that accelerated in 2000 as US natural gas prices rose and imports increased.

A lot of these imports are from China, a country that values its manufacturing sector. And now, the US is dependent upon China to finance its burgeoning debt. Improving the competitive health of our manufacturing sector can help reduce this dependency. Increasing competitiveness of the industrial sector and increasing exports is an important matter of public policy that needs [to be] addressed.

The decision is yours to make. Company CEOs have a responsibility to their shareholders to protect the company's interest and they will. The manufacturing sector is agile and mobile to survive and thrive—it is just a question of where.

Climate Policy and Manufacturing Competitiveness

IECA has not attempted to gain consensus by the industrial sector on what is the best way to regulate GHG emissions for the US economy or for the manufacturing sector. However, there is little question how the majority view policy options.

Every discussion begins and ends with "competitiveness." Manufacturers compete globally and for many, the cost of energy and carbon will determine whether they will successfully compete in domestic and global markets.

The "absolute" cost of energy and carbon does not matter so long as all of our competitors around the world have the same increased costs. What matters to manufacturers is the "relative" cost of energy and carbon compared to our major global competitors regardless of whether they are in Europe or a developing country.

For that reason, US climate policy must not increase our relative costs. This means that manufacturing competitiveness must also be dealt with at the international level. While this presents a challenge for policy makers, it also provides a wonderful opportunity.

Those of us from the industry believe that more GHG emission reductions can be achieved globally when industrial climate policy instruments are focused on productivity that is, increasing production while reducing energy consumption. It's a win-win and recognizes that all players can only manage the energy use inside their plant and often have little control on the type of energy available.

There is general agreement by US manufacturers that other countries will not knowingly sacrifice their manufacturing jobs in response to climate policy. Since China tends to be a policy lynchpin, it is important to note that they especially will not sacrifice their manufacturing competitiveness to address climate change.

It is China's manufacturing sector that has raised its status to a world power by creating jobs and exports that have provided a significant and unequaled trade surplus. Now, the US is dependent upon them to buy our treasury bills and finance our debt. This is not an enviable position for the US nor is it necessary.

To its credit, the Chinese government has a history of emphasizing the importance of the manufacturing sector which is in contrast to the US government. China has also provided export tax credits, subsidies for energy costs and manages its currency. Some US government officials claim that currency control gives China a 40% competitive advantage over US manufacturers. Whether it's the currency or not, China's manufacturing sector is winning and US manufacturing is losing.

Any US climate policy option must hold manufacturing harmless until major competitors in both developed and countries in transition have comparable energy and carbon cost increases. Comparable reduction requirements do not meet the test. Without this protection, US manufacturers will protect their shareholders and move production facilities to countries that offer a competitive environment.

Well intentioned members of Congress have proposed a cap and trade system that would provide manufacturers with "some" free allowances that would decline over time and would cover "some" of the resulting higher energy costs. While appreciated, these provisions are not adequate to allow the industrial sector to compete, and grow domestic production and exports. Many US industries have been working on energy efficiency for decades and simply don't have technology available to make step changes needed to meet these ratcheting targets.

Under these provisions we will still have a declining GHG cap that reduces our production; unpredictable costs for energy, carbon and transaction costs; and un-necessary cost increases. It also does not do anything to help our domestic customers who will be asked to absorb higher costs for our products.

Economy-wide cap and trade is simply the wrong policy platform for the manufacturing sector. IECA wants a climate policy that will allow US manufacturing to: invest in the US; does not create winners and losers; does not penalize those who have already invested in energy efficiency; and transparency so that the system cannot be manipulated or gamed.

Relatively few manufacturers in the industry support cap and trade. The ones that do have either inherent special circumstances that allow them to gain a relative competitive advantage; have already moved their energy intensive manufacturing offshore; will significantly benefit from increased product sales or are simply not energy intensive and are not measurably impacted.

We do not know any manufacturing companies who support carbon cap and trade with auction. This is completely understandable because the manufacturing sector needs predictability over long time horizons for capital investment. The auction of carbon allowances does not give price certainty plus manufacturers are disadvantaged in competing for the auctioned carbon with regulated utilities who can afford to pay any price and then pass the cost on to consumers to pay.

If the government lets Wall Street participate, the auction option gets even worse. In general, manufacturers believe that only companies who are required to reduce GHG emissions should be allowed to purchase carbon allowances or offsets. This leaves Wall Street out.

Auctioning is the quickest way to lose manufacturing jobs and they will go silently, one at a time and without an announcement. Each manufacturing production unit has a cost break—even that varies significantly from plant to plant and from company to company. As the cost of carbon rises, the manufacturer will not have any choice but to shut it down.

Very few companies support cap and trade even if allowances are initially provided free of charge because they recognize that these temporary allowances are not a safety net and their economic viability is in jeopardy long term. The engineering limitations of their manufacturing facilities leave little room for imagination—just realism.

A carbon tax is better than a cap and trade program because it does not constrict our ability to increase the volume of product produced, it is superior in transparency, and more easily adjusted at the border. Nonetheless, it is a cost that is not welcomed and un-necessary for the industrial sector to reduce carbon intensity. Clearly, a high carbon tax will be just as effective in putting us out of business.

There are about 350,000 manufacturing facilities in the US. It is estimated that about 7,800 facilities would emit 10,000 tons of CO_2 per year. By itself, regulating the industrial sector presents a significant regulatory challenge for the federal government. While only 7,800 would be regulated, the other 342,200 facilities and the American consuming public would be asked to absorb higher resulting product costs. . . .

Aaron Ezroj

How Cap and Trade Will Fuel the Global Economy

Cap-and-trade programs and related measures will spark a wide range of business opportunities. The financial sector will grow to facilitate hundreds of billions of dollars worth of climate change-related exchanges. By 2050, markets for low-carbon technologies are likely to be worth at least $500 billion annually, and possibly much more. Numerous carbon capture-and-storage projects will emerge. Plug-in and other advanced technology vehicles will become the norm. Moreover, a plethora of legal and nonlegal consulting agencies will be advising government agencies and companies on climate change. Countries and companies should position themselves now in order to take full advantage of these opportunities.

The European Union (EU) has done the most to position itself and companies within it. Although it initially disfavored cap and trade, the EU has implemented the world's most expansive cap-and-trade program: the EU Emission Trading System (EU ETS). Following the implementation of the EU ETS, financial markets in London are overseeing the trading of billions of dollars in carbon allowances. Diplomats in Brussels are negotiating guidelines for offset projects in China. Moreover, Copenhagen is becoming the global center for the development of wind turbine technology. These efforts have provided the EU and companies within it with a significant head start in positioning themselves for the transition to a low-emissions global economy.

Eying the success of cap and trade in Europe, New Zealand is moving forward with the New Zealand Emissions Trading Scheme (NZ ETS), Australia is moving forward with the Australian Carbon Pollution Reduction Scheme (CPRS), and individual states and provinces within the United States and Canada are moving forward with their programs and joining larger regional collectives, such as the Regional Greenhouse Gas Initiative (RGGI) in the northeastern United States and the Western Climate Initiative (WCI) in the western United States and Canadian Provinces. The U.S. federal government is also contemplating putting together a program of enormous proportions, involving agencies such as the U.S. Environmental Protection Agency (EPA), the U.S. Department of Agriculture, the Federal Energy Regulatory Commission, and the Commodities Futures Trading Commission, among others. As more countries and regions enact cap-and-trade programs, these programs will fuel the global economy.

From *Environmental Law Reporter,* July 2010, pp. 10696–10705. Copyright © 2010 by Environmental Law Institute, Washington, DC. Reprinted by permission of ELI.

Financial Markets

In 2008, transactions on the global carbon market amounted to $92 billion. There were over three billion spot, future, and option contracts traded for a variety of reasons, including compliance, risk management, and arbitrage. Between 2007 and 2008, the value of transactions nearly doubled for the EU ETS. As current cap-and-trade programs expand to cover more sectors of the economy and other countries and regions develop cap-and-trade programs, the global carbon market will continue to grow rapidly. While it is impossible to forecast the growth of this emerging sector with exact accuracy, it has been projected that if all developed countries had carbon markets covering all fossil fuels, the global carbon market would grow by 200%. Moreover, if markets were established in all the top 20 emitting countries, the global carbon market would grow by 400%. According to Louis Redshaw, the head of environmental markets at Barclays Capital: "Carbon will be the world's biggest commodity market, and it could become the world's biggest market overall."

Allowances are the basic unit traded within the global carbon market. A single allowance provides a compliance entity with the right to emit one ton of carbon dioxide (CO_2) or CO_2 equivalent. Allowances are introduced into the market through a distribution made by a government agency or an auction, and resulting revenue funds policy mandates. Allowances are then traded between compliance entities through market exchanges or over-the-counter transactions. At the end of each compliance period, compliance entities surrender allowances to a designated regulator for each ton of CO_2 or CO_2 equivalent that they emitted during the period.

Emissions trading enables a compliance entity to emit more than permitted by its current holding of allowances if it can obtain spare allowances from another compliance entity. The overall environmental outcome is the same as if both compliance entities used their allowances exactly, but with the important difference that both buying and selling companies benefited from the flexibility allowed by trading. Moreover, emissions trading encourages compliance entities to find cost-effective ways to reduce their emissions, which allows compliance entities to purchase fewer allowances.

However, the price for allowances is not static. The price of an allowance can spike upward as a result of weather fluctuations. For instance, a low water year affects the generation of hydroelectricity. The price of an allowance can spike downward if there is a recession, there is less demand for energy, and in turn compliance entities require less allowances to emit greenhouse gases (GHGs). Moreover, the price of an allowance can change as a result of market speculation from investment banks who themselves have no compliance obligations but are still active in carbon markets. Or, the price of allowances can change as a result of environmental nongovernmental organizations (NGOs) purchasing allowances to retire, and thus restricting the ability of compliance entities to emit GHGs.

Furthermore, if compliance entities do not have enough allowances to surrender at the end of a compliance period, they will be forced to pay heavy penalties. The EU ETS fines compliance entities 100 euros for each ton of

CO_2 or CO_2 equivalent emitted for which the operator has not surrendered an allowance. RGGI has a 3x allowance penalty for each ton of CO_2 or CO_2 equivalent emitted for which the operator has not surrendered an allowance. Additionally, each day and each excess ton of emissions is considered separate violations of state law for which the source can be subject to administrative or civil fines and proceedings.

Because the price for allowances fluctuates and compliance entities need to make sure that they will have enough allowances or else they will face heavy penalties, carbon markets currently include or are likely to include a number of financial instruments to manage risk. These instruments, referred to as derivatives, include forward contracts, futures contracts, option contracts, and swaps. Forward contracts allow buyers and sellers to agree upon the delivery of allowances at a specified date. Futures contracts give the holder the right to sell a specified quantity of allowances at a specific price within a specified time, regardless of the market price for allowances. Option contracts give the holder the right to buy a specified quantity of allowances at a specific price within a specified period of time, regardless of the market price for allowances. Lastly, swaps allow for the exchange of one asset or liability for another asset or liability.

In the EU ETS, the majority of allowance-based instruments are traded as derivatives, rather than allowances. Concerns over allowance price volatility or a low volume of allowances may also make this the case in a U.S. carbon market.

Robust carbon markets, involving allowances and derivatives, will lead to numerous employment and investment opportunities. Large investment firms are well aware of the enormous potential that cap and trade presents for the financial industry and are already active in markets for carbon emissions and other climate-related commodities. In 2006, Goldman Sachs made a minority equity investment in Climate Exchange PLC, which owns the European Climate Exchange, the Chicago Climate Exchange, and the newly created California Climate Exchange.

Alternative Energy

In 2007, the size of the market for renewable energy products was approximately $38 billion and employed approximately 1.7 million people. Overall, the market grew by 25% in 2005. Within the overall total, some renewable energy technologies grew at an even faster rate. The global install-capacity of solar photovoltaic rose by 55% in 2005. The market for wind power grew by nearly 50%. In the year prior to August 2006, the market capitalization of solar companies grew 38-fold to $27 billion. Growth in the biofuel sector only rose by 15% in 2005, but still the total market for the sector is worth over $15 billion. It has been predicted that by the year 2050, the annual market for low-carbon technologies could be worth hundreds of billions of dollars and employ over 25 million people.

A study conducted by the United Kingdom's (U.K.'s) Secretary of State for Energy and Climate Change, explained: "Climate change is not only one of the most significant challenges of our generation; it also presents a huge opportunity. Supplying the demands of a low-carbon economy offers a significant potential contribution to the economic growth and job creation in the U.K." Further, the study projected that whole new industrial sectors may emerge and will provide around 100 billion pound sterling worth of investment opportunities and up to 500,000 U.K. renewable energy jobs.

Moreover, the study emphasized:

> The current economic difficulties make this even more important: now is the not the time to scale back our ambitions on tackling climate change and securing our energy supplies. The increased levels of investment in renewable energy in the U.K. and across Europe over the next decade and beyond will involve significant adjustment costs, but the high investment in renewable energy has the potential to boost our economy in the short term and will help kick-start our long-term transition to a low-carbon economy.

The best way for a country and companies within it to capitalize on opportunities in the renewable energy market is to adopt a cap-and-trade program and related measures. Doing so allows for: (A) auction revenue to be allocated to alternative energy projects; (B) offsets enabling projects that would otherwise not be feasible; and (C) renewable energy certificates that subsidize the development of alternative energy projects.

Auction Revenue Will Be Allocated to Alternative Energy Projects

When allowances are auctioned, the revenue from the auction is used to fund specific policy mandates. In H.R. 2454, which details the U.S. House of Representative's proposal for a cap-and-trade program, allowance revenue is used to provide rebates for low- and moderate-income families; to offset increased costs faced by consumers of electricity, natural gas, and heating oil; to subsidize GHG capture and storage; to support other domestic and international technology programs; to safeguard the competitiveness of energy-intensive, trade-exposed industries; and to support domestic and international adaptation programs. From 2012 to 2050, 15% of allowance revenue will go to renewables, efficiency, GHG capture and storage, autos, and other green technologies.

California's cap-and-trade program may also devote auction revenue to carbon reduction technologies, such as alternative energy projects. The Economic and Technology Advancement Advisory Committee, which has been advising the California Air Resources Board, recommended using allowance revenue to fund research and development and to support a green technology workforce training program.

Offset Credits Enable Projects That Would Otherwise Not Be Feasible

Carbon offset credits are awarded for GHG reductions from renewable energy projects that are "additional," meaning that they would not have been financially viable without the prospect of revenue from the sale of offsets.

Renewable energy projects are unlikely to qualify for offset credits within a U.S. cap-and-trade program because project developers would have difficulty demonstrating that a renewable energy project would not have been financially viable without the prospect of offset revenues. In a U.S. cap-and-trade program, energy sector emissions will likely be capped. This will make fossil fuels more expensive and thus make renewable energy sources more attractive.

However, while domestic renewable energy projects are unlikely to qualify for offset credits in a national emissions reduction program, renewable energy projects will still qualify for offset credits in countries without GHG emissions controls on their energy sectors. The EU encourages companies within it to support capacity-building activities in developing countries to help them take advantage of the Clean Development Mechanism (CDM) in a manner that supports sustainable development in the host's country. Indeed, the Kyoto Protocol reads:

> The purpose of the CDM *shall be to assist Parties not included in Annex I in achieving sustainable development* and in contributing to the ultimate objective of the Convention, and to assist Parties included in Annex I in achieving compliance with their quantified emission limitation and reduction commitments under Article 3.

Currently, a large percentage of offset credits are generated from renewable energy projects in countries without GHG emissions controls. Thirty-five percent of the projects in the CDM pipeline are renewable energy projects. There are 399 wind CDM projects in China, producing 22,209 megawatts (MW) of electricity, and 320 wind CDM projects in India, producing 5,915 MW of electricity. There are 819 hydro CDM projects in China, producing 25,896 MW of electricity, and 130 hydro CDM projects in India, producing 5,737 MW of electricity. Moreover, 30% of the projects in the Joint Implementation (JI) pipeline are renewable energy projects.

H.R. 2454 lays out a framework for international offset-crediting and emphasizes that credits can only be issued for projects in developing countries. Thus, assuming that a large percentage of offset credits are generated from renewable energy projects in countries without GHG emissions controls, offset credits will continue to enable projects that would otherwise not be feasible.

Renewable Energy Certificates Will Subsidize the Development of Alternative Energy Projects

Renewable Energy Certificates (RECs) are awarded for each MW-hour of renewable energy generated from qualifying renewable energy projects, such as wind, solar, geothermal, and certain hydropower projects. Typically, RECs are unbundled and sold separately from the underlying electricity generated by renewable energy projects, allowing renewable energy generators to sell both RECs and the wholesale electricity they produce. Overall, RECs act as a subsidy for the development of alternative energy projects.

Currently, REC registries are being set up in many states, regions, and countries. California requires utilities to meet part of their electricity demand through renewable energy sources. The California Energy Commission estimates that renewable energy sources generate 12% of California's retail electricity load. California S.B. 107 requires investor-owned utilities to increase the share of renewables in their electricity portfolios to 20% by 2010. At the same time, public-owned utilities are encouraged to meet the same target. Recently, Gov. Arnold Schwarzenegger called for renewables to make up 33% of electricity portfolios and accordingly, it is anticipated that renewable energy sources will generate 33% of California's electricity by 2020.

H.R. 2454 amends the Public Utility Regulatory Policies Act of 1978 to require retail electricity suppliers to meet 20% of their electricity demand through renewable energy sources and energy efficiency by 2020. Each retail electricity supplier that annually sells four million MW-hours of electricity or more would need to submit RECs equal to at least three-quarters of their allotted requirement.

Other countries are also enacting systems involving mechanisms similar to RECs. For example, a Renewables Obligation (RO) was introduced in the U.K. in 2002. Under the RO, generators receive Renewable Obligation Certificates (ROCs) for renewable electricity. Electricity suppliers are incentivized to buy ROCs from generators, and ROCs provide renewable generators with financial support in addition to what they receive from selling their electricity. The RO has so far increased RO-eligible renewable electricity generation in the U.K. from 1.8% of the country's electricity load in 2002 to 5.3% of the country's electricity load in 2008.

It is unclear whether RECs will be part of a cap-and-trade program or be a related measure in a larger climate change legislative package. Principally, RECs serve as proof that one MW-hour of electricity was generated and delivered to the grid from a qualifying renewable energy source, but the definition of RECs has been extended to imply or explicitly claim that RECs also offset GHG emissions and should be treated as offset credits. There are, however, serious problems with treating RECs as offset credits. First, it is difficult to prove ownership with RECs. Operation of the electric power grid is complex, and it is difficult to establish linkage between renewable energy generation and changes in generation at other power plants on the grid. Second, REC programs have eligibility requirements that do not necessarily consider additionality. While some renewable energy projects may not have been implemented without RECs, other projects may have been implemented without them.

However, whether RECs are directly incorporated into a cap-and-trade program or whether they are supplemental as part of a larger climate change legislative package, they will certainly encourage the development of renewable energy projects.

GHG Capture and Storage

Cap and trade will also lead to the development of numerous GHG capture-and-storage projects. Domestic and international offset projects are rapidly increasing in number. In 2007, the value of transactions in the primary market for offset projects grew 34% to $8.2 billion. Currently, the market is dominated by the main offset mechanisms of the Kyoto Protocol: the CDM and the JI. In 2007, CDM transactions accounted for 87% of project-based transaction volumes and JI transactions doubled in volume and tripled in value. The remaining market activity was split among other compliance mechanisms and voluntary purchases.

The offset market is highly sophisticated, involving a number of government, quasi-government, and private-sector participants. At its 21st meeting, the CDM Executive Board discussed work on the registration of CDM project activities as part of the CDM Management Plan. The CDM Executive Board decided to: "Make publicly available relevant information, submitted to it for this purpose, on proposed CDM project activities in need of funding and on investors seeking opportunities, in order to assist in arranging funding of CDM project activities, as necessary." Subsequently, the United Nations Framework Convention on Climate Change (UNFCCC) established the UNFCCC CDM Bazaar, which is a web-based facility serving as a platform for the exchange of information on CDM project opportunities.

There are offset retailers, such as Climate Trust, TerraPass, NativeEnergy, and Myclimate. Offset prices vary by factors, such as project type, location, and stringency of offset program requirements. Certified emission reductions (CERs), awarded for CDM projects, and emission reduction units (ERUs), awarded for JI projects, can be valued at upward of 80% of the trading price of EU allowances. Prices for voluntary offset credits vary significantly based on project types, project locations, standards used, offset quality, delivery guarantees, and contract terms.

Recently, JP Morgan agreed to purchase EcoSecurities, an offset aggregator, for $204 million. EcoSecurities has been involved in the carbon market for over 10 years and has offices and representatives in more than 20 countries and five continents. It sources, develops, and trades emissions reductions credits and manages a diverse portfolio of credits, including different project types, project locations, volumes, technologies, methodologies, risk profiles, contract terms, volumes, and sustainability co-benefits.

As more countries and regions enact cap-and-trade programs and the demand for offsets increases, two project types are likely to expand rapidly: (1) methane capture and destruction projects; and (2) biological sequestration projects.

Methane Capture and Destruction Projects Will Become Extremely Popular

Methane capture and destruction projects are likely to become extremely popular because methane has 25 times the heat-trapping ability of CO_2 and the global warming potential of GHGs influences the volume of offsets generated by a project. A developer who reduces one ton of methane gas receives 25 times the credits that they would receive for reducing one ton of CO_2.

Current practices could be changed to curb emissions. For instance, in the United States, coal mines account for about 10% of all man-made methane emissions. Because the gas can present a safety risk for miners, methane released during the extraction of coals is removed through ventilation fans and vented into the atmosphere. Through an offset project, the methane could instead be recovered and burned to produce energy or flared to reduce its heat-trapping ability when it is released into the atmosphere.

In 2007, about 49% of the U.S. offset supply was produced from projects that capture and destroy methane from coal mines, agricultural operations, or landfills. Ninety-three of the 211 projects that produced U.S. offsets were methane projects. The Clean Energy Jobs and American Power Act considers: "methane collection and combustion projects at active underground coal mines"; "methane collection and combustion projects at landfills"; and, "nonlandfill methane collection, combustion and avoidance projects involving organic waste streams that would have otherwise emitted methane in the atmosphere, including manure management and biogas capture and combustion."

Biological Sequestration Will Result in Numerous Forestry Projects

Forestry and other land use projects aimed at sequestering GHGs will also increase in popularity. These projects can reduce and avoid the atmospheric buildup of GHGs in a number of ways. First, tree biomass and soils can act as carbon sinks, removing and storing CO_2. Statistics released in an EPA study explained that afforestation can sequester 2.2-9.5 tons of CO_2 per acre per yearj and reforestation can sequester 1.1-7.7 tons of CO_2 per acre per year for 90 to more than 120 years before saturation occurs. Second, CO_2 emissions can be avoided by using biofuels rather than fossil fuels. Third, agricultural emissions from fertilizers can be reduced by changing livestock management and fertilizer applications.

The Kyoto Protocol vaguely promised to include emissions reductions for forestry and other land use projects because these projects have the potential to sequester CO_2. However, there are concerns about how much CO_2 these projects actually remove from the atmosphere, how to measure the CO_2 that they remove, and whether the removal from the atmosphere is permanent. For these reasons, the use of forestry and other land use projects in meeting emissions targets has been controversial. However, the United States and other economies with high energy-intensity and population growth, Australia and Canada, have pushed for a maximum of flexibility in achieving emissions reductions. The United States, specifically, has insisted on the inclusion of sinks from forestry and other land use projects.

In 2007, 17% of the U.S. offset supply was generated from forestry and other land use projects. This includes 52 forestry projects that produced about 7% of the total U.S. supply. The Clean Energy Jobs and American Power Act considers awarding offset credits for: "projects involving afforestation or reforestation of acreage not forested as of January 1, 2009"; "forest management resulting in an increase in forest carbon stores, including harvested wood products"; "agricultural, grassland, and rangeland sequestration and management practices"; and, "changes in carbon stocks attributed to land use change and forestry activities." Moreover, H.R. 2454 gives financial incentives to farmers and ranchers to plant trees. According to an EPA analysis of H.R. 2454, about 18 million acres of new trees would be planted by 2020. With the implementation of a U.S. cap-and-trade program, afforestation efforts would be even greater than those carried out by the Civilian Conservation Corps between 1933 and 1942, which planted 3 billion trees.

Plug-in and Other Advanced Technology Vehicles

Transportation is one of the largest sources of GHG emissions. In California, it is the largest source. Cap and trade and related measures are taking steps to decrease emissions in this sector by encouraging the production and purchasing of plug-in and other advanced technology vehicles.

Auction Revenue Will Be Allocated to Plug-in and Other Advanced Technology Vehicles

Fuel providers are compliance entities in the proposed U.S. cap-and-trade program, California's cap-and-trade program, and New Zealand's cap-and-trade program. Within these programs, auction revenue may be allocated to plug-in and other advanced technology vehicles. For example, under H.R. 2454, the U.S. cap-and-trade program contains significant incentives for automakers to produce plug-in and other advanced technology vehicles. In the beginning of the program, 3% of allowances would be allocated to the automotive sector to provide grants to refit or establish plants to build plug-ins and other advanced vehicles. Depending on the price of allowances, this allocation could end up being worth billions of dollars each year.

Consumers Will Switch to Plug-in and Other Advanced Technology Vehicles Because of Increased Fuel Costs

Because fuel providers will have to purchase allowances in certain cap-and-trade programs, in these programs, their costs will increase and these costs will be passed on to consumers who will have to pay more for gasoline at the pump. Facing increased costs in gasoline, consumers will be incentivized to purchase plug-in and other advance technology vehicles. This trend was demonstrated in the 1980s, for instance, when consumers responded to high gasoline prices by driving smaller, more fuel-efficient cars.

A study conducted by the Center for the Study of Energy Markets postulates that this trend may not be incredibly strong, especially with today's consumers, who may be less likely to curb their gas consumption with increased fuel prices than consumers in earlier decades. This could be because incomes have grown and consumers are now less sensitive to price increases because gasoline consumption is a smaller share of their budget. The study, however, looks at short-run rather than long-run gasoline price increases, and acknowledges that consumers may respond to higher gasoline prices in the long run by purchasing more fuel-efficient vehicles.

Low-Carbon Fuel Standards Will Encourage the Production of Alternative Energy Vehicles

Additionally, programs aimed at reducing the carbon intensity of transportation fuels, such as a low-carbon fuel standard (LCFS), will encourage the production of alternative energy vehicles. The California Air Resources Board developed an LCFS that requires fuel providers to track the average life-cycle GHG intensity of their products, including production, transportation, storage, and fuel use, and reduce the average life-cycle GHG intensity of transportation fuels they sell in California by at least 10% by 2020.

Following California's lead, the EU, several other U.S. states, and some Canadian provinces are developing LCFS proposals. The U.K. has also introduced the U.K. Renewable Transport Fuel Obligation Programme, which includes reporting requirements and methodologies for calculating life-cycle GHG emissions and requires fuel providers to ensure biofuels constitute 2.5% of total road transport fuels in 2008-2009, 3.75% in 2009-2010, and 5% after 2009–2010.

It is unclear how LCFS and other programs aimed at reducing the carbon intensity of transportation fuels will fit into a cap-and-trade program. A study prepared by the Center for the Study of Energy Markets recommended that California's LCFS should be kept separate from California's cap-and-trade program for at least the first 10 years to ensure innovation and investment in low global warming-intensive fuel technologies. However, whether these standards are directly incorporated into a cap-and-trade program or whether they are supplemental as part of a larger climate change legislative package, they will certainly encourage the production of alternative energy vehicles.

Legal and Nonlegal Consulting

Because of the introduction of cap-and-trade programs and related measures, a plethora of legal and nonlegal consulting agencies will be advising companies and government agencies on climate change [and related issues]. . . .

Conclusion

Cap and trade and related measures are not just environmental efforts that will curb the effects of global warming. They also present a wide range of business opportunities that will fuel the global economy. This includes growth in the financial sector, the development of low-carbon technologies, numerous carbon capture-and-storage projects, increased production of plug-in and other advanced technology vehicles, and a plethora of legal and nonlegal consulting opportunities. Countries and companies should position themselves now in order to take full advantage of these opportunities.

EXPLORING THE ISSUE

Will Restricting Carbon Emissions Damage the Economy?

Critical Thinking and Reflection

1. If the regulatory bodies in charge of assigning permitted pollution limits or caps reduce those caps over time, what will happen to overall pollution levels? What will happen if the caps stay the same over time?
2. How do cap-and-trade systems motivate companies to reduce carbon emissions?
3. How would controlling carbon emissions stimulate the development of advanced technology vehicles?
4. Should important decisions—such as whether to restrict emissions of greenhouse gases—be made primarily in terms of near-term effects on the economy, or in terms of long-term effects?

Is There Common Ground?

Both sides in this issue agree that restricting carbon emissions will increase costs for some businesses. One important question is whether the benefits to other businesses will compensate. Another is whether any business has the right to expect the conditions under which it operates not to change.

1. What businesses or industries seem most likely to experience increased costs if carbon emissions are restricted?
2. What businesses or industries seem most likely to benefit economically if carbon emissions are restricted?
3. When the conditions under which a business operates change, what should the business do?

Internet References . . .

350.org

350.org is an international campaign dedicated to building a movement to unite the world around solutions to the climate crisis—the solutions that justice demands. Its goal is to inspire the world to reduce the level of carbon dioxide in the atmosphere to 350 parts per million, the level scientists have identified as the safe upper limit for CO_2 in our atmosphere.

www.350.org/

Climate Change

The United Nations Environmental Program maintains this site as a central source for substantive work and information resources with regard to climate change.

www.unep.org/themes/climatechange/

Intergovernmental Panel on Climate Change

The Intergovernmental Panel on Climate Change (IPCC) was formed by the World Meteorological Organization (WMO) and the United Nations Environment Programme (UNEP) to assess the scientific, technical, and socio-economic information relevant for the understanding of the risk of human-induced climate change.

www.ipcc.ch/

The National Renewable Energy Laboratory

The National Renewable Energy Laboratory is the nation's primary laboratory for renewable energy and energy efficiency research and development. Among other things, it works on wind power and biofuels.

www.nrel.gov/

Nuclear Energy

The U.S. Department of Energy's Office of Nuclear Energy leads U.S. efforts to develop new nuclear energy generation technologies; to develop advanced, proliferation-resistant nuclear fuel technologies that maximize energy from nuclear fuel; and to maintain and enhance the national nuclear technology infrastructure.

www.ne.doe.gov/

Energy Issues

*H*umans cannot live and society cannot exist without producing environmental impacts. The reason is very simple: Humans cannot live and society cannot exist without using resources (e.g., soil, water, ore, wood, space, plants, animals, oil, sunlight), and those resources come from the environment. Many of these resources (e.g., wood, oil, coal, water, wind, sunlight, uranium) have to do with energy. The environmental impacts come from what must be done to obtain these resources and what must be done to dispose of the wastes generated in the process of obtaining and using them. The issues that arise are whether and how we should obtain these resources, whether and how we should deal with the wastes, and whether alternative answers to these questions may be preferable to the answers that experts think they already have.

In 2007, the Intergovernmental Panel on Climate Change released its fourth assessment report, summarizing the scientific consensus as the climate is warming, human activities are responsible for it, and the impact on human well-being and ecosystems will be severe. This brought energy issues to the fore with unprecedented urgency. The six issues presented here are by no means the only issues related to energy, but they will serve to demonstrate the vigor and variety of the current energy debates.

- Is Global Warming a Catastrophe That Warrants Immediate Action?
- Should We Drill for Offshore Oil?
- Is Shale Gas the Solution to Our Energy Woes?
- Is Renewable Energy Really Green?
- Are Biofuels a Reasonable Substitute for Fossil Fuels?
- Is It Time to Revive Nuclear Power?

ISSUE 7

Is Global Warming a Catastrophe That Warrants Immediate Action?

YES: Global Humanitarian Forum, from *Climate Change—The Anatomy of a Silent Crisis* (May 2009)

NO: Bjørn Lomborg, from "Let's Keep Our Cool About Global Warming," *Skeptical Inquirer* (March/April 2008)

Learning Outcomes

After reading this issue, you should be able to:

- Describe the effects of climate change to date on people.
- Describe how climate change affects different socioeconomic classes of people in different ways.
- Describe what can be done in the near future to prepare for future effects of global warming.
- Explain how population size and growth interact with carbon emissions and global warming.

ISSUE SUMMARY

YES: The Global Humanitarian Forum argues that global warming due to human activities, chiefly the emission of greenhouse gases such as carbon dioxide, is now beyond doubt. Impacts on the world's poorest people are already severe and will become much worse. Immediate action is essential to tackle climate change, increase funding for adaptation to its effects, and end the suffering it causes.

NO: Bjørn Lomborg argues that although global warming has genuine impacts on people, the benefits of continuing to use fossil fuels are so much greater than the costs that the best approach to a solution is not to demand draconian cuts in carbon emissions but to invest globally in research and development of non-carbon-emitting energy technologies and thereby "recapture the vision of delivering both a low-carbon and a high-income world."

Exactly what will global warming do to the world and its people? Projections for the future have grown steadily worse; see Eli Kintisch, "Projections of Climate Change Go from Bad to Worse, Scientists Report," *Science* (March 20, 2009). Effects include rising sea level, more extreme weather events, reduced global harvests (Constance Holden, "Higher Temperatures Seen Reducing Global Harvests," *Science*, January 9, 2009), and threats to the economies and security of nations (Michael T. Klare, "Global Warming Battlefields: How Climate Change Threatens Security," *Current History*, November 2007; and Scott G. Bergerson, "Arctic Meltdown: The Economic and Security Implications of Global Warming," *Foreign Affairs*, March/April 2008). With change in rainfall patterns and the rise in sea level, millions of people will flee their homelands; see Alex de Sherbinin, Koko Warner, and Charles Erhart, "Casualties of Climate Change," *Scientific American* (January 2011). Perhaps worst of all, even if we somehow stopped emitting greenhouse gases today, the effects would continue for 1000 years or more, during which sea level rise may exceed "several meters." See Susan Solomon et al., "Irreversible Climate Change due to Carbon Dioxide Emissions," *Proceedings of the National Academy of Sciences* (February 10, 2009).

It seems clear that something must be done, but what? How urgently? And with what aim? Should we be trying to reduce or prevent human suffering? Or to avoid political conflicts? Or to protect the global economy—meaning standards of living, jobs, and businesses? The humanitarian and economic approaches are obviously connected, for protecting jobs certainly has much to do with easing or preventing suffering. However, these approaches can also conflict. In October 2009, the Government Accountability Office (GAO) released "Climate Change Adaptation: Strategic Federal Planning Could Help Government Officials Make More Informed Decisions" (GAO-10-113; www.gao.gov/products/GAO-10-113), which noted the need for multiagency coordination and strategic (long-term) planning, both of which are often resisted by bureaucrats and politicians. Robert Engelman, *Population, Climate Change, and Women's Lives* (Worldwatch Institute, 2010), notes that addressing population size and growth would help but "despite its key contribution to climate change, population plays little role in current discussions on how to address this serious challenge."

The U.S. Climate Change Science Program's "Scientific Assessment of the Effects of Global Change on the United States, A Report of the Committee on Environment and Natural Resources, National Science and Technology Council" (May 29, 2008; available at http://www.whitehouse.gov/files/documents/ostp/NSTC%20Reports/Scientific%20Assessment%20FULL%20Report.pdf) describes the current and potential impacts of climate change. In sum, it says that the evidence is clear and getting clearer that global warming is "very likely" due to greenhouse gases largely released by human activity and there will be consequent changes in precipitation, storms, droughts, sea level, food production, fisheries, and more. Dealing with these effects may require changes in many areas, particularly relating to energy use, but "significant uncertainty exists about the potential impacts of climate change on energy production and distribution, in part because the timing and magnitude of climate impacts are uncertain." See Susan Milius, "Already Feeling the Heat," *Science News* (Web edition) (May 29, 2008).

Consonant with the Bush administration's previous position that imposing restrictions on fossil fuel use or carbon emissions is a bad idea both because of uncertainty about global warming and because it would harm the economy, mitigation of the risks—which the report says are likely or very likely—is barely mentioned. The Bush administration's position was that there are wrong ways and right ways to go about solving the global warming problem. The wrong way is raising energy taxes and prices (which hurts consumers and business), imposing restrictions, and abandoning the use of nuclear power and coal. The right way is setting realistic goals, adopting policies that spur investment in new technologies, encouraging the use of nuclear power and "clean" coal, and enhancing international cooperation through free trade in clean energy technologies. Such measures rely on the market to stimulate voluntary reductions in emissions. It's possible that increasing funding for alternative energy technologies would lead to solutions that greatly ease the problem, but many observers think it folly to rely on future breakthroughs.

President Barack Obama has indicated that his administration will take global warming more seriously. In June 2009, the U.S. House of Representatives passed an energy and climate bill that promised to cap carbon emissions and stimulate use of renewable energy. The Senate version of the bill failed to pass; see Daniel Stone, "Who Killed the Climate and Energy Bill?" *Newsweek* (September 15, 2010). The Obama administration also said it was committed to negotiating seriously at the Copenhagen Climate Change Conference in December 2009. Unfortunately, the Copenhagen meeting ended with little accomplished except agreements to limit global temperature increases to two degrees Celsius by 2100, but only through voluntary cuts in carbon emissions; to have developed nations report their cuts; to have developed nations fund mitigation and adaptation in developing nations; and to continue talking about the problem (see Elizabeth Finkel, "Senate Looms as Bigger Hurdle after Copenhagen," *Science*, January 1, 2010). There were few signs that the world is ready to take the extensive actions deemed necessary by many; see, e.g., Janet L. Sawin and William R. Moomaw, "Renewing the Future and Protecting the Climate," *World Watch* (July/August 2010).

In May 2010, the National Research Council released three books, *Advancing the Science of Climate Change* (www.nap.edu/catalog.php?record_id=12782), *Limiting the Magnitude of Future Climate Change* (www.nap.edu/catalog.php?record_id=12785), and *Adapting to the Impacts of Climate Change* (www.nap.edu/catalog.php?record_id=12783). Together, they stress the reality of the problem, the need for immediate action to keep the problem from getting worse, and the need for advance planning and preparation to deal with the impacts. However, their message may not be enough to convince Congress to pass necessary legislation; see Richard A. Kerr and Eli Kintisch, "NRC Reports Strongly Advocate Action on Global Warming," *Science* (May 28, 2010). One reason is reports that make the reports sound less alarming. For an example, see Ayami Hayashi, Keigo Akimoto, Fuminoro Sano, Shunsuke Mori, and Toshimasa Tomoda,

"Evaluation of Global Warming Impacts for Different Levels of Stabilization as a Step Toward Determination of the Long-Term Stabilization Target," *Climate Change* (January 2010), which says that "global warming will reduce the world's total number of deaths caused by thermal stress owing to the large decrease in the cold-related deaths; it will increase the water stress in some regions, while it will decrease the stress in other regions; . . . and it will enhance an increase in the wheat production potential." However, computer simulations suggest that since 1980, climate changes have reduced maize and wheat harvests by 3.8–5.5 percent; see D. B. Lobell, W. Schlenker, and J. Costa-Roberts, "Climate Trends and Global Crop Production Since 1980," *Science* (published online May 5, 2011). At a meeting of the International Emissions Trading Association, Christiana Figueres, Executive Secretary of the United Nations framework convention on climate change, said that the situation is urgent and the world must immediately agree to change its goal from limiting global warming to 2.0°C to limiting it to 1.5°C, or "we are in big trouble"; see Fiona Harvey, "UN Chief Challenges World to Agree Tougher Target for Climate Change," *The Guardian* (June 1, 2011).

In the following selections, the Global Humanitarian Forum argues that global warming due to human activities, chiefly the emission of greenhouse gases such as carbon dioxide, is now beyond doubt. Impacts on the world's poorest people are already severe and will become much worse. Immediate action is essential to tackle climate change, increase funding for adaptation to its effects, and end the suffering caused by it. Economist Bjørn Lomborg argues that although global warming has genuine impacts on people, the benefits of continuing to use fossil fuels are so much greater than the costs that the best approach to a solution is not to demand draconian cuts in carbon emissions but to invest globally in research and development of non-carbon-emitting energy technologies and thereby "recapture the vision of delivering both a low-carbon and a high-income world."

Climate Change—The Anatomy of a Silent Crisis

Science is now unequivocal as to the reality of climate change. Human activities, in particular emissions of greenhouse gases like carbon dioxide are recognized as its principle cause. This report clearly shows that climate change is already causing widespread devastation and suffering around the planet today. Furthermore, even if the international community is able to contain climate change, over the next decades human society must prepare for more severe climate change and more dangerous human impacts.

This report documents the full impact of climate change on human society worldwide today. It covers in specific detail the most critical areas of the global impact of climate change, namely on food, health, poverty, water, human displacement, and security. The third section of this report highlights the massive socio-economic implications of those impacts, in particular, that the worst affected are the world's poorest groups, who cannot be held responsible for the problem. The final section examines how sustainable development and the millennium development goals are in serious danger, the pressures this will exert on humanitarian assistance, and the great need to integrate efforts in adapting to climate change.

Based on verified scientific information, established models, and, where needed, the best available estimates, this report represents the most plausible narrative of the human impact of climate change. It reports in a comprehensive manner the adverse effects people already suffer today due to climate change within a single volume, encompassing the full spectrum of the most important impacts evidenced to date.

The findings of the report indicate that every year climate change leaves over 300,000 people dead, 325 million people seriously affected, and economic losses of US$125 billion. Four billion people are vulnerable, and 500 million people are at extreme risk. These figures represent averages based on projected trends over many years and carry a significant margin of error. The real numbers could be lower or higher. The different figures are each explained in more detail and in context in the relevant sections of the report. Detailed information describing how these figures have been calculated is also included in the respective sections and in the end matter of the report.

These already alarming figures may prove too conservative. Weather-related disasters alone cause significant economic losses. Over the past five years this toll has gone as high as $230 billion, with several years around

From *Climate Change—The Anatomy of a Silent Crisis,* May 2009. Published by Global Humanitarian Forum.

$100 billion and a single year around $50 billion. Such disasters have increased in frequency and severity over the past 30 years in part due to climate change. Over and above these costs are impacts on health, water supply, and other shocks not taken into account. Some would say that the worst years are not representative and they may not be. But scientists expect that years like these will be repeated more often in the near future.

Climate Change Through the Human Lens

Climate change already has a severe human impact today, but it is a silent crisis—it is a neglected area of research as the climate change debate has been heavily focused on physical effects in the long-term. This human impact report on climate change, therefore, breaks new ground. It focuses on human impact rather than physical consequences. It looks at the increasingly negative consequences that people around the world face as a result of a changing climate. Rather than focusing on environmental events in 50–100 years, the report takes a unique social angle. It seeks to highlight the magnitude of the crisis at hand in the hope to steer the debate towards urgent action to overcome this challenge and reduce the suffering it causes.

The human impact of climate change is happening right now—it requires urgent attention. Events like weather-related disasters, desertification and rising sea levels, exacerbated by climate change, affect individuals and communities around the world. They bring hunger, disease, poverty, and lost livelihoods—reducing economic growth and posing a threat to social and, even, political stability. Many people are not resilient to extreme weather patterns and climate variability. They are unable to protect their families, livelihoods and food supply from negative impacts of seasonal rainfall leading to floods or water scarcity during extended droughts. Climate change is multiplying these risks.

Today, we are at a critical juncture—just months prior to the Copenhagen summit where negotiations for a post-2012 climate agreement must be finalized. Negotiators cannot afford to ignore the current impact of climate change on human society. The responsibility of nations in Copenhagen is not only to contain a serious future threat, but also to address a major contemporary crisis. The urgency is all the more apparent since experts are constantly correcting their own predictions about climate change, with the result that climate change is now considered to be occurring more rapidly than even the most aggressive models recently suggested. The unsettling anatomy of the human impact of climate change cannot be ignored at the negotiating tables.

Climate Change Is a Multiplier of Human Impacts and Risks

Climate change is already seriously affecting hundreds of millions of people today and in the next twenty years those affected will likely more than double—making it the greatest emerging humanitarian challenge of our time. Those seriously affected are in need of immediate assistance either following a weather-related disaster, or because livelihoods have been severely compromised by climate change. The number of those severely affected by climate change is more than ten times greater than for instance those injured in traffic accidents each year, and more than the global annual number of new malaria cases. Within the next 20 years, one in ten of the world's present population could be directly and seriously affected.

Already today, hundreds of thousands of lives are lost every year due to climate change. This will rise to roughly half a million in 20 years. Over nine in ten deaths are related to gradual environmental degradation due to climate change—principally malnutrition, diarrhoea, [and] malaria, with the remaining deaths being linked to weather-related disasters brought about by climate change.

Economic losses due to climate change currently amount to more than one hundred billion US dollars per year, which is more than the individual national GDPs of three quarters of the world's countries.

This figure constitutes more than the total of all Official Development Assistance in a given year. Already today, over half a billion people are at extreme risk to the impacts of climate change, and six in ten people are vulnerable to climate change in a physical and socio-economic sense. The majority of the world's population does not have the capacity to cope with the impact of climate change without suffering a potentially irreversible loss of wellbeing or risk of loss of life. The populations most gravely and immediately at risk live in some of the poorest areas that are also highly prone to climate change—in particular, the semi-arid dry land belt countries from the Sahara to the Middle East and Central Asia, as well as sub-Saharan Africa, South Asian waterways and Small Island Developing States.

A Question of Justice

It is a grave global justice concern that those who suffer most from climate change have done the least to cause it. Developing countries bear over nine-tenths of the climate change burden: 98 percent of the seriously affected and 99 percent of all deaths from weather-related disasters, along with over 90 percent of the total economic losses. The 50 Least Developed Countries contribute less than 1 percent of global carbon emissions.

Climate change exacerbates existing inequalities faced by vulnerable groups, particularly women, children, and the elderly. The consequences of climate change and poverty are not distributed uniformly within communities. Individual and social factors determine vulnerability and capacity to adapt to the effects of climate change. Women account for two-thirds of the world's poor and comprise about seven in ten agricultural workers. Women and children are disproportionately represented among people displaced by extreme weather events and other climate shocks.

The poorest are hardest hit, but the human impact of climate change is a global issue. Developed nations are also seriously affected, and increasingly so. The human impact of recent heat waves, floods, storms and forest fires in rich countries has been alarming. Australia is perhaps the developed nation most vulnerable to the direct impacts of climate change and also to the indirect impact from neighbouring countries that are stressed by climate change.

The Time to Act Is Now

Climate change threatens sustainable development and all eight Millennium Development Goals. The international community agreed at the beginning of the new millennium to eradicate extreme hunger and poverty by 2015. Yet, today, climate change is already responsible for forcing some fifty million additional people to go hungry and driving over ten million additional people into extreme poverty. Between one-fifth and one-third of Official Development Assistance is in climate-sensitive sectors and thereby highly exposed to climate risks.

To avert the worst outcomes of climate change, adaptation efforts need to be scaled up by a factor of more than 100 in developing countries. The only way to reduce the present human impact is through adaptation. But funding for adaptation in developing countries is not even one percent of what is needed. The multilateral funds that have been pledged for climate change adaptation funding currently amount to under half a billion US dollars.

Despite the lack of funding, some cases of successful adaptation do provide a glimmer of hope. Bangladesh is one such example. Cyclone Sidr, which struck Bangladesh in 2007, demonstrates how well adaptation and prevention efforts can pay off. Disaster preparation measures, such as early warning systems and storm-proof houses, minimized damage and destruction. Cyclone Sidr's still considerable death toll of 3,400, and economic damages of $1.6 billion, nevertheless compare favourably to the similar scale cyclone Nargis, which hit Myanmar in 2008, resulting in close to 150,000 deaths and economic losses of around $4 billion.

Solutions do also exist for reducing greenhouse gas emissions, some even with multiple benefits. For instance, black carbon from soot, released by staple energy sources in poor communities, is likely causing as much as 18 percent of warming. The provision of affordable alternative cooking stoves to the poor can, therefore, have both positive health results, since smoke is eliminated, and an immediate impact on reducing emissions, since soot only remains in the atmosphere for a few weeks. Integrating strategies between adaptation, mitigation, development, and disaster risk reduction can and must be mutually reinforcing. Climate change adaptation, mitigation, humanitarian assistance and development aid underpin each other, but are supported by different sets of institutions, knowledge centres, policy frameworks, and funding mechanisms. These policies are essential to combat the human impact of climate change, but their links to one another have received inadequate attention.

A key conclusion of this report is that the global society must work together if humanity is to overcome this shared challenge: nations have to realize their common interest at Copenhagen, acting decisively with one voice; humanitarian and development actors of all kinds have to pool resources, expertise, and efforts in order to deal with the rapidly expanding challenges brought about by climate change; and in general, people, businesses, and communities everywhere should become engaged and promote steps to tackle climate change and end the suffering it causes.

Bjørn Lomborg

 **NO**

Let's Keep Our Cool About Global Warming

There is a kind of choreographed screaming about climate change from both sides of the debate. Discussion would be on much firmer ground if we could actually hear the arguments and the facts and then sensibly debate long-term solutions.

Man-made climate change is certainly a problem, but it is categorically not the end of the world. Take the rise in sea levels as one example of how the volume of the screaming is unmatched by the facts. In its 2007 report, the United Nations estimates that sea levels will rise about a foot over the remainder of the century. While this is not a trivial amount, it is also important to realize that it is not unknown to mankind: since 1860, we have experienced a sea level rise of about a foot without major disruptions. It is also important to realize that the new prediction is lower than previous Intergovernmental Panel on Climate Change (IPCC) estimates and much lower than the expectations from the 1990s of more than two feet and the 1980s, when the Environmental Protection Agency projected more than six feet.

We dealt with rising sea levels in the past century, and we will continue to do so in this century. It will be problematic, but it is incorrect to posit the rise as the end of civilization.

We will actually lose very little dry land to the rise in sea levels. It is estimated that almost all nations in the world will establish maximal coastal protection almost everywhere, simply because doing so is fairly cheap. For more than 180 of the world's 192 nations, coastal protection will cost less than 0.1 percent GDP and approach 100 percent protection.

The rise in sea level will be a much bigger problem for poor countries. The most affected nation will be Micronesia, a federation of 607 small islands in the West Pacific with a total land area only four times larger than Washington, D.C. If nothing were done, Micronesia would lose some 21 percent of its area by the end of the century. With coastal protection, it will lose just 0.18 percent of its land area. However, if we instead opt for cuts in carbon emissions and thus reduce both the sea level rise and economic growth, Micronesia will end up losing a *larger* land area. The increase in wealth for poor nations is more important than sea levels: poorer nations will be less able to defend themselves against rising waters, even if they rise more slowly. This is the same for other vulnerable nations: Tuvalu, the Maldives, Vietnam, and Bangladesh.

From *Skeptical Inquirer*, March/April 2008, pp. 42–45. Copyright © 2008 by Skeptical Inquirer. Reprinted by permission. www.csicop.org

The point is that we cannot just talk about CO_2 when we talk about climate change. The dialogue needs to include *both* considerations about carbon emissions and economics for the benefit of humans and the environment. Presumably, our goal is not just to cut carbon emissions, but to do the best we can for people and the environment.

We should take action on climate change, but we need to be realistic. The U.K has arguably engaged in the most aggressive rhetoric about climate change. Since the Labour government promised in 1997 to cut emissions by a further 15 percent by 2010, emissions have *increased* 3 percent. American emissions during the Clinton/Gore administration increased 28 percent.

Look at our past behavior: at the Earth Summit in Rio in 1992, nations promised to cut emissions back to 1990 levels by 2000. The member countries of the Organisation for Economic Cooperation and Development (OECD) overshot their target in 2000 by more than 12 percent.

Many believe that dramatic political action will follow if people only knew better and elected better politicians. Despite the European Union's enthusiasm for the Kyoto Protocol on Climate Change—and a greater awareness and concern over global warming in Europe than in the United States—emissions per person since 1990 have remained stable in the U.S. while E.U. emissions have *increased* 4 percent.

Even if the wealthy nations managed to reign in their emissions, the majority of this century's emissions will come from developing countries—which are responsible for about 40 percent of annual carbon emissions; this is likely to increase to 75 percent by the end of the century.

In a surprisingly candid statement from Tony Blair at the Clinton Global Initiative, he pointed out:

> I think if we are going to get action on this, we have got to start from the brutal honesty about the politics of how we deal with it. The truth is no country is going to cut its growth or consumption substantially in the light of a long-term environmental problem. What countries are prepared to do is to try to work together cooperatively to deal with this problem in a way that allows us to develop the science and technology in a beneficial way.

Similarly, one of the top economic researchers tells us: "Deep cuts in emissions will only be achieved if alternative energy technologies become available at reasonable prices." We need to engage in a sensible debate about how to tax CO_2. If we set the tax too low, we emit too much. If we set it too high, we end up much poorer without doing enough to reduce warming.

In the largest review of all of the literature's 103 estimates, climate economist Richard Tol makes two important points. First, the really scary, high estimates typically have not been subjected to peer review and published. In his words: "studies with better methods yield lower estimates with smaller uncertainties." Second, with reasonable assumptions, the cost is very unlikely to be higher than $14 per ton of CO_2 and likely to be much smaller. When I specifically asked him for his best guess, he wasn't too enthusiastic about shedding his cautiousness—as few true researchers invariably are—but gave the estimate of $2 per ton of CO_2.

Therefore, I believe that we should tax CO_2 at the economically feasible level of about $2/ton, or maximally $14/ton. Yet, let us not expect this will make any major difference. Such a tax would cut emissions by 5 percent and reduce temperatures by 0.16° F. And before we scoff at 5 percent, let us remember that the Kyoto protocol, at the cost of 10 years of political and economic toil, will reduce emissions by just 0.4 percent by 2010.

Neither a tax nor Kyoto nor draconian proposals for future cuts move us closer toward finding better options for the future. Research and development in renewable energy and energy efficiency is at its lowest for twenty-five years. Instead, we need to find a way that allows us to "develop the science and technology in a beneficial way," a way that enables us to provide alternative energy technologies at reasonable prices. It will take the better part of a century and will need a political will spanning parties, continents, and generations. We need to be in for the long haul and develop cost-effective strategies that won't splinter regardless of overarching ambitions or false directions.

This is why one of our generational challenges should be for *all nations to commit themselves to spending 0.05 percent of GDP in research and development of noncarbon emitting energy technologies.* This is a tenfold increase on current expenditures, yet would cost a relatively minor $25 billion per year (seven times cheaper than Kyoto and many more times cheaper than Kyoto II). Such a commitment could include all nations, with wealthier nations paying the larger share, and would let each country focus on its own future vision of energy needs, whether that means concentrating on renewable sources, nuclear energy, fusion, carbon storage, conservation, or searching for new and more exotic opportunities.

Funding research and development globally would create a momentum that could recapture the vision of delivering both a low-carbon and high-income world. Lower energy costs and high spin-off innovation are potential benefits that possibly avoid ever stronger temptations to free-riding and the ever tougher negotiations over increasingly restrictive Kyoto Protocol-style treaties. A global financial commitment makes it plausible to envision stabilizing climate changes at reasonable levels.

I believe it would be the way to bridge a century of parties, continents, and generations, creating a sustainable, low-cost opportunity to create the alternative energy technologies that will power the future.

To move toward this goal we need to create sensible policy dialogue. This requires us to talk openly about priorities. Often there is strong sentiment in any public discussion that we should do *anything* required to make a situation better. But clearly we don't actually do that. When we talk about schools, we know that more teachers would likely provide our children with a better education. Yet we do not hire more teachers simply because we also have to spend money in other areas. When we talk about hospitals, we know that access to better equipment is likely to provide better treatment, yet we don't supply an infinite amount of resources. When we talk about the environment, we know tougher restrictions will mean better protection, but this also comes with higher costs.

Consider traffic fatalities, which are one of the ten leading causes of deaths in the world. In the U.S., 42,600 people die in traffic accidents and 2.8 million people are injured each year. Globally, it is estimated that 1.2 million people die from traffic accidents and 50 million are injured every year.

About 2 percent of all deaths in the world are traffic-related and about 90 percent of the traffic deaths occur in third world countries. The total cost is a phenomenal $512 billion a year. Due to increasing traffic (especially in the third world) and due to ever better health conditions, the World Heath Organization estimates that by 2020, traffic fatalities will be the second leading cause of death in the world, after heart disease.

Amazingly, we have the technology to make all of this go away. We could instantly save 1.2 million humans and eliminate $500 billion worth of damage. We would particularly help the third world. The answer is simply lowering speed limits to 5 mph. We could avoid almost all of the 50 million injuries each year. But of course we will not do this. Why? The simple answer that almost all of us would offer is that the benefits from driving moderately fast far outweigh the costs. While the cost is obvious in terms of those killed and maimed, the benefits are much more prosaic and dispersed but nonetheless important—traffic interconnects our society by bringing goods at competitive prices to where we live and bringing people together to where we work, and lets us live where we like while allowing us to visit and meet with many others. A world moving only at 5 mph is a world gone medieval.

This is not meant to be flippant. We really could solve one of the world's top problems if we wanted. We know traffic deaths are almost entirely caused by man. We have the technology to reduce deaths to zero. Yet, we persist in exacerbating the problem each year, making traffic an ever-bigger killer.

I suggest that the comparison with global warming is insightful; we have the technology to reduce it to zero, yet we seem to persist in going ahead and exacerbating the problem each year, causing temperatures to continue to increase to new heights by 2020. Why? Because the benefits from moderately using fossil fuels far outweigh the costs. Yes, the costs are obvious in the "fear, terror, and disaster" we read about in the papers every day.

But the benefits of fossil fuels, though much more prosaic, are nonetheless important. Fossil fuels provide us with low-cost electricity, heat, food, communication, and travel. Electrical air-conditioning means that people in the U.S. no longer die in droves during heat waves. Cheaper fuels would have avoided a significant number of the 150,000 people that have died in the UK since 2000 due to cold winters.

Because of fossil fuels, food can be grown cheaply, giving us access to fruits and vegetables year round, which has probably reduced cancer rates by at least 25 percent. Cars allow us to commute to city centers for work while living in areas that provide us with space and nature around our homes, whereas communication and cheap flights have given ever more people the opportunity to experience other cultures and forge friendships globally.

In the third world, access to fossil fuels is crucial. About 1.6 billion people don't have access to electricity, which seriously impedes human development. Worldwide, about 2.5 billion people rely on biomass such as wood and waste (including dung) to cook and keep warm. For many Indian women, searching for wood takes about three hours each day, and sometimes they walk more than 10 kilometers a day. All of this causes excessive deforestation. About 1.3 million people—mostly women and children—die each year due to heavy indoor air pollution. A switch from biomass to fossil fuels would dramatically improve 2.5 billion lives; the cost of $1.5 billion annually would be greatly superseded by benefits of about $90 billion. Both for the developed and the developing world, a world without fossil fuels—in the short or medium term— is, again, a lot like reverting back to the middle ages.

This does not mean that we should not talk about how to reduce the impact of traffic and global warming. Most countries have strict regulation on speed limits—if they didn't, fatalities would be much higher. Yet, studies also show that lowering the average speed in Western Europe by just 5 kilometers per hour could reduce fatalities by 25 percent—with about 10,000 fewer people killed each year. Apparently, democracies in Europe are not willing to give up on the extra benefits from faster driving to save 10,000 people.

This is parallel to the debate we are having about global warming. We can realistically talk about $2 or even a $14 CO_2 tax. But suggesting a $140 tax, as Al Gore does, seems to be far outside the envelope. Suggesting a 96 percent carbon reduction for the OECD by 2030 seems a bit like suggesting a 5 mph speed limit in the traffic debate. It is technically doable, but it is very unlikely to happen.

One of the most important issues when it comes to climate change is that we cool our dialogue and consider the arguments for and against different policies. In the heat of a loud and obnoxious debate, facts and reason lose out.

EXPLORING THE ISSUE

Is Global Warming a Catastrophe That Warrants Immediate Action?

Critical Thinking and Reflection

1. How is climate change killing people today?
2. In what sense is dealing with climate change a matter of social justice?
3. What measures should be taken in the near future to best prepare for the long-term impacts of global warming?
4. How do population size and growth affect carbon emissions and climate change?

Is There Common Ground?

Both sides in this issue agree that global warming is already affecting people. They differ in how concerned we should be, or in how much we should be worrying about "social justice." Visit the Social Justice Training Institute at http://www.sjti.org/ to find out more.

1. What is social justice?
2. How does the concept apply to environmental problems such as global warming?
3. What other aspects of life can you apply the concept to?

ISSUE 8

Should We Drill for Offshore Oil?

YES: **Stephen L. Baird**, from "Offshore Oil Drilling: Buying Energy Independence or Buying Time?" *The Technology Teacher* (November 2008)

NO: **Mary Annette Rose**, from "The Environmental Impacts of Offshore Oil Drilling," *The Technology Teacher* (February 2009)

Learning Outcomes

After reading this issue, you should be able to:

- Explain the case against offshore drilling.
- Explain the case for offshore drilling.
- Describe how oil and gas development affect human and environmental health.
- Describe what oil spilled during offshore drilling does to marine organisms.

ISSUE SUMMARY

YES: Stephen L. Baird argues that the demand for oil will continue even as we develop alternative energy sources. Drilling for offshore oil will not give the United States energy independence, but the nation cannot afford to ignore energy sources essential to maintaining its economy and standard of living.

NO: Mary Annette Rose argues that the environmental impacts of exploiting offshore oil—including toxic pollution, ocean acidification, and global warming—are so complex and far-reaching that any decision to expand U.S. oil drilling must be based on more than public opinion driven by consumer demands for cheap energy, economic trade imbalances, and politics.

Petroleum was once known as "black gold" for the wealth it delivered to those who found rich deposits. Initially those deposits were located on land, in places such as Pennsylvania, Texas, Oklahoma, California, and Saudi Arabia.

As demand for oil rose, so did the search for more deposits, and it was not long before they were being found under the waters of the North Sea and the Gulf of Mexico, and even off the California beaches. For the "History of Offshore Oil," see the *Congressional Digest* (June 2010).

In 1969, a drilling rig off Santa Barbara, California, suffered a blowout, releasing more than three million gallons of oil and fouling 35 miles of the coast with tarry goo. John Bratland, "Externalities, Conflict, and Offshore Lands," *Independent Review* (Spring 2004), calls this incident the origin of the modern conflict over offshore drilling for oil. He notes that since then most accidental oil releases have been related to transportation (as when an oil tanker runs into rocks; the *Exxon Valdez* spill was a striking example; see John Terry, "Oil on the Rocks—the 1989 Alaskan Oil Spill," *Journal of Biological Education,* Winter 1991). Underwater oil releases have largely been prevented by the development of blowout-prevention technology. But Santa Barbara has not forgotten, and residents do not trust blowout prevention technology. See William M. Welch, "Calif.'s Memories of 1969 Oil Disaster Far from Faded," *USA Today* (July 14, 2008).

Some people do not seem disturbed by the prospect of oil blowouts or spills. Ted Falgout, director of the port at Port Fourchon, Louisiana, looks at the forest of oil rigs in the Gulf of Mexico and "sees green: the color of money that comes from the nation's busiest haven of offshore drilling. 'It's OK to have an ugly spot in your backyard,' Falgout says, 'if that spot has oil coming out of it.'" See Rick Jervis, William M. Welch, and Richard Wolf, "Worth the Risk? Debate on Offshore Drilling Heats Up," *USA Today* (July 13, 2008).

In 2008, oil and gasoline prices reached record highs. Many people were concerned that prices would continue to rise, with the result being rapid investment in alternative energy sources such as wind. At the same time, those who favored increased drilling, both onshore and offshore, began to call for the government to open up more land for exploration. In its last few months in office, the Bush administration issued leases for lands near national parks and monuments in Utah and lifted an executive order banning offshore drilling. Both measures were considered justified because they would reduce dependence on foreign sources of oil, ease a growing balance of payments problem, and ensure a continuing supply of oil. Critics pointed out that any oil from new wells, on land or at sea, would not reach the market for a decade or more. They also stressed the risks to the environment and called for more attention to alternative energy technologies.

The debate over whether to expand drilling for offshore oil is by no means over. In the wake of President Bush's lifting of the executive order banning offshore drilling, the federal Minerals Management Service (MMS) proposed 31 oil and gas lease sales in areas of the nation's Outer Continental Shelf. These areas are estimated to contain at least 86 billion barrels of oil and 420 trillion cubic feet of natural gas, although specific deposits had not yet been discovered. According to the MMS press release, MMS director Randall Luthi said, "We're basically giving the next Administration a two-year head start. This is a multi-step, multi-year process with a full environmental review and several opportunities for input from the states, other government agencies and interested parties, and the general public."

Reactions from states off whose shores these areas lie have been largely negative. Shortly after President Obama took office, the new Department of the Interior Secretary, Ken Salazar, canceled the Bush administration's Utah oil and gas leases and announced a new strategy for offshore energy development that would include oil, gas, and renewable resources such as wind power. The press release (www.mms.gov/ooc/press/2009/press0210.htm) said that Salazar's strategy calls for extending the public comment period on a proposed 5-year plan for oil and gas development on the U.S. Outer Continental Shelf by 180 days, assembling a detailed report from the Department of Interior agencies on conventional and renewable offshore energy resources, holding four regional conferences to review these findings, and expediting renewable energy rulemaking for the Outer Continental Shelf. "To establish an orderly process that allows us to make wise decisions based on sound information, we need to set aside the Bush Administration's midnight timetable for its OCS drilling plan and create our own timeline," Salazar said.

Unfortunately, the MMS subsequently issued many permits without adequate review. One of those covered the BP Deepwater Horizon drilling rig's Macondo well in the Gulf of Mexico, which exploded on April 20, 2010, resulting in the release of 200 million gallons of oil into the Gulf of Mexico with environmental impacts that are still being assessed. This record-setting disaster is reviewed in Joel K. Bourne, Jr., "The Gulf of Oil: The Deep Dilemma," *National Geographic* (October 2010). See also Mac Margolis, "Drilling Deep," *Discover* (September 2010). As one result, many people and organizations are saying that the government's approach to regulating offshore drilling needs to be changed; see "National Environmental Health Association (NEHA) Position on Offshore Oil Drilling," *Journal of Environmental Health* (September 2010). Some changes have already been announced. In response to the disaster, the U.S. government banned offshore oil drilling for the time being. In September 2010, the Department of Interior (of which the MMS is part) announced that "new rules and the aggressive reform agenda we have undertaken are raising the bar for the oil and gas industry's safety and environmental practices on the Outer Continental Shelf. Under these new rules, operators will need to comply with tougher requirements for everything from well design and cementing practices to blowout preventers and employee training. They will also need to develop comprehensive plans to manage risks and hazards at every step of the drilling process, so as to reduce the risk of human error." See "Salazar Announces Regulations to Strengthen Drilling Safety, Reduce Risk of Human Error on Offshore Oil and Gas Operations" (www.doi.gov/news/pressreleases/ Salazar-Announces-Regulations-to-Strengthen-Drilling-Safety-Reduce-Risk-of-Human-Error-on-Offshore-Oil-and-Gas-Operations.cfm). On May 17, 2011, Salazar testified before the U.S. Senate Committee on Energy and Natural Resources that progress is being made and Congress must "provide the tools for the federal government to effectively oversee offshore oil and gas development, including codifying [new] safety and environmental standards" (www.doi.gov/news/ pressreleases/Salazar-Hayes-Bromwich-Testify-on-Safe-Responsible-Domestic-Oil-and-Gas-Production.cfm).

In the following selections, from before the 2010 BP Deepwater Horizon spill in the Gulf of Mexico, Stephen L. Baird argues that the demand for oil will continue even as we develop alternative energy sources. Drilling for offshore oil will not give the United States energy independence, but the nation cannot afford to ignore energy sources essential to maintaining its economy and its standard of living. He claims the environmental objections just do not add up. In a direct response to Baird's essay, Mary Annette Rose argues that the environmental impacts of exploiting offshore oil—including toxic pollution, ocean acidification, and global warming—are so complex and far-reaching that any decision to expand U.S. oil drilling must be based on more than public opinion driven by consumer demands for cheap energy, economic trade imbalances, and politics.

YES

Stephen L. Baird

Offshore Oil Drilling: Buying Energy Independence or Buying Time?

Skyrocketing fuel prices, unprecedented home foreclosures, rising unemployment, escalating food prices, increasing climate disasters, and the continued war on two fronts have prompted greater public support for renewed offshore drilling for oil. A Gallup poll conducted in May of 2008 found that 57 percent of respondents favored such drilling, while 41 percent were opposed. [. . .] The political landscape is also being changed in favor of offshore drilling, with the results of a Zogby poll (Zogby International has been tracking public opinion since 1984) showing that three in four likely voters—74 percent—support offshore drilling for oil in U.S. coastal waters, and more than half (59 percent) also favor drilling for oil in the Alaska National Wildlife Refuge. [. . .] The tide is turning in favor of offshore drilling, with environmental concerns given less thought because of the increasing financial strain being realized by a majority of the American public. The debate on offshore drilling has captured headlines in newspapers, stirred debate on talk radio, and has been at the forefront on the nightly news.

The rising tide for support of offshore drilling recently gathered momentum when, on July 14, 2008, President George W. Bush lifted a 1990 executive order by the first President Bush banning offshore drilling, while at the same time calling for drilling in the Arctic National Wildlife Refuge. As of August 2008, however, a 1982 congressional ban is still in place, making Bush's action a symbolic gesture, and now the congressional ban is being debated in terms of both environmental issues and U.S. energy independence. In an almost complete reversal of policy, on July 30, 2008, the U.S. Department of the Interior released a news report saying that the nation's energy situation has dramatically changed in the past year. Secretary of the Interior, Dirk Kempthorne, said, "Areas that were considered too expensive to develop a year ago are no longer necessarily out of reach based on improvements to technology and safety." Kempthorne went on to say that, "The American people and the President want action, and a new initiative (the development of a new oil and natural gas leasing program for the U.S. Outer Continental Shelf) can accelerate an offshore exploration and development program that would increase production from additional domestic energy resources." President Bush is urging Congress

From *The Technology Teacher*, November 2008, pp. 13–17. Copyright © 2008 by International Technology Education Association. Reprinted by permission.

to enact legislation that would allow states to have a say regarding operations off their shores and to share in the resulting revenues. [. . .] Shortly after the Interior Department released plans for jumpstarting new offshore oil exploration, on August 16, 2008 the Speaker of The House, Nancy Pelosi, dropped her opposition to a vote on coastal oil exploration and expanded offshore drilling (with appropriate safeguards and without taxpayer subsidies to big oil) as part of broad energy legislation to be addressed when Congress returned in September. [. . .] Today, with the high price of oil and a widening gap between U.S. energy consumption and supply, the ban on offshore oil drilling is being rethought by the general public, politicians, and the oil industry.

The energy stalemate between environmentalists and industry that has inhibited U.S. offshore oil production since the late 1960s is being broken, environmental arguments no longer add up, and working Americans are now taking energy policy inaction personally. According to a Pew Research Center poll conducted in July 2008, 60 percent of respondents considered energy supplies more important than environmental protection, and a majority of young Americans, 18–29, now consider energy exploration more important than conservation. [. . .]

Addressing Environmental and Safety Concerns

Though offshore drilling conjures up fears of catastrophic spills, (such as the 80,000 barrels that spilled six miles off Santa Barbara, California, inundating beaches and aquatic life in January 1969), the petroleum industry rightly argues that safety measures have improved considerably in recent years. According to the U.S. Minerals Management Service, since 1975, 101,997 barrels spilled from among the 11.855 billion barrels of American oil extracted offshore. This is a 0.001 percent pollution rate. That equates to 99.999 percent clean—compare that with Mother Nature herself, as 620,500 barrels of oil ooze organically from North America's ocean floors each year. [. . .]

The United States has been a leader in the creation of the modern offshore oil industry and has pioneered many new safety technologies, ranging from blowout preventers to computer-controlled well data designed to help oil companies' efforts to prevent disasters. Sensors and other instruments now help platform workers monitor and handle the temperatures and pressures of subsea oil, even as drilling is occurring. Hurricanes have become manageable, with oil lines now being capped at or beneath the ocean floor. Even if oil platforms snap loose and blow away, industrial seals restrain potentially destructive petroleum leaks from hundreds or even thousands of feet below the ocean's surface. In August and September of 2005, the 3,050 offshore oil structures endured the wrath of Hurricanes Katrina and Rita without damaging petroleum spills. While 168 platforms and 55 rigs were destroyed or seriously damaged, the oil they pumped remained safely encased, thanks to heavy underwater machinery. The U.S. Minerals Management Service concluded, "Due to the prompt evacuation and shut-in preparations made by operating and service personnel, there was no loss of life and no major oil spills attributed to either storm." [. . .] If it can be done in an environmentally friendly fashion—and with oil companies themselves footing the bill—increasing opportunities for new offshore drilling might be worthwhile.

Offshore territories and public lands like the Alaska National Wildlife Refuge (ANWR) that don't allow drilling have been estimated to contain up to 86 billion barrels of oil according to the U.S. government's Energy Information Administration. Although analysts say that amount of oil will not greatly affect the price of oil, and that renewed offshore drilling would have little impact on gas prices anytime soon, in the short term, oil prices could go down slightly if Congress lifts its moratorium on new offshore drilling because the market would factor in the prospect of additional oil supplies later on. A spokeswoman for the American Petroleum Institute said that, "If we had new territory, we could hypothetically make a big find." [. . .] Offshore drilling might not be the end-all solution to our oil dependence, but any serious energy proposal has to be comprehensive and should include more oil supply and production from the outer continental shelf.

How Dependent Are We on Foreign Oil?

Although the United States is the third largest oil producer (the U.S. produces 10 percent of the world's oil and consumes 24 percent), most of the oil we use is imported. The U.S. imported about 60 percent of the oil consumed in 2006. [. . .] About half of the oil we import comes from the Western Hemisphere (North, South, Central America, and the Caribbean including U.S. territories). [. . .] We imported only 16 percent of our crude oil and petroleum products from the Persian Gulf countries of Bahrain, Iraq, Kuwait, Qatar, Saudi Arabia, and the United Arab Emirates. During 2006, our five biggest suppliers of crude oil were: Canada (17.2%), Mexico (12.4%), Saudi Arabia (10.7%), Venezuela (10.4%), and Nigeria (8.1%). It is usually impossible to tell whether the petroleum products that you use came from domestic or imported sources of oil once they are refined. [. . .] According to the United States Energy Information Administration, the United States spends more than $20 billion, on average, per month to purchase oil, gasoline, and diesel fuel from abroad.

The negative aspects of this dependency are fairly obvious, and they have been well documented. First, oil imports contribute heavily to the United States' trade deficit, which is at record levels. Second, the United States is forced to make political decisions that it might not make otherwise (invading Iraq, cooperating with hostile governments such as Venezuela and Nigeria, looking the other way at Saudi Arabia's reactionary regime, etc.) because it needs their oil. Third, up to now the availability of oil at a fairly reasonable price has left the United States to continue down a path of using more and more energy. [. . .] From Nixon to now, every sitting President has promised to make sure that we wouldn't have a future energy problem . . . though we certainly do now.

> Richard Nixon, 1974: "We will lay the foundation for our future capacity to meet America's energy needs from America's own resources."

Gerald Ford, 1975: "I am proposing a program, which will begin to restore our country's surplus capacity in total energy. In this way, we will be able to assure ourselves reliable and adequate energy and help foster a new world energy stability for other major consuming nations."

Jimmy Carter, 1980: "We must take whatever actions are necessary to reduce our dependence on foreign oil—and at the same time reduce inflation."

Ronald Reagan, 1982: "We will ensure that our people and our economy are never again held hostage by the whim of any country or oil cartel."

George H. W. Bush, 1990: "The Congress should, this month, enact measures to increase domestic energy production and energy conservation in order to reduce dependence on foreign oil."

George W. Bush, 2008: "And here we have a serious problem, America is addicted to oil." [. . .]

It is somewhat misleading when politicians talk about "America's addiction to oil" because there are some mitigating factors that make this situation a lot less dire than it might seem. The two largest foreign suppliers of oil to the United States are friendly to us: Canada and Mexico. These countries have increased their exports to the United States for the past decade and are well-positioned to continue doing so. Thus, fears that the United States will be dependent on "enemy" regimes are overblown, and the price of oil is a world-market price, so the United States is not being gouged. The United States can buy oil from anywhere (except where it imposes sanctions, like Iran), and it doesn't really matter if the oil comes from internal sources or imports. In fact, the United States exports some oil from Alaska, because it is more efficient to send that oil to Japan than it is to send it down to refineries in California. [. . .] The world oil markets are very competitive, and the locating, capturing, refining, and selling of oil is a very complex system. Saying that the United States is too dependent on foreign oil and that this will spell disaster in the near future is not an accurate statement. A more accurate statement would be to say that we have many wasteful energy habits, and that we need to focus on how to reduce our energy use and to expand alternative energy sources without drastically affecting our lifestyles and our economy. But oil is essential to our country's normal functioning, and therefore more American oil must be part of an American energy solution.

Why Drill Offshore for New Oil?

Is more drilling for American oil an essential part of lowering energy costs and freeing us from dependence on foreign sources of energy? Opening up new areas for exploration in the Outer Continental Shelf and the Alaska National Wildlife Refuge in the United States, even if new supplies won't actually reach our gas tanks for several years, would immediately impact the amount of upward speculation on long-term commodity investment in oil. Oil speculators would see a greater supply ahead and that the future of oil would be less constrained on the supply side. Also, fears of Middle Eastern turmoil or South American unrest that could disrupt supply shipments would be much less of a reason to drive up the price of crude if a stable United States could supply additional millions of barrels of oil.

Today, oil drilling is prohibited in all offshore regions along the North Atlantic coast, most of the Pacific coast, parts of the Alaska coast, and most of the eastern Gulf of Mexico. The central and western portions of the Gulf of Mexico therefore account for almost all current domestic offshore oil production, providing 27 percent of the United States' domestic oil production. The areas under the congressional ban contain an estimated additional 18 billion barrels of oil. This estimate is considered conservative since little exploration has been conducted in most of those areas during the past quarter of a century due to the congressional ban. Estimates tend to increase dramatically as technology improves and exploration activities occur. [. . .] Major advances in seismic technology and deep-water drilling techniques have already led the Interior Department's Minerals Management Service to increase its original estimate of untapped Gulf of Mexico oil from 9 billion barrels to 45 billion barrels. In short, there could be much more oil under the sea than previously thought.

The Interior Department has already taken steps for new offshore oil exploration, announcing plans for a lease program that could open up new areas off the coasts of Florida, Georgia, Texas, North Carolina, Virginia, and other coastal states to drilling if Congress lifts the ban. Randall Luthi, director of the department's Minerals Management Service, which handles offshore oilfield regulations and leases, said, "The technology has improved . . . the safety systems we now require have greatly improved . . . and the industry has a good record." According to Luthi, new tools such as high-tech computers that make exploration easier and tougher building materials on platforms are making offshore drilling safer. Seismic technology and directional drilling techniques let oil companies drill 100 exploratory wells from a single offshore platform, reducing the number of derricks and therefore the potential for problems. Automatic shut-off valves underneath the seabed can cut the flow of oil immediately if there's a problem or a storm coming. Blowout prevention equipment can automatically seal off pipes leading to the surface in the case of an unexpected pressure buildup, and undersea pipelines and wellheads can be monitored with special equipment such as unmanned, camera-equipped, and sensor-laden underwater vehicles. [. . .] New drilling technologies and the industry's track record in the Gulf of Mexico show that offshore drilling for oil is safer than it ever has been, proving that you can drill and still be environmentally friendly.

Conclusion

The demand for energy is going up, not down, and for a long time, even as alternative sources of energy are developed, more oil will be needed. The strongest argument against drilling is that it could distract the country from the pursuit of alternative sources of energy. The United States cannot drill its way to energy independence. But with the developing economies of China and India steadily increasing their oil needs in their latter-day industrial revolutions, the United States can no longer afford to turn its back on finding all the sources of fuel necessary to maintain its economy and standard of living. What is required is a long-term, comprehensive plan that includes wind, solar, geothermal, biofuels, and nuclear—and that acknowledges that oil and gas will be instrumental to the United States' well-being for many years to come.

Mary Annette Rose  **NO**

The Environmental Impacts of Offshore Oil Drilling

Stephen L. Baird's article in the November 2008 issue of *The Technology Teacher* describes a contemporary debate about opening more U.S. land and coastal regions to oil and gas exploration and production (E&P). While Baird's thesis—"informed and rational decisions can be reached through the understanding of how complex technological systems can impact the environment, our economy, our politics, and ultimately our culture" [. . .] epitomizes the goal of a technologically literate citizen, his article is a stark contradiction to this call for understanding. His one-sided argument is built upon public opinion driven by consumer demand for cheap energy, economic trade imbalance, and politics. Baird fails to connect the offshore oil and gas E&P to the toxins and greenhouse gases these technological processes release to the marine environment and the atmosphere. Decades of empirical evidence indicates that all stages of offshore oil and gas activity have consequences for the health and survivability of marine plants and animals, humans, and our planet.

In the following, I counter Baird's proposition that "environmental arguments no longer add up" [. . .] by identifying a few of the impacts of offshore oil E&P on the environment. My hope is that this analysis will better prepare teachers to foster the development of *critical-thinking* skills in their students. These skills are prerequisite to assessing the impacts of technology upon the environment and society [. . .] and essential for making environmentally sustainable choices.

Technology Assessment

When we ask our students to assess the impacts of technology, we ask them to engage in a process of inquiry, a cognitive journey of questioning assumptions, hypothesizing, gathering and reviewing evidence and trends, and testing their hypotheses against the body of evidence. This process, known as Technology Assessment (TA), refers to an examination of the potential or existing risks and consequences of developing, adopting, or using a technology. TA begins by bounding the study to identify time horizons, impact zones, stakeholders, and a host of relevant technical and environmental information. Tools, such as cross-impact analysis, mathematical models, and regression analysis, are used to analyze this data and to predict outcomes and risks associated with

From *The Technology Teacher,* February 2009, pp. 27–31. Copyright © 2009 by International Technology Education Association. Reprinted by permission.

possible decisions. TA results in a list of "if/then" statements, options, trade-offs, or alternative future scenarios, which decision makers use to inform policies, make investments, and plan for the future.

Nature of Petroleum (Hydrocarbons)

To examine environmental impacts, we should begin by looking at the nature of crude oil (petroleum). Petroleum is a fossil fuel that forms from the remains of prehistoric vegetation and animals as a result of millions of years of heat and pressure. It is a complex mixture of hydrocarbons, several minor constituents (e.g., sulfur), and trace metals (e.g., chromium). The chemical composition of crude oil varies by the age of the geologic formation from which it came. When crude oil is released into the environment, biological, physical, and chemical processes (referred to as weathering) alter the oil's original characteristics. [. . .] Lighter oils tend to be volatile, reactive, and highly flammable, while heavier crudes tend to be tarry and waxy and contain cancer-causing polycyclic aromatic hydrocarbons (PAH) and other toxic substances.

Offshore Oil Exploration and Production

One challenge for offshore (waters beyond three miles from the shoreline) E&P operators is to control the dynamic changes in temperature and pressures when drilling into rock formations located deep beneath the ocean. As depths increase, the pressure of drilling fluids (muds) is used to counter the deep-sea pressures related to depth and the pockets of high-pressure and high-temperature (HPHT) gases. If not contained, these HPHT gases result in dangerous oil-well blowouts that emit a buoyant plume of oil, produced water, and pressurized natural gas (methane).

The 1969 blowout off the Santa Barbara, California coast that spilled 80-100,000 gallons of crude oil and inundated local beaches was an ecological disaster, killing thousands of birds, fish, and marine mammals. [. . .] However, this pales in comparison to the 1979 blowout at the Ixtoc 1 offshore oil rig in the Gulf of Mexico, which spewed an estimated 140 million gallons of crude oil until that well was capped nine months later. [. . .]

Modern technology and the research and monitoring systems of the Department of Interior's Minerals Management Service (MMS), which manages E&P in the outer continental shelf (OCS), have reduced the frequency of these ecological catastrophes. However, weather, tectonic events, equipment failure, transportation accidents, human error, and deliberate unethical choices continue to make oil spills a reality.

Today, there are nearly 4,000 active platforms in the OCS. [. . .] MMS [. . .] indicates that 115 platforms were destroyed and 600 offshore pipelines were damaged by Hurricanes Katrina and Rita in 2005. For these hurricanes, "124 spills were reported with a total volume of roughly 17,700 barrels of total petroleum products, of which about 13,200 barrels were crude oil and condensate from platforms, rigs, and pipelines, and 4,500 barrels were refined products from platforms and rigs." [. . .] In 2008, 60 platforms were destroyed

during Hurricanes Gustav and Ike [. . .]; data on oil spills and damage to pipelines has not yet been released.

Offshore E&P also requires transportation vessels and terrestrial storage.
As a direct result of Hurricane Katrina, an above-ground storage tank in St.
Bernard Parish, Louisiana, spilled over 25,110 barrels of mixed crude oil. [. . .]
The oil inundated about 1,700 homes and several canals. Testing conducted in
2006 confirmed that contaminants, including PAHs, arsenic, and other toxics
were above acceptable risk standards. [. . .]

Wastes from Offshore E&P

The less dramatic, yet more disturbing impacts of offshore oil E&P relate to
the volume and type of wastes these processes generate. Wastes include produced water, drilling fluids (muds), cuttings (crushed rock), diesel emissions,
and chemicals associated with operating mechanical, hydraulic, and electrical
equipment, such as biocides, solvents, and corrosion inhibitors.

Produced Water.   By volume, 98% of the waste from E&P is *produced water,*
with estimates at 480,000 barrels per day in 1999. [. . .] Produced water is a
water mixture consisting of hydrocarbons (e.g., PAH, organic acids, phenols,
and volatiles), naturally occurring radioactive materials, dissolved solids, and
chemical additives used during drilling. Glickman [. . .] concluded that "hydrocarbons are likely contributors to produced water toxicity, and their toxicities
are additive, so that although individually the toxicities may be insignificant,
when combined, aquatic toxicity can occur." [. . .] Furthermore, studies document that sediments become contaminated with these toxins and that the
concentration has a direct correlation with produced water discharges. [. . .]

Citing "relief to coastal waters, which support spawning grounds,
nurseries, and habitats for commercial and recreational fisheries: reducing
documented aquatic 'dead zone' impacts; reduction of potential cancer
risks to anglers from consuming seafood contaminated by produced water
radionuclides; and reducing potential exposure of endangered species to toxic
contaminants" the EPA [. . .] banned the release of produced waters to inland
and coastal waters (extending to three miles from shore). However, offshore
E&P operations in U.S. waters may legally discharge treated produced water
directly into the ocean or inject it into underground wells.

Drilling Fluids.   Drilling muds and cuttings are of environmental concern
because of their potential toxicity and the large volume that are discharged
during drilling. Three types occur, including oil-based (OBM; diesel or mineral
oil serves as base fluid), water-based (WBM), and synthetic-based muds (SBM).
The EPA [. . .] requires zero discharge of OBM—to dispose, OBM is shipped to
onshore oil field waste sites or injected into disposal wells at sea. WBM and
SBM typically contain arsenic, barium, cadmium, chromium, copper, iron,
lead, mercury, and zinc. Barium and barium compounds are used in drilling
muds because they act as lubricants and increase the density of mud, thereby
sealing gases in the well.

Drilling in deep water (>1,000 ft) uses rotary bits that chip through thousands of feet of rock to access oil and gas deposits. Diesel-powered engines provide the power to operate the drilling rig and drive the drill. As paraphrased from Continental Shelf Associates, Inc. [. . .], the general sequence of events entails initial "open hole" drilling where a drill bit positioned within a drill pipe chips away at the ocean floor. As the drill bit spins, mud (WBM) and water are forced at high velocity around the drill bit to force rock chips (cuttings) up and out to the seabed. After a known distance, the drill bit and pipe are removed and a wellhead is installed. WBM is typically discharged and replaced with SBM. Additionally, a marine riser system is connected to the wellhead to return fluids, muds, and cuttings to the drill rig where they are separated. Cuttings are discharged in a plume from the platform, and mud is recycled back to the drill bit. [. . .]

Continental Shelf Associates, Inc. [. . .] examined the impact of synthetic-based drilling fluids (SBF: mixtures of organic isomers) by comparing indicators from near and far distances from four E&P drilling sites in the Gulf of Mexico. [. . .] Cuttings and SBM extended several hundred meters from the well site and up to 45 cm in thickness. Analyses indicated that near-field sediments were toxic to amphipods (crustaceans). Chemicals associated with both WBM and SBM waste solids in near-field sediments contributed to sediment toxicity. Significantly higher mercury and lead concentrations were found in near-field sediments than in far-field sediments for some sites. Red crabs had high concentrations of toxins, such as arsenic, barium, chromium, and mercury.

Ethical Climate

Offshore E&P is conducted by a small number of operators, most of which are multinational oil corporations with single-year income larger than the GDP of entire nations; e.g., ExxonMobil Corporation [. . .] reported $40.610 billion in net income for 2007. The isolated nature of offshore drilling operations fosters a climate conducive to environmentally irresponsible behaviors. For instance, as reported by the EPA [. . .], two large electrical transformers located on Platform Hondo, Exxon's Santa Ynez Unit, leaked nearly 400 gallons of fluids contaminated with PCBs into the Pacific Ocean. In one instance, a transformer leaked for almost two years before repairs were made. Cleanup workers were not provided protective equipment to protect themselves against direct contact with and inhalation of PCBs. Exxon agreed to a settlement of $2.64 M in violation of the federal *Toxic Substances Control Act.*

Environmental Impacts

There are known detrimental impacts upon the marine environment for all phases of offshore E&P. [. . .] While natural seepages contribute more hydrocarbons to the marine environment by volume, the quick influx and concentration of oil during a spill makes them especially harmful to localized marine organisms and communities. Plants and animals that become coated in oil perish from mechanical smothering, birds die from hypothermia as their feathers lose their waterproofing, turtles die after ingesting oil-coated food, and animals become disoriented and exhibit other behavior changes after breathing volatile organic compounds.

When emitted into the marine environment, oil, produced water, and drilling muds may adversely impact an entire population by disrupting its food chain and reproductive cycle. Marine estuaries are especially susceptible, as hydrocarbons and other toxins tend to persist in the sediments where eggs and young often begin life. However, the severity and effects of oil exposure vary by concentration, season, and life stage. The oil spill from the Ixtoc 1 blowout threatened a rare nesting site of the Kemp's Ridley sea turtle, an endangered species. Field and laboratory data on the nests of turtle eggs found a significant decrease in survival of hatchlings, and some hatchlings had developmental deformities. [. . .]

Marine organisms that live near an existing or sealed wellhead or an oil spill area experience persistent exposure to a complex web of hydrocarbons, petroleum-degrading microbes, and toxic substances associated with drilling muds and produced water. Abundance and diversity of marine life, especially those living near or in the seabed, decline. The growth and reproduction rates of entire populations that live in the water column may decline for months after a spill [. . .], natural defense mechanisms necessary to deal with disease (immune suppression) become compromised [. . .], and genetic mutations may occur. Many of these toxins (e.g., arsenic, chromium, mercury, and PAH) move up the food chain and biomagnify, i.e., increase in concentration.

One of the most disturbing trends is the evidence that common hydrocarbon contaminants (e.g., PAH) act as endocrine disrupters. Endocrine disrupters are chemicals that can act as hormones or anti-hormones in aquatic ecosystems, thus disrupting normal reproductive and developmental patterns. [. . .] Evidence also suggests that polychlorinated biphenyls (PCBs), a known carcinogen, also exhibit these endocrine effects. [. . .]

Climate Change and Ocean Acidification.   Greenhouse gases (GHG) are generated directly by offshore oil E&P and indirectly by enabling future emissions of oil as it is refined, distributed, and consumed, primarily by the transportation sector. In the U.S., 2007 emissions of methane from petroleum E&P was estimated at 22 MMTCO$_2$e (million metric tons of carbon dioxide equivalents) in addition to total contributions of petroleum at 2,579.9 MMTCO$_2$e. [. . .] These GHGs are primary drivers that disrupt the carbon cycle involving the biosphere, atmosphere, sediments (including fossil fuels), and the ocean. The consequences of this disruption include climate change (global warming, melting of ice at the poles, and ocean acidification). The ocean acts as a carbon sink, absorbing CO$_2$. As CO$_2$ increases, the acidity of the ocean increases and carbonate becomes less available to marine organisms that need it to build shells and skeletal material. Corals, calcareous phytoplankton, and mussels are especially susceptible to acidosis, which "can lead to lowered immune response, metabolic depression, behavioral depression affecting physical activity and reproduction, and asphyxiation." [. . .]

Human Health Impacts

Workers, victims of oil spills, and rescue workers are exposed to a host of chemical hazards. When people come in dermal contact with drilling fluids, muds, and cuttings, they can experience dermatitis; as exposure increases, impacts can include hypokalemia, renal toxicity, and cardiovascular and neuromuscular effects. [. . .] Exposure to volatile aromatic hydrocarbons (e.g., benzene) results in respiratory distress and unconsciousness. Long-term exposure can cause anemia, leukemia, reproductive problems, and developmental disorders. [. . .] Exposure to fine particulate matter, nitrogen oxides, sulphur, and dozens of hydrocarbons (e.g., PAH) emitted from diesel and gasoline engines, is linked to a variety of health impacts, including asthma attacks, cancer, endocrine disruption, and cardiopulmonary ailments. Because toxins bioaccumulate in fish, people who eat fish and shellfish from affected waters may experience nervous system effects, such as impairment of peripheral vision and seizure. Children and fetuses are especially vulnerable; exposure to toxins impairs physical and cognitive development.

Conclusion

The environmental impacts of offshore oil exploration and production are complex and far-reaching. Petroleum, produced water, and the chemicals used to extract petroleum from under the ocean have mechanical, toxic, carcinogenic, and mutagenic impacts on marine life and the humans who eat its marvelous bounty. But more profoundly, the combustion of petroleum is a major contributor of carbon dioxide to the atmosphere which, in turn, drives global warming and ocean acidification.

Therefore, the decision to open additional U.S. lands and oceanic territories to oil exploration and production should be based on more than public opinion and some unquestioned assumption that the U.S. has a right to consume about 25% of the world's energy and contribute 21% of the worlds' CO_2 emissions. Engaging students in a process of technology assessment is a viable pedagogy to help students develop dispositions and critical-thinking skills. These skills will enable students to not only recognize narrow, one-sided perspectives, but be better prepared to seek and apply valid data and analytical strategies to the critically important decisions that could impact life on this planet. In an age of complexity, these skills are essential for making environmentally sustainable choices. [. . .]

EXPLORING THE ISSUE

Should We Drill for Offshore Oil?

Critical Thinking and Reflection

1. List three strong arguments against offshore drilling.
2. List three strong arguments for offshore drilling.
3. In what ways does oil and gas development (onshore or offshore) affect human well-being?
4. How does nature handle oil spills, including both natural seeps and human-caused spills?

Is There Common Ground?

One cannot help but wonder whether, in the wake of the 2010 BP Deepwater Horizon spill in the Gulf of Mexico, Stephen L. Baird has changed his mind about whether environmental objections to offshore drilling hold up. Read Joel K. Bourne, Jr., "The Gulf of Oil: The Deep Dilemma," *National Geographic* (October 2010), and Michael Grunwald, "BP Oil Spill: Has the Damage Been Exaggerated?" Time (July 29, 2010), and answer the following questions:

1. How bad has the environmental impact of the spill been?
2. Is the environmental impact as bad as people initially expected?
3. If your answer to the previous question is "No," describe at least three factors that help to explain the smaller impact.
4. Should we continue to worry about the environmental impacts of offshore drilling?

ISSUE 9

Is Shale Gas the Solution to Our Energy Woes?

YES: Diane Katz, from "Shale Gas: A Reliable and Affordable Alternative to Costly 'Green' Schemes," *Fraser Forum* (July/August 2010)

NO: Deborah Weisberg, from "Fracking Our Rivers," *Fly Fisherman* (April/May 2010)

Learning Outcomes

After reading this issue, you should be able to:

- Explain how shale gas threatens water supplies and even human health.
- Discuss the relative merits of putting energy-related decision making in the hands of private investors or government.
- Explain how an influx of cheap fossil fuel will affect the development of alternative (e.g., wind and solar) energy systems.
- Discuss whether increasing prices through regulation of an industry, in order to protect public health, is justifiable.

ISSUE SUMMARY

YES: Diane Katz argues that new technology has made it possible to release vast amounts of natural gas from shale far underground. As a result, we should stop spending massive sums of public money to develop renewable energy sources. The "knowledge and wisdom of private investors" are more likely to solve energy problems than government policymakers.

NO: Deborah Weisberg argues that the huge amounts of water and chemicals involved in "fracking"—hydraulic fracturing of shale beds to release natural gas—pose tremendous risks to both ground and surface water, and hence to public health. There is a need for stronger regulation of the industry.

Fossil fuels have undeniable advantages. They are compact and easy to transport. In the form of petroleum and natural gas and their derivatives, they are well suited to powering automobiles, trucks, and airplanes. They are also abundant and relatively inexpensive, although the end of the era of oil abundance is in sight, and prices are rising. However, fossil fuels also have disadvantages, for their use puts carbon dioxide in the air, which threatens us with global warming. Oil is associated with disastrous oil spills such as the one that resulted from the failure of a British Petroleum drilling rig in the Gulf of Mexico in 2010. Coal mining leaves enormous scars on the landscape, and coal burning emits pollutants that must be controlled. Natural gas alone seems relatively benign, for although it emits carbon dioxide when burned, it emits less than oil or coal. It produces fewer air pollutants, it cannot be spilled (if released, it can cause explosions and fires, but outdoors it mixes with air and blows away), and obtaining it has not meant huge damage to the environment. Much of the United States' demand for natural gas is met by domestic production, but demand is rising and imports—now at about 15 percent of demand—will have to rise to keep up, just as they have with oil. Lacking new sources of natural gas or a shift to coal, nuclear power, or alternatives such as wind and solar power, the nation must inevitably become more dependent on foreign energy suppliers.

It has long been known that large amounts of "unconventional" natural gas reside in deep layers of sedimentary rock such as shale. However, this gas could not be extracted with existing technology, at least not at a price that would permit a profit once the gas was sold. In recent years, this has changed, for drilling technology now allows drillers to bend drill holes horizontal to follow rock layers. Injecting millions of gallons of water and chemicals at extraordinarily high pressure can fracture (or "frack") the rock surrounding a drill hole and permit trapped gas to escape. See Richard A. Kerr, "Natural Gas from Shale Bursts onto the Scene," *Science* (June 25, 2010). Mark Fischetti, "The Drillers Are Coming," *Scientific American* (July 2010), notes that the Marcellus shale formation, which stretches from upstate New York through Pennsylvania to Tennessee, may contain enough gas to meet U.S. needs for 40 years. There are other shale formations in the United States, Canada, and Europe. The total U.S. supply may be enough to meet needs for a century; see Steve Levine, "Kaboom!" *New Republic* (May 13, 2010). See also Paul Stevens, "Cheap Gas Coming?" *World Today* (August–September 2010).

Not surprisingly, many people are concerned about the environmental impacts of "fracking" and disposing of used water, chemicals, and drilling wastes. Richard A. Kerr describes threats to groundwater in "Not Under My Backyard, Thank You," *Science* (June 25, 2010). But the industry insists that it will deal responsibly with its wastes and hastens to reassure people living near drilling sites. Alex Halperin, "Drill, Maybe Drill?" *American Prospect* (May 2010), describes the debate over shale-gas drilling in upstate New York. The area has suffered large job losses, something the shale-gas industry may remedy. Many landowners—including farmers—see the potential for huge boosts to their income. But the industry has reportedly persuaded people to lease drilling rights on their property by making promises that cannot be kept.

Environmental impacts are a huge concern. So far, regulations are slowing development of the industry in New York—a moratorium on drilling permits has been proposed; see Theresa Keegan, "Controversy Rages in Hydro-Fracking Debate," *Hudson Valley Business Journal* (June 28, 2010). In neighboring Pennsylvania, the drillers are already producing large quantities of natural gas and spilled fracking fluid has caused problems. The potential problems are discussed in Brian Colleran, "The Drill's About to Drop," *E Magazine* (March/April 2010). James C. Morriss, III, and Christopher D. Smith, "The Shales and Shale-Nots: Environmental Regulation of Natural Gas Development," *Energy Litigation Journal* (Summer 2010), contend that if companies in the industry act to prevent problems before regulators require such action, this both demands a better understanding of the technology and prevents future litigation.

According to Marc Levy and Mary Esch, "EPA Takes New Look at Gas Drilling, Water Issues," *AP* (www.google.com/hostednews/ap/article/ALeqM5jrnCodm MZhIWjyiXJP_JYQUd77BgD9H33C703) (July 20, 2010), in 2004 an Environmental Protection Agency (EPA) study said that fracking was little or no threat to drinking water and Congress exempted fracking from federal regulation. Now, however, the EPA is holding hearings and will conduct a $1.9 million study to reevaluate fracking technology; see Tom Zeller, Jr., "E.P.A. Considers Risks of Gas Extraction," *The New York Times* (July 23, 2010). Preliminary results of the study are expected in 2012. The EPA's Web page on fracking and this study is at www.epa.gov/ogwdw000/uic/wells_hydrofrac.html.

Concern over such problems has prompted the New York State Department of Environmental Conservation to revisit its 1992 Generic Environmental Impact Statement (GEIS), noting that "After a comprehensive review of all the potential environmental impacts of oil and gas drilling and production in New York, the Department found in the 1992 GEIS that issuance of a standard, individual oil or gas well drilling permit anywhere in the state, when no other permits are involved, does not have a significant environmental impact. A separate finding was made that issuance of an oil- and gas-drilling permit for a surface location above an aquifer is also a nonsignificant action. . . ." However, the new fracking technology warrants further review based on "required water volumes in excess of GEIS descriptions, possible drilling in the New York City Watershed, in or near the Catskill Park, and near the federally designated Upper Delaware Scenic and Recreational River, and longer duration of disturbance at multi-well drilling sites." The Department's "Draft Supplemental Generic Environmental Impact Statement on the Oil, Gas and Solution Mining Regulatory Program" is available at www.dec.ny.gov/energy/58440.html.

In the YES selection, Diane Katz argues that the new "fracking" technology has made it possible to release vast amounts of natural gas from deep shale deposits. As a result, we should stop spending massive sums of public money to develop renewable energy sources. The "knowledge and wisdom of private investors" are more likely to solve energy problems than government policymakers. In the NO selection, Deborah Weisberg argues that the huge amounts of water and chemicals involved in "fracking" pose tremendous risks to both ground and surface water, and hence to public health. There is a need for stronger regulation of the industry.

YES

Diane Katz

Shale Gas: A Reliable and Affordable Alternative to Costly "Green" Schemes

Governments at every level across North America are collectively showering billions of tax dollars on "green energy" schemes in an effort to avert global warming and end our "dependence on foreign oil." But in the political arena, there is precious little attention being paid to a far more affordable alternative energy source with great potential to reduce both fossil fuel emissions and imports of Middle Eastern oil.

In contrast to government tax breaks, preferential loans, grants, and other forms of subsidies to wind and solar projects, private investors are moving capital into the production of "shale gas." Trapped within dense sedimentary rock, this "unconventional" natural gas was for decades considered too costly to retrieve. But advances in drilling technologies, along with the rising cost of conventional natural gas, have transformed the economics of shale gas extraction. Consequently, the vast stores of shale gas buried a thousand meters or more below the surface of North America (and beyond) have the potential to dramatically alter both environmental politics and geopolitics.

The actual volume of recoverable shale gas remains imprecise as supplies are still being mapped and evaluated. The National Energy Board estimates Canada's volume to be 1,000 trillion cubic feet, with similar reserves in the United States. Europe also may be home to nearly 200 trillion cubic feet of shale gas.

In Canada, there are major shale gas "plays" in the Horn River Basin and the Montney Formation, both in British Columbia. Major exploration for shale gas is also occurring in the Colorado Group in Alberta and Saskatchewan, the Utica Shale in Quebec, and the Horton Bluff Shale in New Brunswick and Nova Scotia.

When burned, shale gas emits just half the carbon dioxide of coal. Unlike wind and solar power, which produce power intermittently, natural gas is continuously available to produce the steam that powers turbines in the production of electricity. In addition, distribution networks for natural gas already exist, meaning that there is less need to build costly infrastructure. These and other advantages of shale gas call into question the massive public outlays for more problematic "renewable" power sources.

From *Fraser Forum*, July/August 2010, pp. 18–20. Copyright © 2010 by Fraser Institute. Reprinted by permission. www.fraserinstitute.org

According to energy analyst Amy Myers Jaffe, shale gas "is likely to upend the economics of renewable energy. It may be a lot harder to persuade people to adopt green power that needs heavy subsidies when there's a cheap, plentiful fuel out there that's a lot cleaner than coal, even if [natural] gas isn't as politically popular as wind or solar."

That very dynamic stymied energy mogul T. Boone Pickens in his plan to build the world's largest wind farm in the Texas Panhandle. The plan called for the construction of a wind farm with 687 turbines, driving the production of 1,000 megawatts of electricity—the equivalent of a nuclear power plant.

Shortly after the debut of the project in 2008, natural gas prices declined, making wind energy not competitive enough to attract the $2 billion needed in financing. As Pickens told the *Dallas Morning News,* "You had them standing in line to finance you when natural gas was $9 [per million Btu] . . . Natural gas at $4 [per million Btu] doesn't have many people trying to finance you." The lack of a transmission line to move the wind power to urban centers also contributed to his decision to kill the project, Pickens said.

But governments across Canada have virtually unlimited financing at their disposal in the form of tax revenues, and thus are forcing taxpayers to subsidize costly "renewable" energy projects and transmission build-outs, even though more efficient alternatives exist. The government of Ontario, for example, is forcing utilities (read consumers) to buy "green" power at more than double the market rate for conventional electricity.

In the past, the fine grain of shale rock made tapping the natural gas within particularly difficult. The National Energy Board describes shale as "denser than concrete" and thus virtually impermeable. But from the tenacity of a lone Texan, a productive method to set the gas flowing has emerged. As the *Sunday Times* reports:

> It all began in 1981 when Mitchell Energy & Development, a Texas gas producer, was, quite literally, running out of gas. [George] Mitchell, who founded the firm, ordered his engineers to look into tapping shale, which drillers usually passed through to get to the oil and gas fields below them. . . . For years, [the shale] had been ignored, but Mitchell had a hunch about their potential. "I thought there had to be a way to get at it," he said. "My engineers were always adamant. They would say, 'Mitchell, you're wasting your money.' And I said, 'Let me.'" It took 12 years, more than 30 experimental wells and millions of dollars before he came up with the technical solution.

That technical solution is known as "hydraulic fracturing" (or "fracking"), which involves injecting at high pressure a mixture of water, sand, and chemicals into the shale to fracture the rock and allow the release of the natural gas therein. In conjunction with fracking, horizontal drilling is used to maximize the surface area of the borehole through which the gas is collected.

Some environmentalists complain that the chemical compounds used in fracking threaten to pollute soil and groundwater, and they decry the volumes of water used in the production process. In addition, some global warming alarmists oppose the development of new stores of fossil fuel. But in many instances, fracking is conducted thousands of feet below aquifers, and the strata are separated by millions of tons of impermeable rock. Moreover, ever larger quantities of the water used in fracking are recycled. The industry also maintains that stringent regulatory standards are in place to protect the environment. And, as detailed in another article in this edition of *Fraser Forum*, all sources of energy—"renewables" included—involve environmental trade-offs.

Initially, fracking and horizontal drilling were too costly for widespread adoption. But as oil prices rose, these techniques became more cost-effective. Since then, economies of scale and technological innovations have "halved the production costs of shale gas, making it cheaper even than some conventional sources."

Energy analysts expect further cost reductions in shale gas production as major oil and gas companies invest in new technologies. For example, production costs have fallen to $3 per million Btu at the Haynesville Formation, which encompasses much of the US Gulf Coast, down from $5 or more at the Barnett Shale in the 1990s.

The turnabout in shale gas fortunes is all the more remarkable given predictions in the past decade that Canada and the United States were running low on natural gas. US Federal Reserve Chairman Alan Greenspan, for example, declared in 2003 that the United States would have to import liquid natural gas to meet demand.

Doing so would have increased reliance on supplies from Russia and Iran, hardly an appealing prospect for anyone intent on "energy independence." Before the shale gas boom, both countries were thought to control more than half of the known conventional gas reserves in the world. Now, however, Canada and the United States have access to huge domestic stores.

This could cause dramatic shifts in global petro-politics. As energy analyst Amy Myers Jaffe notes, "Consuming nations throughout Europe and Asia will be able to turn to major US oil companies and their own shale rock for cheap natural gas, and tell the Chavezes and Putins of the world where to stick their supplies—back in the ground."

The new accessibility to shale gas will also moderate the influence of OPEC and any potential natural gas cartel by providing affordable and reliable alternative sources of energy. Indeed, US production of natural gas in March hit an historical monthly high of 2.31 trillion cubic feet, topping Russia to become the largest producer in the world. Consequently, natural gas exports once headed to North America are instead heading to Europe, thereby forcing Russia to lower prices for its once-captive customers.

Illustrating the new political tectonics is the recent agreement between Chevron and Poland for natural gas development and production. According to Dr. Daniel Fine of the Mining and Minerals Resources Institute at MIT, "When Chevron announces that they have gas [in Poland], then Russia is shut out" from having a monopoly in Eastern Europe.

Canada will also feel the effects of the energy market shifts. For example, the expansion of US supplies means that Canada will need to find new export opportunities for its natural gas. However, this should not cause problems, analysts say, because supplies of conventional natural gas are declining else-where while fuel demands for transportation and electricity are growing.

The private sector is adept at adjusting to shifting trends. For example, a shipping terminal for natural gas imports to be built by Kitimat LNG Inc. was redesigned for exports to the Pacific Rim due to "increases in supply throughout North America—including in the US, Canada's traditional export market."

Unfortunately, federal and provincial governments remain wedded to energy policies that lack the knowledge and wisdom of private investors and fail to account for the dynamic nature of the market. Vast infusions of subsi-dies obscure the true costs of various energy sources, while disparate regula-tions and mandates inhibit the unfettered competition that would otherwise determine the most efficient and beneficial fuels. Policy makers and politicians could dramatically improve energy policy by releasing their ham-fisted grip on the energy market.

Deborah Weisberg  **NO**

Fracking Our Rivers

On Christmas Day 2007, George Watson returned home from a family dinner to find one of his prized Black Angus cows dead alongside Hargus Creek, a stream that runs through his southwestern Pennsylvania farm.

Over the next three months, Watson lost 16 more cattle—all of which had been bred—making it, as he said, "a double loss." Up to three in one day were found lying near the water. A series of calves died soon after birth.

"I've been raising cattle for 22 years and never had anything like that," said Watson, a Vietnam veteran, who also was having problems with discolored, sludgy well water. A local vet tested the dead cows, but failed to find anything abnormal. Looking back now, Watson wishes he'd had someone test the water in the creek.

Although natural gas wells were being developed all around him, rumors of illegal wastewater dumping in local streams, and a 43-mile fish kill on Dunkard Creek in the same Monongahela River watershed two years later, fueled his darkest fears.

"After my cows died, I suspected it was from brine and waste being dumped, although I can't prove it now," said Watson, who later leased the mineral rights on his farm to Range Resources for $3,000 an acre plus 15 percent production royalties. Drilling hadn't begun as of late last year.

Range is one of 40 companies driving the boom in hydraulic fracturing for natural gas in Pennsylvania, where 53,000 wells are turning pastures and woods into industrial sites. Although hundreds of thousands more have changed the landscape in at least 31 states, Pennsylvania and New York have an abundance of Marcellus Shale wells and, unlike out West, they are close to end users. While vertical drilling and "hydrofracking" for gas has existed for decades, new technologies enable extractors to go more than a mile deep and a mile horizontally to fracture the Marcellus—and release embedded gas—using millions of gallons of sandy, chemical-laden water.

Dunkard Fish Kill

CONSOL Energy's Morris Run borehole and other sources in the Dunkard watershed are under investigation by several federal and state agencies, including the Pennsylvania Attorney General's Office, over possible illegal discharges of hydraulic fracturing fluid, since the level of total dissolved solids, including chlorides, in Dunkard Creek was higher than anything previously associated with coal bed methane wastewater, the only discharge permitted at Morris Run.

"There's pretty strong evidence there was more than coal bed methane water going down that borehole," said Charlie Brethauer of Pennsylvania DEP's water management section. "As far as allegations of illegal activity, I think there's something to it, although to what extent, we don't have any idea yet. We haven't ruled out 'fracking' fluid."

Ed Pressley and his wife Verna live along Dunkard Creek in Brave, Pennsylvania, and watched in horror as fish began going belly up in September 2009 in what would become a massive loss of wildlife that continued for a month. The shells of rare mussels popped open, said Verna, and muskellunge and smallmouth bass bled to death from their gills.

"Kids were putting fish into buckets trying to save them—the tears were running down their cheeks—but there was nowhere to take the fish to," said Verna, a retired science teacher. "We counted 600 dead fish—the stench was overwhelming—just below our dam. It was one of the most devastating emotional experiences of my life."

What made it especially heartbreaking for the Pressleys is that their dream was to turn their property into a living classroom, where children could study kingfishers, blue herons, mudpuppies, turtles, and other forms of wildlife sustained by the water. They were negotiating a conservation easement agreement with the US Department of Agriculture that would protect their land against development for generations to come, and with American Rivers to have a relic industrial dam removed from their section of the stream.

"The folks at Agriculture and American Rivers say they're going to stick with it," Verna said as she stood along Dunkard and peered into the eerily empty water last fall. "But it's going to be years before you'll see fish in here again. I know it's not going to happen in my lifetime."

An EPA interim report about Dunkard's demise cites the presence of golden algae, a toxic organism indigenous to southern U.S. coastal waters, but never before documented in Pennsylvania. Whether it got to Dunkard on migratory birds' feet, drilling equipment that originated in Texas, or by some other means may never be known, but the EPA confirmed that excessive levels of total dissolved solids turned Dunkard so salty the algae were able to thrive.

Golden algae was later found on Whitely Creek, a stocked trout fishery in the same watershed, said Brethauer, who indicated it is likely to spread to other streams.

While the gas drilling industry touts hydraulic fracturing as America's path to energy independence—the Natural Gas Supply Association claims there are enough reserves to meet the nation's needs for a century—some watchdogs say weak regulations and poor enforcement are fueling an environmental nightmare.

The 2005 Energy Policy Act exempts injection of hydraulic fracturing fluids from a key provision in the Safe Drinking Water Act, and federal regulations governing wastewater disposal are limited, according to Deborah Goldberg of Earthjustice, a nonprofit environmental law firm. "Gas wastewater treatment is mostly left to states to regulate and monitor, and most states are way behind the curve."

Ron Bishop, a biochemistry lecturer at SUNY College at Oneonta and a nationally certified chemical hazards management expert, put it this way: "You have to go through more permitting hoops to put a new garage on your property than to drill for gas."

Pennsylvania is in the process of tightening limits on total dissolved solids that can be discharged in rivers and streams, and New York is considering new permitting requirements—generating a de facto moratorium on drilling—although many environmental stakeholders, including New York City, say they aren't strong enough to protect watersheds such as the Delaware River, which provides drinking water to 17 million people.

"Government has to ramp up its regulations tremendously," said Jeff Zimmerman, an attorney for Damascus Citizens for Sustainability and Friends of the Upper Delaware River, groups which formally have protested the New York proposal. "Until an environmentally infallible extraction system can be assured without qualification, the gas drilling industry should not be allowed to operate. It must be failsafe. A single mistake or uncontrolled accident can wipe out, for years and years, important resources, such as those of Dunkard Creek."

Federal lawmakers are also considering legislation—the Fracturing Responsibility and Awareness of Chemicals (FRAC) Act—that would reverse the Clean Water Act exemption and force industry to disclose the names of all of the hundreds of chemicals used in the hydrofracking process, including those traditionally guarded as proprietary information. Pennsylvania makes the names of chemicals available, but not the proportions.

"Some of them are really nasty, like toluene and benzene, which are known to cause cancer," said Bishop. "Others are harmful to wildlife. DB-NPA is a biocide commonly added to fracking water to kill bacteria and algae. Even in amounts too tiny to show up on chemical tests, it's lethal to bay oysters, water fleas, and brown trout."

The staggering volume of fracking fluid used in each horizontal well—up to 6 million gallons of water and 50,000 pounds of chemicals—means environmental impacts can occur on a massive scale, Bishop said. Spills at drill sites and well casing failures—the two most common problems associated with hydrofracking—can cause escaping fluids to contaminate ground and surface water, and gas to migrate underground.

Violations

PADEP cited drillers for more than 450 violations last year. Cabot Oil Co. was charged with a series of spills that polluted a wetlands and killed fish in Stevens Creek, a Susquehanna River tributary in northeast Pennsylvania.

In a separate matter, Cabot is being sued by 15 Dimock residents who claim drilling operations contaminated their drinking water and caused them to suffer neurological and gastrointestinal ills. They are seeking a halt to drilling plus establishment of a trust fund to cover their medical care.

[On Jan. 9, 2010, PADEP also announced it had fined Atlas Resources $85,000 for violations at 13 different well sites in Greene, Fayette, and Washington counties. The violations included failure to restore well sites after drilling, failure to prevent discharges of silt-laden runoff, and for discharging industrial waste including production fluids onto the ground at 7 of the 13 sites.   THE EDITOR.]

Among the many environmental threats or impacts associated with hydrofracking—including huge withdrawals of water from lakes and streams and erosion and sedimentation from truck traffic on rural roads—one of the more concerning is disposal of wastewater, since about half the liquid used in fracking flows back with additional toxins, including brine six times saltier than ocean water, Bishop said. "This hazardous, industrial waste must be disposed of, but there's no good answer as to how or where. Texas and Oklahoma allow deep well injection, but it doesn't work in Pennsylvania and New York because our rock 10,000 feet down isn't porous enough to absorb the waste."

Before water can be discharged into streams it must be strained, desalinated and restored to an acceptable pH level, but few sewage treatment plants are equipped to deal with the volume and chemical composition of fracking water, and many streams have reached their capacity for assimilating more total dissolved solids, Bishop said. "It's a gigantic problem."

Chris Tucker of Energy In-Depth, a coalition of trade groups managed by the Independent Petroleum Association of America, agrees wastewater disposal is one of the industry's biggest bugaboos.

"Everyone knows we have to get on top of it," he said. "Producers are taking a lot of the wastewater from Pennsylvania into Ohio for deep-well injection. The industry is also looking at mobile recycling facilities, but they're getting quoted one cent a gallon. Consider what that would cost when you're dealing with 3 or 4 million gallons of water."

Although Range Resources' CEO John Pinkerton insists that his company's wastewater poses no threat to freshwater streams, Range has turned to recycling in Pennsylvania, where one-acre impoundments and miles of aboveground pipes circulate frack wastewater among several wells. Range also is exploring additional technologies, including crystallization and evaporation—essentially boiling wastewater and skimming off the salt which could be sold for road de-icing.

"We don't know how much Marcellus play there will be but wastewater disposal will keep pace. If it doesn't, the drilling will cease," Pinkerton said. "We have millions of dollars invested in each well. We have to know where every gallon coming out of the ground will go. It's in our best interest to do it right. To do otherwise would be business suicide."

As a fly fisher, Pinkerton considers himself an environmentalist, and he said natural gas extraction is the only practical alternative to foreign oil and coal. "The idea that we can go to 100 percent renewables before you and I pass away is ludicrous. We need a portfolio of energy solutions—a balanced energy policy—so if oil goes to $300 a barrel, we're not stuck. If we don't figure this out, we're dead meat."

He said every industry has risks and impacts—"you've got to cut down trees to print your magazine," he said—"but temporary inconveniences are necessary for tremendous, long-term gain, unless we all want to walk or ride horses to work."

Both Pinkerton and Tucker decry direct EPA permitting, which the FRAC Act would require. "It wouldn't just slow us down, it would bring us to a stop for four or five years," said Tucker, who points to the job growth he claims his industry has spawned. "We put 48,000 people to work in Pennsylvania and zero in New York because of the de facto moratorium. Where I come from, Wilkes-Barre/Scranton, gas is a godsend for folks who are economically depressed."

Fly fishing guide Glenn McConnell said he felt better about leasing the mineral rights to his land in the Pennsylvania Wilds after Range agreed to address Trout Unlimited.

"The drillers are just as concerned about the environment as you and me," McConnell said. "They don't want to make a bad name for themselves. If something isn't right, they'll correct it immediately."

PA Council Trout Unlimited environmental chair Greg Grabowicz is more focused on problem prevention. "We want assurances that operations will be fail-safe. Our immediate concern is whether DEP can enforce even existing regulations, with such a small staff and so many wells," said Grabowicz, a professional forester. "There's no doubt Pennsylvania's watersheds will change dramatically over the next 30 years from new roads and pipelines, but only time will tell if drillers run into problems that cause catastrophes."

Others, though, already have seen impacts to their favorite coldwater fisheries, including Sam Harper, the DEP water management program chief monitoring Dunkard, who has a camp in the Allegheny National Forest. "There's been a dramatic change in the South Branch of Tionesta Creek, where I fish," he said. "We're seeing a lot fewer brook trout and a lot more roads leading to wells."

And there are likely to be more impacts to woodland streams as ozone from diesel-powered trucks and drilling equipment cause leaf burn and deforestation, according to Al Appleton, a former New York City Department of Environmental Protection commissioner, who serves as technical advisor to Damascus Citizens.

Appleton said too little is also made of the millions of gallons of water sucked from lakes and streams for each hydrofracking operation.

"They may not impact flow during certain times of the year, but drilling isn't a seasonal business," he said. "These companies are withdrawing significant amounts of water constantly."

While PADEP raised drilling permit fees last year to help pay for more site inspections, it also streamlined the permit approval process to 28 days with completion of a basic application—even though the agency admits the need to put more teeth into existing regulations. "Environmentalists focus on wastewater, but the biggest issue for us is what happens at the site," said PADEP spokesman Tom Rathbun. "Is the well 'cased' properly? Are the water pipes built properly?

What about how trucks are crossing streams? That's where our focus needs to be."

In the meantime, lawmakers expect to hold hearings on hydrofracking and to request an EPA study on its effects on the environment, according to Kristopher Eisenla, an aide to FRAC Act co-sponsor Congresswoman Diana DeGette (D-Colorado). "The industry has had a free ride for so long, if greater oversight costs it a few more bucks, in the interest of public health, it's worth it."

[In the Sept. 2009 issue, John Randolph in his page 2 article "The Threats Posed by Marcellus Drilling" identified "the single largest threat to Pennsylvania (also new York, West Virginia, and Ohio) wild-trout streams since the coal/steel era of the Industrial Revolution." After that issue went to subscribers, a "total" fish kill on 43 miles of Dunkard Creek in Pennsylvania raised the question again: "Is the Commonwealth of Pennsylvania protecting its waterways?"   THE EDITOR.]

EXPLORING THE ISSUE

Is Shale Gas the Solution to Our Energy Woes?

Critical Thinking and Reflection

1. Do we need energy so badly that we should ignore risks to the water supply and human health?
2. In what sense is the knowledge and wisdom of private investors preferable to that of government policymakers when it comes to deciding what to do about energy?
3. How will ample supplies of cheap natural gas affect development of renewable energy supplies such as wind and solar power?
4. Is it true that if greater oversight of an industry (such as the shale gas industry) costs a few more bucks, in the interest of public health, it is worth it?

Is There Common Ground?

Even if the proponents of unrestrained exploitation of shale gas by fracking are right when they say it solves our energy problems, the supply of shale gas will not last forever. The public—and its health—will remain, as will concern over carbon emissions and the need for ample amounts of energy to run our civilization.

1. Should we, as suggested by Diane Katz, stop investing public money in developing alternative energy sources? If not, why not?
2. Is government regulation essential to protect public health? Visit the Public Health Service at www.usphs.gov/aboutus/mission.aspx to explore one agency's approach.
3. Another kind of fossil fuel we have not yet tapped in any major way is shale oil. (See the Bureau of Land Management's oil shale information at http://ostseis.anl.gov/guide/oilshale/.) Discuss the potential benefits (and environmental costs) of exploiting this resource.

ISSUE 10

Is Renewable Energy Really Green?

YES: Andrea Larson, from "Growing U.S. Trade in Green Technology," testimony before the U.S. House Committee on Energy and Commerce, Subcommittee on Commerce, Trade and Consumer Protection (October 7, 2009)

NO: Senator Lamar Alexander (R-TN), from "The Perils of 'Energy Sprawl'," *Resources for the Future* (October 5, 2009)

Learning Outcomes

After reading this issue, you should be able to:

- Explain what "green" means.
- Describe the environmental impacts of renewable energy technologies.
- Explain how renewable energy technologies benefit the economy.
- Explain why nuclear power may be preferable to renewable energy technologies.

ISSUE SUMMARY

YES: Andrea Larson argues that "green" technologies include, among other things, renewable energy technologies and these technologies are essential to future U.S. domestic economic growth and to international competitiveness.

NO: Senator Lamar Alexander (R-TN) argues that the land use requirements of solar and wind power threaten the environment. We must therefore be very careful in how we implement these "green" energy technologies. He also believes the best way to address climate change (by cutting carbon emissions) is with nuclear power.

"**G**reen" has long been understood to mean environmentally friendly. A green energy technology is sustainable. It is not based on fossil fuels—which is not quite the same as saying it is based on renewable energy sources—so it does

179

not add carbon dioxide or other greenhouse gases to the atmosphere. It does not pollute air or water. It does not diminish biodiversity or harm ecosystems.

Examples of green energy technologies include solar power, wind power, biomass power, hydropower, tidal power, geothermal power, and wave power. The first two—solar and wind—are the ones most frequently discussed in magazines and newspapers, perhaps because they can be installed in units small enough to fit in a family's backyard. John Gulland and Wendy Milne describe in "Choosing Renewable Energy," *Mother Earth News* (April/May 2008), the process—and difficulties—of converting their home to run mostly on solar and wind power, with a wood stove (biomass power) for heat. "[U]sing renewables," they write, "has increased our independence and sense of security, and lessened our carbon emissions . . . another benefit is in knowing that we are contributing, even in a small way, to the health and sustainability of the local community."

Green energy technologies can also be scaled up to provide much larger amounts of electricity. An Australian study says they could provide 60 percent of Australia's electricity by 2040 and reduce associated carbon dioxide emissions by 78 percent; at present, Australia has more than 50 wind farms with about a thousand wind turbines. In Iceland, green energy technologies account for three quarters of the installed electrical generation capacity. In California, the corresponding figure is one quarter; in Sweden, one third; and in Norway, one half. Wind alone meets a fifth of Denmark's needs. See Rachel Sullivan and Mary-Lou Considine, "Hastening Slowly in the Global Renewables Race," *Ecos* (April–May 2010).

In the United States, green energy technologies play much less of a role. Hydropower, at 7 percent (77,000 MW) of national electrical generating capacity, is an important contributor; unfortunately, potential growth is very limited. Wind power is growing rapidly (see *Vital Signs 2010,* Worldwatch (2010)) and the potential is huge, but at the end of 2009 it provided only 34,863 MW of electrical generating capacity, about 3 percent of the total (and only about 0.2 percent of actual electricity produced). Solar power plays an even smaller role, with just over 2000 MW installed; it too is growing rapidly. Wind and solar thus promise continued growth in manufacturing of equipment, jobs for installation, and displacement of fossil fuels.

Unfortunately, wind and solar power take up a lot of land. Wind farms cover miles of high or windy terrain with towers and spinning blades. One major trend is to move them offshore, where they are out of sight to most people (and the wind is steadier). Solar electric power requires large expanses of solar panels or concentrators, preferably in locations where the sky is rarely cloudy, meaning arid lands or deserts. There is in fact a plan afoot to develop solar power in the Sahara desert to supply Europe with up to 15 percent of its electricity as early as 2015. See Ashley Seager, "Solar Power from Sahara a Step Closer," *The Guardian* (November 1, 2009; www.guardian.co.uk/business/2009/nov/01/solar-power-sahara-europe-desertec), and Daniel Clery, "Sending African Sunlight to Europe, Special Delivery," *Science* (August 13, 2010). It has been noted that if just 1 percent of the Sahara were covered with solar concentrators, it could meet the electrical needs of the entire world.

Power lines can be run across or under the Mediterranean Sea to bring electricity from the Sahara to Europe. Getting it to the rest of the world is a more difficult problem. It thus seems sensible to install green energy technologies closer to where the energy will be used. In Ontario, Canada, the provincial government has embarked on a major effort to increase the green component of its energy supply. As one component of the effort, the Hay Solar company is offering to provide farmers with free barns—whose roofs are covered with solar cells. Hay calculates that the barns will provide a large area for solar energy collection and—despite many more cloudy days in Ontario than in the Sahara—pay for themselves over 20 years, after which they belong to the farmers. See Chris Sorensen, "Absolute Power?" *Maclean's* (June 14, 2010). Rooftop solar power has been discussed in the United States as well; it has a number of advantages—it does not interfere with other uses of land, has minimal red tape, and is close to end users. Southern California Edison has proposed generating 500 MW of electricity by scattering solar power units on rooftops in the region (see David Anthony, "Where Will Solar Power Plants Be Built—Deserts or Rooftops?" *Greentechmedia* (February 22, 2010; www.greentechmedia.com/articles/read/where-will-solar-power-plants-be-built-deserts-or-rooftops/).

It thus seems clear that green energy technologies can be deployed without necessarily having huge environmental impacts. The real question is whether we can deploy *enough* green energy technology to replace fossil fuels, and one major obstacle is the sheer scale of the need for energy; see Richard A. Kerr, "Do We Have the Energy for the Next Transition?" *Science* (August 13, 2010). Many people contend that nuclear power is much better suited to meeting future needs; see Charles Forsberg, "The Real Path to Green Energy: Hybrid Nuclear-Renewable Power," *Bulletin of the Atomic Scientists* (November/December 2009), and John Bradley, "The Nuclear Revivalist," *Popular Science* (July 2010). Allan Sloan and Marilyn Adamo, "If You Believe in Magic, Green Energy Will Be Our Salvation," *Fortune* (July 26, 2010), add that part of the problem in shifting to green energy is that we must start from such a small base. Another part is that we use so much energy for transportation.

Testifying at the October 7, 2009, hearing on "Growing U.S. Trade in Green Technology" before the U.S. House Committee on Energy and Commerce, Subcommittee on Commerce, Trade and Consumer Protection, Mary Saunders, Acting Assistant Secretary for Manufacturing and Services, International Trade Administration, U.S. Department of Commerce, says she sees increasing demand for green energy technologies providing great export opportunities and adds, "Policies that support the early development and commercialization of green technologies are critical to the competitiveness of U.S. firms and improve their competitive edge in the global marketplace." If we do not support such technologies, other nations will claim the lead and the economic benefits.

The current rapid growth of wind and solar power supports, in the YES selection, professor Andrea Larson's argument that green technologies—including, among other things, renewable energy technologies such as wind and solar power—are essential to future U.S. domestic economic growth and to international competitiveness. However, in the NO selection, Senator Lamar Alexander (R-TN) argues that the land use requirements of solar and wind power threaten the environment. We must therefore be very careful in how we implement these "green" energy technologies. He also believes the best way to address climate change (by cutting carbon emissions) is with nuclear power.

YES

Andrea Larson

Growing U.S. Trade in Green Technology

Green technology and clean commerce are the future. Green technology has become, and will increasingly be, a major economic growth area for the U.S. and world trade. There is no reason the U.S. cannot be a world leader through export of clean technology and clean commerce innovation, and U.S. leadership should be a strategic goal.

Why? Because:

1. Investing in clean energy and clean materials is essential for intelligent economic development, human health protection, and ecosystem preservation.
2. U.S. leadership in clean energy and materials (green technology) creates jobs, stimulates innovation, drives exports, and differentiates U.S. technology, education, and skills in global markets.
3. The U.S. could have an advantage in world trade, but on the current path the U.S. will continue to fall behind.

Green Tech and Clean Commerce Is the Future

Population and economic development pressures are colliding with the ability of nature to deliver clean air, water, and soil. Yet the design of the industrial system that brought us to this point in history was based on assumptions of limitless resources and limitless capacity for natural system regeneration, even in the face of our waste streams. Responding to climate change and green tech opportunities are just the beginning of a major shift in this century for business. New design for business is imperative because the forces of change are accelerating.

It is not just the current economic downturn that confounds us. We face unacceptable income and opportunity disparities at home and poverty worldwide as global population grows from 6.5 to 9 billion in the next few decades. Worldwide over 2 billion people are moving rapidly into the middle class, and they will want all the opportunities and material wealth that the richest populations in western societies now view as normal. Today we concurrently face an economic downturn, a climate crisis, an energy security crisis, energy price volatility, new environmental health challenges, and ecological systems in dramatic decline.

U.S. House of Representatives, October 7, 2009.

If that were not enough, the U.S. also faces a competitiveness crisis as it loses ground to other countries that are already strategically committed to mobilizing state resources behind domestic businesses that will produce solutions to these problems. Other countries have mounted national efforts to reach clean commerce goals (e.g. renewable energy, domestic "green" companies, dramatic efficiencies, accelerating advances in PV solar design innovation, advancing clean public transportation, protecting consumers from toxic materials, and providing subsidies and incentives to advance their industries in global markets).

The larger picture shows capitalism as currently designed is at a crossroads. It must deliver on its promise of broad prosperity, yet its very design appears to undermine the ecological systems and healthy communities on which it depends. It needs an overhaul: clean energy and materials provide an answer. The U.S. should be leading this change, not following. . . .

Defining Green Technology

Green technology is one term of several used today to encompass a range of activity and innovation to simultaneously address economic development needs, health protection, and preservation of ecosystem services (e.g. the natural systems that provide us with clean air, water, soil, and food). Other terms include sustainability, clean commerce, cleantech, sustainable business, and sustainability innovation. The activities these terms reference challenge existing ways of designing and delivering not just energy, but the entire set of interdependent systems and supply chains that provide food, shelter, consumer products, and transportation modes.

We will use the abbreviation GT/CC throughout this testimony to refer to green tech and clean commerce, two terms that represent the ideas under discussion.

GT/CC refers to technology innovation, but also non-technical innovation, the latter represented by innovative supply chain management or innovative financing mechanisms to install urban PV solar installations that pay residents to sell excess electricity back to the grid. The non-technical innovative frontier must also be a focus for green tech and clean commerce innovation and U.S. competitiveness.

Furthermore, GT/CC is not just about energy. The fundamental basis of commerce and trade is energy AND materials. Both must be managed and designed to meet human needs and optimize ecological system functions. Thus green chemistry and green engineering practices are equally as important to green tech and clean commerce (GT/CC) as renewable energy technologies. PV solar systems that expose their production workers to toxins, are thrown away in landfills after use, then pollute water supplies, are not the solutions we need. "Fresh" vegetables and fruit grown with agricultural chemicals, processed, and transported thousands of miles and lacking fundamental nutrients that urban garden-grown food provides are not the solutions we need. More efficient lighting replacements that create mercury waste may save energy but are still poor designs. In other words, poorly thought out, so-called green technology improvements focused on today's hot topics (climate and energy are the focus today) are common. But a deeper design perspective is needed. First, a systems view is required. One that understands every "green" energy solution, in fact every energy AND product selection by a company or a consumer, reflects materials choices and embedded energy decisions that must be made visible, examined and evaluated for their life cycle implications. Fortunately this is now happening, led by innovative entrepreneurs. But it must be expanded and accelerated.

Nor is green technology just about efficiency. It is about that, but more importantly it is about innovation. Efficiency just allows us to do the same old things at lower cost and using less energy and fewer materials. A laudable improvement, but not the solution. Innovation creates fundamentally new solutions, preferably systems-oriented solutions that prevent and eliminate the problems we face now with climate alteration and unsafe products.

The concept that ties together innovation and both clean energy and materials is the notion of cradle to cradle design. Our current commercial practices extract raw materials, make products, generate waste streams that impact air and water, expose production workers, sell to consumers who use the products and throw them away, and leave the materials to decompose and contaminate our air and water from the landfill, incinerator or Third World country dumping destination. Think about how the costs and benefits are allocated in this linear system. This is called a cradle to grave product life cycle. The alternative is cradle to cradle design derived from systems thinking, that reduces or eliminates energy and material inputs, including toxicity *by design from the outset* to avoid employee, user/consumer, and ecosystem contamination. Under a cradle to cradle design, selected materials can be safely returned to the earth or maintained within closed recycling systems that use waste from one production and use process, as the feedstock for another.

The "greentech" issues or what I am calling the green technology and clean commerce issues (GT/CC) constitute a central challenge for governments. Providing ever growing volumes of products and services (under current design parameters) to support economic development also gives us pollution and costs that are externalized (and inequitably so) onto the population in one form or another (higher taxes for regulation, disease, and more expensive health insurance for chronic illnesses). Examples are air pollution (excessive concentrations of toxins in the air contributing to the asthma epidemic, among other respiratory problems), unsafe foods (linked to diabetes, obesity, and food contamination), excessive carbon dioxide concentration in the atmosphere (climate change and volatility), and water supply threats and shortages due to industrial contamination.

As world population rises to 9 billion in the next few decades and capitalism as currently designed stumbles in its promise of greater prosperity and results instead in wealth creation accompanied by income disparities, climate change, and waste streams increasingly tied to chronic human health challenges, a clean commerce solution is emerging. This is an alternative approach to business that we call green technology and clean commerce. This movement is obvious in the current emphasis on clean energy alternatives in response to climate change. . . .

U.S. Competitiveness

Transformation in the next decade to an alternative mindset about energy and materials is key to U.S. competitiveness and mandatory if global society is to handle the challenges of population growth, energy demands, and material throughput volumes required to provide prosperity for billions more people. We can choose to let others lead or we can mobilize and combine all the elements we have in this country to lead.

This discussion acknowledges that the U.S. has declared 25% renewable energy goals by 2025 with the February 2009 ARRA legislation. The clean technology stimulus accounts for about $66 billion, just ahead of China's stimulus investment. The important fact, nonetheless, is that we come to the table late. By way of example, according to the U.S. International Trade Commission, "Denmark, Germany, India, Japan, and Spain accounted for a combined 91 percent of global exports of wind-powered generating sets in 2008."

Globally, investments in GT/CC have been growing rapidly. For instance, new investments in sustainable energy increased between 25% and 73% annually from 2002 to 2007, until growth fell to only 5% in 2008 following the 2007–08 recession. Nonetheless, even in 2008, total investments in sustainable energy projects and companies reached $155 billion, with wind power representing the largest share at $51.8 billion. Meanwhile, the world's 12 major economic stimulus packages proposed to invest another $180 billion collectively in coming years. Also in 2008, sustainability-focused companies as identified by the Dow Jones Sustainability Index or Goldman Sachs SUSTAIN list outperformed their industries by 15% over a six-month period. Longer horizon analyses indicate companies screened for sustainability factors match or exceed the performance of conventional firms. These are companies that focus not only on renewable energy sources but also energy conservation, environmentally safer products, and improved corporate governance.

Despite being a leader in some areas, however, the U.S. was not an overall leader in GT/CC. From 2000 to 2008, venture capital investments in U.S.-based renewable energy companies increased from 0.6% of all VC investments to 11.84%, and in 2008, venture capital and private equity made new investments in energy efficiency and renewable energy worth $7.72 billion in North America and $3.05 billion in Europe. Moreover, the U.S. had the most GT/CC business incubators in 2008, with 56. The UK was next in incubators with 21, and 16 were in Germany. Yet Europe as whole was home to 46% of the global total of incubators, versus 40% for the U.S. Furthermore, North American investments in sustainable energy shrank 8% in 2008 to $30.1 billion, while in Europe they increased 2% to $49.7 billion. Many other major emerging economies also saw investments in their renewable energy sectors increase: Brazil's increased 76% to $10.8 billion (mainly due to ethanol), China's increased 18% to $15.6 billion, and India's increased 12% to $3.7 billion. Even in Spain investments reached $17.4 billion in 2008, or $430 per capita compared to North America's $57 per capita. For investments specifically in publically traded renewable energy and efficiency companies, Chinese companies led in 2008 with $2.8 billion, followed by Portugal ($2.6 billion), the U.S.

($2.1 billion), and Germany ($1.5 billion). In fact, in 2008, China became the world's largest manufacturer of photovoltaic panels, with 95% of them destined for export. This output means China may soon surpass both German and American manufacturers.

Indeed, China has recently made massive moves toward a CT/CC economy. For instance, China now has 60% of the total global capacity for solar thermal water heaters. Even such a relatively minor innovation saved 3 million tons of oil equivalent in 2006 according to the International Energy Agency. China is also nurturing and protecting its domestic wind power producers, reserving contracts for them and restricting foreign firms. The size of China's market for GT/CC creates significant opportunities for development of domestic innovators and mass producers. Nonetheless, China has a way to go: other countries have put themselves into leadership positions over the past two decades through a series of policies. Those world leaders have been Japan, Denmark, Spain, and Germany.

In 1996, Japan set a target by 2010 of using 3% (roughly 19 gigaliters oil equivalent) of primary energy supply from renewable sources excluding hydropower and geothermal energy. In 2008, the target was amended to represent an upper bound while 15.1 Gl was established as a lower bound. That goal plus grants for residential solar PV installations allowed Japan to lead the world in installed solar capacity from 1999 to 2005, which also allowed Japanese companies such as Sharp to gain an early manufacturing lead. Sharp and other Japanese companies remain competitive in the U.S. market to this day, even though Germany overtook Japan in installed capacity in 2006. In 2007, Japan established Renewable Portfolio Standards that required utilities to use renewable sources of electricity generation, to reach 16 TWh by 2014. The RPS also set prices for solar PV rates, and in December 2008, Japan allocated another $9 billion for solar subsidies, which is less than California's current solar subsidy program but reaches more eligible people. Japan continues to invest in solar research, including space-based solar energy.

Denmark began to shape its lead in GT/CC in 1976, when its Energy Research Program granted generous subsidies to renewable energies. Danish renewable energy companies turned heavily toward wind power, selling that technology domestically and abroad, especially in California. In 1989, new laws required utilities to buy electricity from renewable sources and co-generation plants, and a series of subsidies and other government support boosted GT/CC through the 1990s. By 2003, Denmark dominated the global market for wind-power generator sets, selling $966 million or 79.5% of the market. Denmark still gets a larger share of its energy from wind than any other country and sold $1.2 billion worth of generator sets in 2008, or 23.4% of the global market. Meanwhile, Danish Vestas controls 17.8% of the wind turbine market, putting Danish companies behind Germany and ahead of the U.S., Spain, and China in that field. In 2008, the Danish government's Agreement on Energy Policy sets goals of 20% of gross energy consumption from renewable sources by 2011, with incentives for de-centralized production, research, and other activity.

On the other side of Europe, Spain had a mere 979 GWh of renewable energy generation, almost all of it hydro-electric, in 1990. Yet in 2007, that same generation had risen 33-fold to 32,714 GWh, with wind accounting for about two-thirds of the total. A series of steps similar to those in Japan and Denmark led to this rapid rise, which has ultimately left Spain a major force in the world's solar and wind energy markets. Spain's 1980 Law for the Conservation of Energy first established subsidies for renewable energy sources feeding into grid. In 1997, the Law of the Electricity Sector guaranteed grid access for renewable sources and later laws set prices as well as targets, such as 12% of energy from renewable sources by 2010. With this support, Spain ranked third globally in 2008 in installed wind capacity with 16.8 GW and controlled 8.8% of the market for wind generator sets and 14.9% for turbines. It has also been a leader in solar thermal plants, building Europe's first in 2007 and continuing to develop others.

Germany, finally, has achieved some of the broadest, most profound changes en route to a GT/CC economy. It reached its Kyoto Protocol emissions target of a 20% reduction of GHG emissions from 1990 levels in 2007, a year early. A series of policies has enabled this progress, such as the 1991 Feed-in Tariff Act that required utilities to purchase electricity from any supplier on the grid. Later laws, such as the 2000 Renewable Energies Act and its subsequent updates, have guaranteed prices for renewable energies and set broad environmental targets. Germany in 2009 set even more ambitious plans for reducing overall emissions and dependence on fossil fuels. . . .

In 2008 in Germany, revenue from construction of renewable energy facilities was 13.1 billion Euros (approximately $19.7 billion) and from operation was 15.7 billion Euros ($23.6 billion), representing approximately 278,000 jobs in all. The total revenue from these two activities increased 188% relative to 2003. Meanwhile, the German government's Market Incentive Program, through grants and other incentives, encourages renewable energies by direct funding, which attracts additional investment. From 2000 to 2008, 1.2 billion Euros of direct funding attracted an additional 8.6 billion Euros of outside investment, with government funding for renewable energy R&D directed mainly to solar and wind. The results have been a near quintupling of electricity generated from renewable sources since 1990. In contrast, U.S. government subsidies totaled $29 billion from 2002–2008 for renewable energies, more than half for corn ethanol, which paled in comparison to $72 billion in subsidies for fossil fuels.

What you see when reviewing different countries' strategies is policy variation customized to local conditions but built upon a consistent pattern of core features that includes protections to control consumer costs and mitigation for windfall profits to any players. Simplicity is important to keep public administration costs low and company and individual transaction costs minimal. Consistent policies, gradual amendments to update, and stable supports (whether direct investments or tax incentives) are essential to encourage equipment manufacturers to innovate and to mass produce. Clear and consistent signals also reassure investors that markets will be relatively predictable within adequate time frames for generating returns. In summary, successful government policies appear to include key stakeholders and set ambitious targets, and then address concerns about price-gouging and the factors that typically drive innovators and companies away: instability, uncertainty, and inconsistency.

The U.S. can catch up, but when other countries are working from 20 year-plus guaranteed grid access for renewable energy producers in Spain and Germany (starting in 1991 in Germany) and well-established Spanish Feed-In Tariffs (TIFs) that built on German and Danish examples established well over a decade ago, it suggests the magnitude of the catch up challenge. These countries jumped in early, learned and adapted, and can now act faster and more effectively to build their CT/CC going forward. For the huge and rapidly growing markets for GT/CC in India and China, the U.S. faces governments quickly moving to protect and support fledgling industries that will produce clean cars and public transportation technologies to address pollution impacts, clean energy production (to offset reliance on dirty coal), and the state of the art green components and systems to address the many development and pollution/health problems they know they must solve.

Final Thoughts

The economic growth paradigm and accompanying common knowledge that told us growth had to come first, followed only much later by investment in environmental and health protection (the path of western industrialized societies) will not be sufficient for India and China. I tell my MBA students that given the pace of innovation in those countries around clean commerce goals, the U.S. will be buying most of its clean technology solutions from Indian and Chinese companies in 10 years.

I would also suggest that the U.S.'s geopolitical decline, should it come to pass, will be reflected in our unwillingness to step up to the GT/CC challenge that current population, resource, pollution, and technology development conditions impose.

I am not an advocate of government regulation unless the private sector lacks the ability to provide for the public good. Unfortunately, companies trying to move toward GT/CC, while admirable, are in a race against the cumulative decisions of firms and individuals that continue to erode the commons that is our ultimate source of all wealth, social and financial.

We tend to think of the commons as natural systems (air, water, or land); we might want to consider adding our children's bodies to that collective commons. The Centers for Disease Control [and Prevention] extensive research on contaminants in human blood, immune, and reproductive systems suggest that this century long industrial experiment that clearly has had decisive negative influences on our ecological systems and atmosphere, is also at work on the human body and children's health. Are we surprised?

The last thing I want to see is unnecessary regulation. I work with private sector innovators and emphasize the amazing capacity of markets and entrepreneurial forces in society to create the changes we need to see. But this activity must be framed with enabling and supporting policy that sets the rules and provides consistent and intelligent guidance so that markets and human ingenuity can do the rest.

In addition, let us keep in mind, in the polarized and ideologically laced discussions that pass for policy debate, that there are no purists. State subsidies and consistent long-term government support for fossil fuels played a large part in giving us the energy and materials system we live with today. Subsidies, just in recent years alone, explain why GT/CC activities remain vulnerable and investment capital moves slowly.

Can the U.S. build a GT/CC strategy? Through insufficient investment and lack of policy leadership the U.S. continues to lose ground in its learning pace and its domestic experience to countries willing to back their companies with capital and create mutually reinforcing incentives to mobilize citizen behavior, corporate investment, education, and state decision making. While the hesitancy of the U.S. to create industrial policy to lead in GT/CC is historically understandable, other countries without our political and ideological history (and gridlock) have put policies in place. First we must get our own house in order. It is only then that we will have built the necessary platform for leadership in world trade.

The challenge is straightforward, if ambitious. Future prosperity depends on economic development solutions that address poverty and extreme disparities in income distribution while simultaneously delivering on job creation, skill development, and education for the future. Industrial and commercial activity that fails to actively support provision of clean, healthy products, and clean air, water, shelter, transport, and food, by definition undermines that prosperity. Fortunately the know-how and tools are now available in the form of GT/CC practices and innovation. . . .

Senator Lamar Alexander (R-TN) # NO

The Perils of "Energy Sprawl"

... **I** believe that a new Nature Conservancy scientific paper titled "Energy Sprawl or Energy Efficiency: Climate Policy Impacts on Natural Habitat for the United States of America" will one day occupy a place among the pioneering actions that we honor in the conservation movement. The paper warns that during the next 20 years new energy production, especially biofuels and wind power, will consume a land mass larger than the state of Nebraska. This "energy sprawl," as the authors termed it, will be the result of government cap and trade and renewable mandate proposals designed to deal with climate change. The paper should serve as a Paul Revere ride for the coming renewable energy sprawl. There are negative consequences, as well as positive effects, from producing energy from the sun, the wind, and the earth. And, unless we are as wise in our response as the authors have been in their analysis, our nation runs the risk of damaging the environment in the name of saving the environment.

The first insight in the *Nature Conservancy* paper is in describing the sheer size of the sprawl. The second insight is in carefully estimating the widely varying amounts of land consumed by different kinds of energy production. Finally, the paper suggests four ways to reduce carbon emissions while minimizing the side effects of energy sprawl on the landscape and wildlife habitat. The first recommendation is energy conservation. Second is generating electricity on already developed sites, as when solar panels are put on rooftops or when a chemical company uses byproducts from its production processes to make heat and power. The third recommendation is to make carbon regulation flexible enough to allow for coal plants that recapture carbon or nuclear power plants that produce no carbon or for international offsets. Fourth, the paper suggests careful site selection.

This makes me think of my own experience as Governor 25 years ago when Tennessee banned new billboards and junkyards on a highway over which 2 million visitors travel each year to the Great Smoky Mountains National Park. Then, that decision attracted little attention. Today, it helps to preserve one of the most attractive gateways to any national park. But, as all of us know, if the billboards had gone up then it would be almost impossible today to take them down today. The same will be true with wind turbines, solar thermal plants, and other new forms of energy production.

From remarks presented at the RFF Policy Leadership Forum, October 5, 2009. Copyright © 2009 by Resources for the Future, an independent research institute focused on environmental, energy, and natural resource policies. All rights reserved. Reprinted by permission. www.rff.org

My purpose here today is to challenge you and the organizations who have traditionally protected our landscape, air and water, and wildlife habitat to do the same with the threat of energy sprawl. To ask you, first, to suggest to governments and policy makers and landowners before it is too late the best choices and the most appropriate sites for low-carbon or carbon-free energy production. And, second, I want to ask you to do something that gives many conservationists a stomachache whenever it is mentioned—and that is to rethink nuclear power, because as the *Nature Conservancy*'s paper details, nuclear power in several ways produces the largest amounts of carbon-free electricity with the least impact.

I learned a long time ago that it helps an audience to know where a speaker is coming from. Well, I grew up hiking and camping in the Great Smoky Mountains where I still live two miles from the park. As a Senator I have fought for strict emission standards for sulfur, nitrogen, and mercury because many of us still breathe air that is too polluted. I have introduced legislation to cap carbon from coal plants because I believe that human production of carbon contributes to global warming. I have helped to create 10,000 acres of conservation easements adjacent to the Smokies because it preserves views of the mountains and wildlife needs the space. I drive one of the first plug-in electric hybrid cars because I believe electrifying half our cars and trucks is the quickest way to clean the air, keep fuel prices down, reduce foreign oil use, and help deal with climate change. And I object to 50-story wind turbines along the Appalachian Trail, for the same reason I am co-sponsor of legislation to end the coal mining practice called mountaintop removal—not because I am opposed to coal plants or wind power in appropriate places, but because I want to save our mountaintops.

* * *

Let me offer a few examples to paint a clearer picture of what this energy sprawl might look like in 20 years.

As the *Nature Conservancy* paper notes, most new renewable electricity production will come from wind power which today provides about 1.5 percent of our country's electricity. Hydroelectric dams produce about 7 percent of our electricity and some of them are being dismantled. Solar and all other forms of renewable electricity produce less than 1 percent today. President Bush first suggested that wind power could grow from 1.5 percent to 20 percent by 2030 and President Obama has set out enthusiastically to get this done. In fact, the combination of presidential rhetoric, taxpayer subsidies and mandates have very nearly turned our national electricity policy into a national windmill policy.

To produce 20 percent of America's electricity from wind turbines would require erecting 186,000 1.5-megawatt wind turbines covering an area the size of West Virginia. According to the American Wind Energy Association, one megawatt of wind requires about 60 acres of land, or in other words, that's one 1.5 megawatt wind turbine every 90 acres. These are not your grandmother's windmills. They are 50 stories high. Or, if you are a sports fan, they are three times as tall as the skyboxes at the University of Tennessee football stadium. The turbines themselves are the length of a football field, they are noisy and their flashing lights can be seen for up to twenty miles. In the eastern U.S., where the wind blows less, turbines would work best along scenic ridge tops and coastlines. National Academy of Sciences says up to 19,000 miles of new high voltage transmission lines would be needed to carry electricity from 186,000 wind turbines in remote areas to and through population centers.

So many wind turbines can create real threats to wildlife. The Governor of Wyoming has expressed concern about protecting the Sage Grouse's diminishing population in his state as a result of possible habitat destruction from wind farms. The American Bird Conservancy estimates that each wind turbine in this country kills as many as seven or eight birds each year. Multiply that by 186,000 wind turbines and you could predict the annual death of close to 1.4 million birds per year. Then there are the solar thermal plants, which use big mirrors to heat a fluid and which can spread over many square miles. Secretary of Interior Ken Salazar recently announced plans to cover 1,000 square miles of federally owned land in Nevada, Arizona, California, Colorado and New Mexico and Utah with such solar collectors to generate electricity. Senator Dianne Feinstein of California, who has spent most of her career trying to make the Mojave desert a national monument, strongly objected to a solar thermal plant in the desert on federal land just outside the Mojave National Preserve that would have covered an area 3 miles by 3 miles. Plans for the plant were recently canceled.

The only wind farm in the southeastern United States is on the 3,300 foot tall Buffalo Mountain in Tennessee. The wind there blows less than 20 percent of the time making the project a commercial failure. Because of the unavailability of wind power, renewable energy advocates suggest that we southeasterners use biomass, a sort of controlled bonfire that burns wood products to make electricity. Biomass has promise, to a point. Paper mills can burn wood byproducts to make energy. And clearing forests of dead wood and then burning it not only produces energy but can help to avoid forest fires. According to the Conservancy's paper, biofuels and biomass burning of energy crops for electricity take the most space per unit of energy produced. For example, the Southern Company is building a new 100 megawatt biomass plant in Georgia. Southern estimates it will keep 180 trucks a day busy hauling a million tons of wood a year to the plant. One hundred megawatts is less than one-tenth the production of a nuclear plant which will fit on one square mile. To produce the same amount of energy as one nuclear plant would require continuously foresting an area one-third larger than the 550,000 acre Great Smoky Mountain National Park. You can make your own estimate of the number of trucks it would take to haul that much wood.

That is the second important insight of the *Nature Conservancy* report: a careful estimate of the widely different amounts of land each energy-producing technique requires. The gold standard for land usage is nuclear power. You can get a million megawatt hours of electricity a year—that's the standard unit the authors chose—per square mile, using nuclear power. The second most compact form of renewable energy is geothermal energy. To generate the same amount of power, coal requires four square miles, taking into account all the land required for mining and extraction. Solar thermal takes six square miles. Natural gas takes seven square miles and petroleum seventeen. Photovoltaic cells that turn sunlight directly into electricity require 14 square miles and wind is even more dilute, taking 28 square miles to produce the same unit of electricity.

These differences in land use are pronounced even though the paper's analysis is conservative. The authors include upstream inputs and waste disposal as part of their estimate of an energy producer's footprint. They add uranium mining and Yucca Mountain's 220 square miles to the area our 104 nuclear reactors actually occupy. If one were to consider only each energy plant's footprint, to produce 20 percent of U.S. electricity would take 100 nuclear reactors on 100 square miles or 186,000 wind turbines on 25,000 square miles. Visualize the difference this way. Thru hikers regularly travel the 2,178 miles of the Appalachian Trail from Springer Mountain in Georgia to Mount Katahdin in Maine. A row of fifty story wind turbines along that entire 2,178-mile trail would produce the same amount of electricity produced by 4 nuclear reactors on four square miles.

So, because of these wide differences, policy makers have the opportunity to choose carefully among the various forms of producing carbon-free electricity as well as to think about where such energy production should or should not go.

These are the four ways that the *Nature Conservancy* suggests we approach those decisions:

First, focus on energy conservation. This is the paper's preferred alternative to energy sprawl—and it is hard to see how anyone could disagree. To cite just one example, my home state of Tennessee leads the nation in residential per capita electricity use. If Tennesseans simply used electricity at the national average, the amount of electricity we would save each year would equal that amount produced by two nuclear power plants. Oak Ridge National Laboratory scientists have said that fuel efficiency standards are the single most important step our country can take to reduce carbon emissions.

The second recommendation for dealing with energy sprawl is end-use generation of electricity which usually occurs on already developed sites. One example of this is the co-generation that occurs at a paper factory that uses waste product to produce electricity and heat to run its facility. The most promising example is likely to be solar power on rooftops. In other words, since rooftops already exist, covering them with hundreds of square miles of solar panels would create no additional sprawl. There still are obstacles to the widespread use of solar power. In the southeast, solar still costs 4–5 times what TVA pays on the average for other electricity. There is the obstacle of aesthetics. But companies are now producing solar film embedded within attractive roofing materials—although this costs more. And there still is the problem that solar power is only available when the sun shines and, like wind, it can't be stored in large quantities. But unlike wind, which often blows at night when there is plenty of unused electricity, the sun shines when most people are at their peak power use. As former Energy Secretary James Schlesinger wrote in the *Washington Post*, because of their intermittency, wind and solar systems have to be backed up by other forms of electricity generation—which adds to cost and land usage.

The third recommendation is to make carbon regulation flexible, allowing for carbon recapture at coal plants, for nuclear power and for international offsets. So far the sponsors of climate and energy bills in the Congress haven't heeded this advice. In fact, both the Waxman Markey bill in the House and the Bingaman energy bill in the Senate contain very narrowly defined "renewable energy" mandates. Instead of allowing states to choose their methods of producing the required amount of carbon-free electricity, the legislation heavily tilts toward requiring wind power. For example, the legislation allows existing and new wind turbines, but only new hydroelectric. It does not count nuclear power, municipal solid waste, or landfill gas as "renewable." In the same way, 75 percent of the so-called "renewable electricity" subsidies enacted since 1978 have gone to wind developers. A study by the Energy Information Administration shows that wind gets a subsidy 31 times that of all other renewables combined. These policies have created a heavy bias toward the form of renewable electricity—wind power—that would consume our treasured mountaintops and can be very destructive to wildlife. And a national policy that also encourages wind power in the southeast where the wind barely blows makes as much sense as mandating new hydroelectric dams in the western desert where there is no water. It is my opinion that if we are truly seeking to reduce our carbon output, the policy that would create the least energy sprawl would would be a "carbon-free electricity standard," allowing for the maximum flexibility for those renewable electricity techniques that consume less land and require fewer new transmission lines.

Finally, the *Nature Conservancy* suggests paying attention to site selection for new energy projects. This is where those of you who represent organizations who have spent a century protecting wildlife and treasured landscapes could be of the greatest help in asking the right questions and providing wise answers. For example, should energy projects be placed in National Parks? In National Forests? If so, which forests and which energy projects? Should there be generous taxpayer subsidies for renewable energy projects within 20 miles of the Grand Tetons or along the Appalachian Trail? What about the large amounts of water needed for solar thermal plants or nuclear plants? Should turbines be concentrated in shallow waters 20 miles or more offshore where they can't be seen from the coast and transmission lines run underwater? Couldn't turbines be located in the center of Lake Michigan instead of along its shoreline? Should there be renewable energy zones, such as the solar zones Secretary Salazar is planning, where most new projects are placed—and where are the most appropriate locations for those zones and their transmission lines? In a recent op-ed in the *New York Times*, the Massachusetts Secretary of Energy and Environmental Affairs asked, wouldn't it make more sense to place wind turbines in the Atlantic and run transmission lines underwater than to build new transmission lines to carry wind power from the Great plains to Boston? Should the subsidies for cellulosic ethanol be larger than those for corn ethanol, or should there be no subsidies at all? Should there be a special effort to encourage conservation easements on private lands that protect treasured viewscapes and habitats? According to the Wall Street Journal, on August 13 Exxon Mobil pleaded guilty in federal court to killing 85 birds that had come into contact with crude oil or other pollutants in uncovered tanks of waste-water facilities on its properties. The birds were protected by the Migratory Bird Treaty Act, which dates back to 1918. The company paid $600,000 in fines and fees.

Should the Migratory Bird law be enforced against developers of other energy projects—for example, renewable electricity and transmission lines? One wind farm near Oakland California estimates that its turbines kill 80 golden eagles a year. The American Bird Conservancy estimates that the 25,000 wind turbines in the United States kill between 75,000 and 275,000 birds per year. "Somebody is getting a get out of jail card free," Michael Fry of the Bird Conservancy told the Journal. And what would be the fine for the almost 1.4 million birds that 186,000 turbines might kill?

This raises the question of whether there should be some parity among all energy companies in the application of laws and policies. For example, oil and gas companies receive taxpayer subsidies but they bid to lease and drill on federal land or waters and then pay a royalty for the privilege. Should taxpayer subsidized renewable energy companies also be required to pay a royalty for the privilege of producing electricity on federal lands or waters? And, if so, could this be a source of permanent funding for the Land and Water Conservation fund or other conservation projects on the theory that if the law allows an environmental burden it ought to require an environmental benefit? Based on estimates from the Joint Committee on Taxation and the Congressional Budget Office, taxpayers will pay wind developers a total of $29 billion in federal subsidies over the next 10 years to increase windpower production from 1.5 to 4 percent of our total electricity. . . .

This brings me to my last point, which is to ask you to rethink nuclear power.

In our country, fears about safety, proliferation, and waste disposal have stymied the "atoms for peace" dream of large amounts of low-cost clean reliable energy from nuclear power. Twelve states even have moratoria against building new nuclear plants. Still, the 104 U.S. reactors built between 1970 and 1990 produce 19 percent of America's electricity and, as I have said, 70 percent of our carbon-free electricity. I believe that what Americans should most fear about nuclear power is this: the rest of the world will use it to create low-cost, carbon-free electricity while we—who invented the technology—will not. That would send our jobs overseas looking for their cheap energy. And it would deprive us of the technology most likely to produce large amounts of carbon free electricity to help deal with climate change—and to do it in the way least likely to harm the landscape and wildlife habitats.

Look at what the rest of the world is doing. Of the top five emitters, who together produce 55 percent of the carbon in the world, only the U.S. has no new nuclear plants under construction. China, the world's largest carbon emitter, recently upped its goal for nuclear reactors to 132. Russia, the number three emitter, plans two new reactors every year until 2030. Of the next two emitters, India has six reactors under construction and ten more planned. Japan already has 55 reactors, gets 35 percent of its electricity from nuclear, has two under construction and plans for ten more by 2018.

According to the International Atomic Energy Agency (IAEA), worldwide there are 53 reactors under construction in 11 countries, mostly in Asia. South Korea gets nearly 40 percent of its electricity from nuclear and plans another eight reactors by 2015. Taiwan gets 18 percent of its power from nuclear and is building two new reactors.

In the West, France gets 80 percent of its electricity from nuclear and has among the lowest electricity rates and carbon emissions in Western Europe (behind Sweden and Switzerland which are both half nuclear.) Great Britain has hired the French electric company EDF to help build reactors. Italy has announced it will go back to nuclear.

So where does this leave the United States? Well, we still know how to run reactors better than anyone else; we just don't build them anymore. Our fleet of 104 plants is up and running 90 percent of the time. We have 17 applications for new reactors pending before the Nuclear Regulatory Commission but haven't started construction on any new ones—and the 104 we currently have in operation will begin to grow too old to operate in twenty years.

That is why I believe the U.S. should build 100 new nuclear plants in 20 years. This would bring our nuclear-produced electricity to more than 40 percent of our total generation. Add 10 percent for hydroelectric dams, 7–8 percent for wind and solar (now 1.5 percent), 25 percent for natural gas (which is low-carbon) and you begin to get a real clean—and low-cost—electricity policy.

According to the National Academy of Sciences, construction costs for 100 nuclear plants are about the same as for 186,000 wind turbines. New reactors could be located mostly on sites with existing reactors. There would be little need for new transmission lines. Taxpayer subsidies for nuclear would be one-tenth what taxpayers would pay wind developers over 10 years. As for so called "green jobs," building 100 nuclear plants would provide four times as many construction jobs as building 186,000 turbines. And, of course, nuclear is a base load source of power operating 90 percent of the time, the kind of reliable power that a country that uses 25 percent of the energy in the world must have. Wind and solar are useful supplements but they are only available, on average, about one-third of the time and can't be stored in large amounts.

And what about the lingering fears of nuclear? Obama Administration Energy Secretary Dr. Steven Chu, the Nobel Prize winning physicist, says nuclear plants are safe and he wouldn't mind living near one. That view is echoed by the thousands of U.S. Navy personnel who have lived literally on top of nuclear reactors in submarines and Navy ships for 50 years without incident. The Nuclear Regulatory Commission agrees and its painstaking supervision and application process is intended to do everything humanly possible to keep our commercial fleet of reactors safe.

On the issue of waste, Dr. Chu says there is a two step solution. Step one is store the waste on site for 40 to 60 years. The Nuclear Regulatory Commission agrees this can be done safely, perhaps even for 100 years. Step two is research and development to find the best way to recycle fuel so that its mass is reduced by 97 percent, pure plutonium is never created, and the waste is only radioactive for 300 years instead of 1 million years.

That kind of recycling would take care of both the waste and the third fear of nuclear power, the threat that other countries might somehow use plutonium to build a bomb. One could argue that because the U.S. failed to lead in developing the safe use of nuclear technology for the last 30 years, we may have made it easier for North Korea and Pakistan to steal or buy nuclear secrets from rogue countries.

Now, let me conclude with this prediction: taking into account these energy sprawl concerns, I believe the best way to reach the necessary carbon goals for climate change with the least damage to our environment and to our economy will prove to be (1) building 100 new nuclear plants in 20 years, (2) electrifying half the cars and trucks in 20 years; we probably have enough unused electricity to plug these vehicles in at night without building one new power plant; (3) putting solar panels on our rooftops. To make this happen, the government should launch mini-Manhattan projects like the one we had in World War II: for recycling used nuclear fuel, for better batteries for electric vehicles, to make solar panels cost competitive, and in addition, to recapture carbon from coal plants. This plan should produce the largest amount of electricity with the smallest amount of carbon at the lowest possible cost thereby avoiding the pain and suffering that comes when high-cost energy pushes jobs overseas and makes it hard for low-income Americans to afford their heating and cooling bills.

My fellow Tennessean Al Gore won a Nobel Prize for arguing that global warming is the inconvenient problem. If you believe he is right, and if you are also concerned about energy sprawl, then I would suggest that nuclear power is the inconvenient solution.

EXPLORING THE ISSUE

Is Renewable Energy Really Green?

Critical Thinking and Reflection

1. What does it mean to be "green"?
2. Which is the better reason for promoting the development of an energy technology—impact on the environment or impact on domestic economic growth?
3. How could society diminish its energy needs to make the problem of supplying enough energy more manageable?
4. If we choose to build huge arrays of solar power collectors, where should we put them—in deserts, on rooftops, or at sides of skyscrapers?

Is There Common Ground?

In the debate over whether renewable energy technologies are truly "green," both sides agree that we need to replace fossil fuels. Unfortunately, there is no easy way to do that, for every energy technology has environmental impacts. But solar and wind power are only two of several renewable energy technologies. Investigate geothermal power, wave power (www.energysavers .gov/renewable_energy/ocean/index.cfm/mytopic=50009), and ocean thermal energy conversion (OTEC; see http://www.nrel.gov/otec/what.html) and answer the following questions:

1. Which of these three energy technologies has the smallest "footprint" on the landscape?
2. In what ways are these three energy technologies likely to interfere with human activities?
3. How large a contribution to society's energy needs can these three energy technologies make?

ISSUE 11

Are Biofuels a Reasonable Substitute for Fossil Fuels?

YES: Keith Kline, Virginia H. Dale, Russell Lee, and Paul Leiby, from "In Defense of Biofuels, Done Right," *Issues in Science and Technology* (Spring 2009)

NO: David Pimentel, Alison Marklein, Megan A. Toth, Marissa N. Karpoff, Gillian S. Paul, Robert McCormack, Joanna Kyriazis, and Tim Krueger, from "Food Versus Biofuels: Environmental and Economic Costs," *Human Ecology* (February 2009)

Learning Outcomes

After reading this issue, you should be able to:

- Explain why there is so much interest in using biofuels to replace fossil fuels for powering vehicles.
- Describe how biofuel production affects food supply and prices.
- Describe how biofuel production affects the environment.
- Explain why some kinds of biofuels are more desirable than others.

ISSUE SUMMARY

YES: Keith Kline, Virginia H. Dale, Russell Lee, and Paul Leiby argue that the impact of biofuel production on food prices is much less than alarmists claim. If biofuel development focused on converting biowastes and fast-growing trees and grasses into fuels, the overall impact would be even better, with a host of benefits in reduced fossil fuel use and greenhouse gas emissions, increased employment, enhanced wildlife habitat, improved soil and water quality, and more stable land use.

NO: David Pimentel, Alison Marklein, Megan A. Toth, Marissa N. Karpoff, Gillian S. Paul, Robert McCormack, Joanna Kyriazis, and Tim Krueger argue that it is not possible to replace more than a small fraction of fossil fuels with biofuels. Furthermore, producing biofuels consumes more energy (as fossil fuels) than it makes available, and because biofuels compete with food production for land, water, fertilizer, and other resources, they necessarily drive up the price of food, which disproportionately harms the world's poor. It might also damage the environment in numerous ways.

$\mathbf{T}$he threat of global warming has spurred a great deal of interest in finding new sources of energy that do not add to the amount of carbon dioxide in the air. Among other things, this has meant a search for alternatives to fossil fuels, which modern civilization uses to generate electricity, heat homes, and power transportation. Finding alternatives for electricity generation (which relies much more on coal than on oil or natural gas) or home heating (which relies more on oil and natural gas) is easier than finding alternatives for transportation (which relies on oil, refined into gasoline and diesel oil). In addition, the transportation infrastructure, consisting of refineries, pipelines, tank trucks, gas stations, and an immense number of cars and trucks that will be on the road for many years, is well designed for handling liquid fuels. It is not surprising that industry and government would like to find nonfossil liquid fuels for cars and trucks (as well as ships and airplanes).

There are many suitably flammable liquids. Among them are the so-called biofuels or renewable fuels, plant oils, and alcohols that can be distilled from plant sugars. According to Daniel M. Kammen, "The Rise of Renewable Energy," *Scientific American* (September 2006), the chief biofuel in the United States so far is ethanol, distilled from corn and blended with gasoline. Production is subsidized with $2 billion of federal funds, and "when all the inputs and outputs were correctly factored in, we found that ethanol" contains about 25 percent more energy (to be used when it is burned as fuel) than was used to produce it. At least one study says the "net energy" is actually less than the energy used to produce ethanol from corn; see Dan Charles, "Corn-Based Ethanol Flunks Key Test," *Science* (May 1, 2009). If other sources, such as cellulose-rich switchgrass or cornstalks, can be used, the "net energy" is supposedly much better. However, generating ethanol requires first converting cellulose to fermentable sugars, which is so far an expensive process (although many people are working on making the process cheaper; see George W. Huber and Bruce E. Dale, "Grassoline at the Pump," *Scientific American* (July 2009). A significant additional concern is the amount of land needed for growing crops to be turned into biofuels; in a world where hunger is widespread, this means land is taken out of food production. If additional land is cleared to grow biofuel crops, this must mean loss of forests and wildlife habitat, increased erosion, and other environmental problems. However, one new study says that greenhouse-gas emissions may be significantly less; see Kevin Bullis, "Do Biofuels Reduce Greenhouse Gases?" *Technology Review* online (May 20, 2011) (www.technologyreview.com/energy/37609/?mod=chfeatured).

Under the Energy Policy Act of 2005, the U.S. Environmental Protection Agency (EPA) requires that gasoline sold in the United States contain a minimum volume of renewable fuel. Under the Renewable Fuel Program (also known as the Renewable Fuel Standard Program, or RFS Program), that volume will increase over the years, reaching 36 billion gallons by 2022. However, some think it is premature and even dangerous to put so much emphasis on biofuels. Robbin S. Johnson and C. Ford Runge, "Ethanol: Train Wreck Ahead," *Issues in Science and Technology* (Fall 2007), argue that the U.S. government's bias in favor of corn-based ethanol rigs the market against more efficient alternatives. It also leads to rising food prices, which particularly affects the world's poor. Donald Mitchell argues in "A Note on Rising Food Prices," The World Bank Development Prospects Group (July 2008), that although many factors contributed to the increase in internationally traded food prices from January 2002 to June 2008, the most important single factor—accounting for as much as 70 percent of the rise in food prices—was the large increase in biofuel production from grains and oilseeds in the United States and the European Union. Without these increases, global wheat and maize stocks would not have declined appreciably and price increases due to other factors would have been moderate.

There are clearly problems with biofuels, and those problems are getting a great deal of attention. David Pimentel, "Biofuel Food Disasters and Cellulosic Ethanol Problems," *Bulletin of Science, Technology & Society* (June 2009), says that because using 20 percent of the U.S. corn crop displaces a mere 1 percent of oil consumption, corn ethanol is a disaster, while using crop wastes and other biological materials poses its own problems. Robert F. Service, "Another Biofuels Drawback: The Demand for Irrigation," *Science* (October 23, 2009), notes that biofuel crops can compete for irrigation water, leading to both water supply and water quality problems. Gernot Stoeglehner and Michael Narodoslawsky, "How Sustainable Are Biofuels? Answers and Further Questions Arising from an Ecological Footprint Perspective," *Bioresource Technology* (August 2009), note that the sustainability of biofuel production depends very much on regional context. Alena Buyx and Joyce Tait, "Ethical Framework for Biofuels," *Science* (April 29, 2011), lay out five ethical principles for biofuels development, including sustainability, human rights, greenhouse-gas reductions, and equitable benefits and note a lack of incentives to support such development.

Laura Venderkam, "Biofuels or Bio-Fools?" *American: A Magazine of Ideas* (May/June 2007), describes the huge amounts of money being invested in companies planning to bring biofuels to market. A great deal of research is also going on, including efforts to use genetic engineering to produce enzymes that can cheaply and efficiently break cellulose into its component sugars, make bacteria or yeast that can turn a greater proportion of sugar into alcohol, and even make bacteria that can convert sugar or cellulose into hydrocarbons that can easily be turned into gasoline or diesel fuel. If these efforts succeed, the price of biofuels may drop drastically, leading investors to abandon the field. Such a price drop would, of course, benefit the consumer and lead to wider use of biofuels. It would also, say C. Ford Runge and Benjamin Senauer, "How Biofuels Could Starve the Poor," *Foreign Affairs* (May/June 2007), ease the impact on food supply. However, David Biello, "The False Promise of Biofuels," *Scientific American* (August 2011), notes that progress in developing alternatives to corn-based ethanol (which competes directly with food production) is very slow and biofuels are not yet commercially competitive; some producers are already shifting from making relatively low-profit biofuels to making high-profit specialty chemicals.

D. A. Walker, "Biofuels—for Better or Worse?" *Annals of Applied Biology* (May 2010), views much biofuel advocacy as based on misinformation. Peter Rosset, "Agrofuels, Food Sovereignty, and the Contemporary Food Crisis," *Bulletin of Science, Technology & Society* (June 2009), says that biofuels are not a prime cause of the 2008 food crisis but they are "clearly contraindicated." On the other hand, Jose C. Escobar et al., "Biofuels: Environment, Technology and Food Security," *Renewable & Sustainable Energy Reviews* (August 2009), considers that increased reliance on biofuels is inevitable due to "the imminent decline of the world's oil production, its high market prices and environmental impacts." According to Amela Ajanovic, "Biofuels versus Food Production: Does Biofuels Production Increase Food Prices," *Energy* (April 2011), biofuels and food production can coexist, especially for second-generation biofuels, but it will never be possible to replace with biofuels all the fossil fuel used for transportation.

Suzanne Hunt, "Biofuels, Neither Saviour Nor Scam: The Case for a Selective Strategy," *World Policy Journal* (Spring 2008), argues that all the concerns are real, but the larger problem is that "our current agricultural, energy, and transport systems are failing." John Ohlrogge et al., "Driving on Biomass," *Science* (May 22, 2009), point to a potential fix for the failure of the transportation system when they show that converting biomass to biofuels is much less efficient than burning it to produce electricity and using the electricity to power electric cars, which are in fact projected to "gain substantial market share in the coming years."

In the following selections, Keith Kline, Virginia H. Dale, Russell Lee, and Paul Leiby argue that the impact of biofuel production on food prices is much less than alarmists claim. If biofuels development focused on converting bio-wastes and fast-growing trees and grasses into fuels, the overall impact would be even better, with a host of benefits in reduced fossil fuel use and greenhouse gas emissions, increased employment, enhanced wildlife habitat, improved soil and water quality, and more stable land use. David Pimentel, Alison Marklein, Megan A. Toth, Marissa N. Karpoff, Gillian S. Paul, Robert McCormack, Joanna Kyriazis, and Tim Krueger argue that it is not possible to replace more than a small fraction of fossil fuels with biofuels. Furthermore, producing biofuels consumes more energy (as fossil fuels) than it makes available, and because biofuels compete with food production for land, water, fertilizer, and other resources, they necessarily drive up the price of food, which disproportionately harms the world's poor. It might also damage the environment in numerous ways.

YES ↵ 　　　　　　　　Keith Kline et al.

In Defense of Biofuels, Done Right

Biofuels have been getting bad press, not always for good reasons. Certainly important concerns have been raised, but preliminary studies have been misinterpreted as a definitive condemnation of biofuels. One recent magazine article, for example, illustrated what it called "Ethanol USA" with a photo of a car wreck in a corn field. In particular, many criticisms converge around grain-based biofuel, traditional farming practices, and claims of a causal link between U.S. land use and land-use changes elsewhere, including tropical deforestation.

Focusing only on such issues, however, distracts attention from a promising opportunity to invest in domestic energy production using biowastes, fast-growing trees, and grasses. When biofuel crops are grown in appropriate places and under sustainable conditions, they offer a host of benefits: reduced fossil fuel use; diversified fuel supplies; increased employment; decreased greenhouse gas emissions; enhanced habitat for wildlife; improved soil and water quality; and more stable global land use, thereby reducing pressure to clear new land.

Not only have many criticisms of biofuels been alarmist, many have been simply inaccurate. In 2007 and early 2008, for example, a bumper crop of media articles blamed sharply higher food prices worldwide on the production of biofuels, particularly ethanol from corn, in the United States. Subsequent studies, however, have shown that the increases in food prices were primarily due to many other interacting factors: increased demand in emerging economies, soaring energy prices, drought in food-exporting countries, cut-offs in grain exports by major suppliers, market-distorting subsidies, a tumbling U.S. dollar, and speculation in commodities markets.

Although ethanol production indeed contributes to higher corn prices, it is not a major factor in world food costs. The U.S. Department of Agriculture (USDA) calculated that biofuel production contributed only 5% of the 45% increase in global food costs that occurred between April 2007 and April 2008. A Texas A&M University study concluded that energy prices were the primary cause of food price increases, noting that between January 2006 and January 2008, the prices of fuel and fertilizer, both major inputs to agricultural production, increased by 37% and 45%, respectively. And the International Monetary Fund has documented that since their peak in July 2008, oil prices declined by 69% as of December 2008, and global food prices declined by 33% during the same period, while U.S. corn production has remained at 12 billion bushels a month, one-third of which is still used for ethanol production.

Reprinted with permission from *Issues in Science and Technology*, by Keith Kline et al, Spring 2009, pp. 75–84. Copyright © 2009 by the University of Texas at Dallas, Richardson, TX.

In another line of critique, some argue that the potential benefits of biofuel might be offset by indirect effects. But large uncertainties and postulations underlie the debate about the indirect land-use effects of biofuels on tropical deforestation, the critical implication being that use of U.S. farmland for energy crops necessarily causes new land-clearing elsewhere. Concerns are particularly strong about the loss of tropical forests and natural grasslands. The basic argument is that biofuel production in the United States sets in motion a necessary scenario of deforestation.

According to this argument, if U.S. farm production is used for fuel instead of food, food prices rise and farmers in developing countries respond by growing more food. This response requires clearing new land and burning native vegetation and, hence, releasing carbon. This "induced deforestation" hypothesis is based on questionable data and modeling assumptions about available land and yields, rather than on empirical evidence. The argument assumes that the supply of previously cleared land is inelastic (that is, agricultural land for expansion is unavailable without new deforestation). It also assumes that agricultural commodity prices are a major driving force behind deforestation and that yields decline with expansion. The calculations for carbon emissions assume that land in a stable, natural state is suddenly converted to agriculture as a result of biofuels. Finally, the assertions assume that it is possible to measure with some precision the areas that will be cleared in response to these price signals.

A review of the issues reveals, however, that these assumptions about the availability of land, the role of biofuels in causing deforestation, and the ability to relate crop prices to areas of land clearance are unsound. Among our findings:

First, sufficient suitably productive land is available for multiple uses, including the production of biofuels. Assertions that U.S. biofuel production will cause large indirect land-use changes rely on limited data sets and unverified assumptions about global land cover and land use. Calculations of land-use change begin by assuming that global land falls into discrete classes suitable for agriculture—cropland, pastures and grasslands, and forests—and results depend on estimates of the extent, use, and productivity of these lands, as well as presumed future interactions among land-use classes. But several major organizations, including the Food and Agriculture Organization (FAO), a primary data clearinghouse, have documented significant inconsistencies surrounding global land-cover estimates. For example, the three most recent FAO Forest Resource Assessments, for periods ending in 1990, 2000, and 2005, provide estimates of the world's total forest cover in 1990 that vary by as much as 470 million acres, or 21% of the original estimate.

Cropland data face similar discrepancies, and even more challenging issues arise when pasture areas are considered. Estimates for land used for crop production range from 3.8 billion acres (calculated by the FAO) to 9 billion acres (calculated by the Millennium Ecosystem Assessment, an international effort spearheaded by the United Nations). In a recent study attempting to reconcile cropland use circa 2000, scientists at the University of Wisconsin-Madison and McGill University estimated that there were 3.7 billion acres of cropland, of which 3.2 billion were actively cropped or harvested. Land-use studies consistently acknowledge serious data limitations and uncertainties, noting that a majority of global crop lands are constantly shifting the location of cultivation, leaving at any time large areas fallow or idle that may not be captured in statistics. Estimates of idle croplands, prone to confusion with pasture and grassland, range from 520 million acres to 4.9 billion acres globally. The differences illustrate one of many uncertainties that hamper global land-use change calculations. To put these numbers in perspective, USDA has estimated that in 2007, about 21 million acres were used worldwide to produce biofuel feedstocks, an area that would occupy somewhere between 0.4% and 4% of the world's estimated idle cropland.

Diverse studies of global land cover and potential productivity suggest that anywhere from 600 million to more than 7 billion additional acres of underutilized rural lands are available for expanding rain-fed crop production around the world, after excluding the 4 billion acres of cropland currently in use, as well as the world's supply of closed forests, nature reserves, and urban lands. Hence, on a global scale, land per se is not an immediate limitation for agriculture and biofuels.

In the United States, the federal government, through the multiagency Biomass Research and Development Initiative (BRDI), has examined the land and market implications of reaching the nation's biofuel target, which calls for producing 36 billion gallons by 2022. BRDI estimated that a slight net reduction in total U.S. active cropland area would result by 2022 in most scenarios, when compared with a scenario developed from USDA's so-called "baseline" projections. BRDI also found that growing biofuel crops efficiently in the United States would require shifts in the intensity of use of about 5% of pasture lands to more intensive hay, forage, and bioenergy crops (25 million out of 456 million acres) in order to accommodate dedicated energy crops, along with using a combination of wastes, forest residues, and crop residues. BRDI's estimate assumes that the total area allocated to USDA's Conservation Reserve Program (CRP) remains constant at about 33 million acres but allows about 3 million acres of the CRP land on high-quality soils in the Midwest to be offset by new CRP additions in other regions. In practice, additional areas of former cropland that are now in the CRP could be managed for biofuel feedstock production in a way that maintains positive impacts on wildlife, water, and land conservation goals, but this option was not included among the scenarios considered.

Yields are important. They vary widely from place to place within the United States and around the world. USDA projects that corn yields will rise by 20 bushels per acre by 2017; this represents an increase in corn output equivalent to adding 12.5 million acres as compared with 2006, and over triple that area as compared with average yields in many less-developed nations. And there is the possibility that yields will increase more quickly than projected in the USDA baseline, as seed companies aim to exceed 200 bushels per acre by 2020. The potential to increase yields in developing countries offers tremendous opportunities to improve welfare and expand production while reducing or maintaining the area harvested. These improvements are consistent with U.S. trends during the past half century showing agricultural output growth averaging 2% per year while cropland use fell by an average of 0.7% per year. Even without large yield increases, cropland requirements to meet biofuel production targets may not be nearly as great as assumed.

Concerns over induced deforestation are based on a theory of land displacement that is not supported by data. U.S. ethanol production shot up by more than 3 billion gallons (150%) between 2001 and 2006, and corn production increased 11%, while total U.S. harvested cropland fell by about 2% in the same period. Indeed, the harvested area for "coarse grains" fell by 4% as corn, with an average yield of 150 bushels per acre, replaced other feed grains such as sorghum (averaging 60 bushels per acre). Such statistics defy modeling projections by demonstrating an ability to supply feedstock to a burgeoning ethanol industry while simultaneously maintaining exports and using substantially less land. So although models may assume that increased use of U.S. land for biofuels will lead to more land being cleared for agriculture in other parts of the world, evidence is lacking to support those claims.

Second, there is little evidence that biofuels cause deforestation, and much evidence for alternative causes. Recent scientific papers that blame biofuels for deforestation are based on models that presume that new land conversion can be simulated as a predominantly market-driven choice. The models assume that land is a privately owned asset managed in response to global price signals within a stable rule-based economy—perhaps a reasonable assumption for developed nations.

However, this scenario is far from the reality in the smoke-filled frontier zones of deforestation in less-developed countries, where the models assume biofuel-induced land conversion takes place. The regions of the world that are experiencing first-time land conversion are characterized by market isolation, lawlessness, insecurity, instability, and lack of land tenure. And nearly all of the forests are publicly owned. Indeed, land-clearing is a key step in a long process of trying to stake a claim for eventual tenure. A cycle involving incremental degradation, repeated and extensive fires, and shifting small plots for subsistence tends to occur long before any consideration of crop choices influenced by global market prices.

The causes of deforestation have been extensively studied, and it is clear from the empirical evidence that forces other than biofuel use are responsible for the trends of increasing forest loss in the tropics. Numerous case studies document that the factors driving deforestation are a complex expression of cultural, technological, biophysical, political, economic, and demographic interactions. Solutions and measures to slow deforestation have also been analyzed and tested, and the results show that it is critical to improve governance, land tenure, incomes, and security to slow the pace of new land conversion in these frontier regions.

Selected studies based on interpretations of satellite imagery have been used to support the claims that U.S. biofuels induce deforestation in the Amazon, but satellite images cannot be used to determine causes of land-use change. In practice, deforestation is a site-specific process. How it is perceived will vary greatly by site and also by the temporal and spatial lens through which it is observed. Cause-and-effect relationships are complex, and the many small changes that enable larger future conversion cannot be captured by satellite imagery. Although it is possible to classify an image to show that forest in one period changed to cropland in another, cataloguing changes in discrete classes over time does not explain why these changes occur. Most studies asserting that the production and use of biofuels cause tropical deforestation point to land cover at some point after large-scale forest degradation and clearing have taken place. But the key events leading to the primary conversion of forests often proceed for decades before they can be detected by satellite imagery. The imagery does not show how the forest was used to sustain livelihoods before conversion, nor the degrees of continual degradation that occurred over time before the classification changed. When remote sensing is supported by a ground-truth process, it typically attempts to narrow the uncertainties of land-cover classifications rather than research the history of occupation, prior and current use, and the forces behind the land-use decisions that led to the current land cover.

First-time conversion is enabled by political, as well as physical, access. Southeast Asia provides one example where forest conversion has been facilitated by political access, which can include such diverse things as government-sponsored development and colonization programs in previously undisturbed areas and the distribution of large timber and mineral concessions and land allotments to friends, families, and sponsors of people in power. Critics have raised valid concerns about high rates of deforestation in the region, and they often point an accusing finger at palm oil and biofuels.

Palm oil has been produced in the region since 1911, and plantation expansion boomed in the 1970s with growth rates of more than 20% per year. Biodiesel represents a tiny fraction of palm oil consumption. In 2008, less than 2% of crude palm oil output was processed for biofuel in Indonesia and Malaysia, the world's largest producers and exporters. Based on land-cover statistics alone, it is impossible to determine the degree of attribution that oil palm may share with other causes of forest conversion in Southeast Asia. What is clear is that oil palm is not the only factor and that palm plantations are established after a process of degradation and deforestation has transpired. Deforestation data may offer a tool for estimating the ceiling for attribution, however. In Indonesia, for example, 28.1 million hectares were deforested between 1990 and 2005, and oil palm expansion in those areas was estimated to be between 1.7 million and 3 million hectares, or between 6% and 10% of the forest loss, during the same period.

Initial clearing in the tropics is often driven more by waves of illegitimate land speculation than agricultural production. In many Latin American frontier zones, if there is native forest on the land, it is up for grabs, as there is no legal tenure of the land. The majority of land-clearing in the Amazon has been blamed on livestock because, in part, there is no alternative for classifying the recent clearings and, in part, because land holders must keep it "in production" to maintain claims and avoid invasions. The result has been the frequent burning and the creation of extensive cattle ranches. For centuries, disenfranchised groups have been pushed into the forests and marginal lands where they do what they can to survive. This settlement process often includes serving as low-cost labor to clear land for the next wave of better-connected colonists. Unless significant structural changes occur to remove or modify enabling factors, the forest-clearing that was occurring before this decade is expected to continue along predictable paths.

Testing the hypothesis that U.S. biofuel policy causes deforestation elsewhere depends on models that can incorporate the processes underlying initial land-use change. Current models attempt to predict future land-use change based on changes in commodity prices. As conceived thus far, the computational general equilibrium models designed for economic trade do not adequately incorporate the processes of land-use change. Although crop prices may influence short-term land-use decisions, they are not a dominant factor in global patterns of first-time conversion, the land-clearing of chief concern in relating biofuels to deforestation. The highest deforestation rates observed and estimated globally occurred in the 1990s. During that period, there was a surplus of commodities on world markets and consistently depressed prices.

Third, many studies omit the larger problem of widespread global mismanagement of land. The recent arguments focusing on the possible deforestation attributable to biofuels use idealized representations of crop and land markets, omitting what may be larger issues of concern. Clearly, the causes of global deforestation are complex and are not driven merely by a single crop market. Additionally, land mismanagement, involving both initial clearing and maintaining previously cleared land, is widespread and leads to a process of soil degradation and environmental damage that is especially prevalent in the frontier zones. Reports by the FAO and the Millennium Ecosystem Assessment describe the environmental consequences of repeated fires in these areas. Estimates of global burning vary annually, ranging from 490 million to 980 million acres per year between 2000 and 2004. The vast majority of fires in the tropics occur in Africa and the Amazon in what were previously cleared, nonforest lands. In a detailed study, the Amazon Institute of Environmental Research and Woods Hole Research Center found that 73% of burned area in the Amazon was on previously cleared land, and that was during the 1990s, when overall deforestation rates were high.

Fire is the cheapest and easiest tool supporting shifting subsistence cultivation. Repeated and extensive burning is a manifestation of the lack of tenure, lack of access to markets, and severe poverty in these areas. When people or communities have few or no assets to protect from fire and no incentive to invest in more sustainable production, they also have no reason to limit the extent of burning. The repeated fires modify ecosystem structure, penetrate ever deeper into forest margins, affect large areas of understory vegetation (which is not detected by remote sensing), and take an ever greater cumulative toll on soil quality and its ability to sequester carbon. Profitable biofuel markets, by contributing to improved incentives to grow cash crops, could reduce the use of fire and the pressures on the agricultural frontier. Biofuels done right, with attention to best practices for sustained production, can make significant contributions to social and economic development as well as environmental protection.

Furthermore, current literature calculates the impacts from an assumed agricultural expansion by attributing the carbon emissions from clearing intact ecosystems to biofuels. If emission analyses consider empirical data reflecting the progressive degradation that occurs (often over decades) before and independently of agriculture market signals for land use, as well as changes in the frequency and extent of fire in areas that biofuels help bring into more stable market economies, then the resulting carbon emission estimates would be worlds apart.

Brazil provides a good case in point, because it holds the globe's largest remaining area of tropical forests, is the world's second-largest producer of biofuel (after the United States), and is the world's leading supplier of biofuel for global trade. Brazil also has relatively low production costs and a growing focus on environmental stewardship. As a matter of policy, the Brazilian government has supported the development of biofuels since launching a National Ethanol Program called Proálcool in 1975. Brazil's ethanol industry began its current phase of growth after Proálcool was phased out in 1999 and the government's role shifted from subsidies and regulations toward increased collaboration with the private sector in R&D. The government helps stabilize markets by supporting variable rates of blending ethanol with gasoline and planning for industry expansion, pipelines, ports, and logistics. The government also facilitates access to global markets; develops improved varieties of sugarcane, harvest equipment, and conversion; and supports improvements in environmental performance.

New sugarcane fields in Brazil nearly always replace pasture land or less valuable crops and are concentrated around production facilities in the developed southeastern region, far from the Amazon. Nearly all production is rainfed and relies on low input rates of fertilizers and agrochemicals, as compared with other major crops. New projects are reviewed under the Brazilian legal framework of Environmental Impact Assessment and Environmental Licensing. Together, these policies have contributed to the restoration or protection of reserves and riparian areas and increased forest cover, in tandem with an expansion of sugarcane production in the most important producing state, Sao Paulo.

Yet natural forest in Brazil is being lost, with nearly 37 million acres lost between May 2000 and August 2006, and a total of 150 million acres lost since 1970. Some observers have suggested that the increase in U.S. corn production for biofuel led to reduced soybean output and higher soybean prices, and that these changes led, in turn, to new deforestation in Brazil. However, total deforestation rates in Brazil appear to fall in tandem with rising soybean prices. This co-occurrence illustrates a lack of connection between commodity prices and initial land clearing. This phenomenon has been observed around the globe and suggests an alternate hypothesis: Higher global commodity prices focus production and investment where it can be used most efficiently, in the plentiful previously cleared and underutilized lands around the world. In times of falling prices and incomes, people return to forest frontiers, with all of their characteristic tribulations, for lack of better options.

Biofuels Done Right

With the right policy framework, cellulosic biofuel crops could offer an alternative that diversifies and boosts rural incomes based on perennials. Such a scenario would create incentives to reduce intentional burning that currently affects millions of acres worldwide each year. Perennial biofuel crops can help stabilize land cover, enhance soil carbon sequestration, provide habitat to support biodiversity, and improve soil and water quality. Furthermore, they can reduce pressure to clear new land via improved incomes and yields. Developing countries have huge opportunities to increase crop yield and thereby grow more food on less land, given that cereal yields in less developed nations are 30% of those in North America. Hence, policies supporting biofuel production may actually help stop the extensive slash-and-burn agricultural cycle that contributes to greenhouse gas emissions, deforestation, land degradation, and a lifestyle that fails to support farmers and their families.

Biofuels alone are not the solution, however. Governments in the United States and elsewhere will have to develop and support a number of programs designed to support sustainable development. The operation and rules of such programs must be transparent, so that everyone can understand them and see that fair play is ensured. Among other attributes, the programs must offer economic incentives for sustainable production, and they must provide for secure land tenure and participatory land-use planning. In this regard, pilot biofuel projects in Africa and Brazil are showing promise in addressing the vexing and difficult challenges of sustainable land use and development. Biofuels also are uniting diverse stakeholders in a global movement to develop sustainability metrics and certification methods applicable to the broader agricultural sector.

Given a priority to protect biodiversity and ecosystem services, it is important to further explore the drivers for the conversion of land at the frontier and to consider the effects, positive and negative, that U.S. biofuel policies could have in these areas. This means it is critical to distinguish between valid concerns calling for caution and alarmist criticisms that attribute complex problems solely to biofuels.

Still, based on the analyses that we and others have done, we believe that biofuels, developed in an economically and environmentally sensible way, can contribute significantly to the nation's—indeed, the world's—energy security while providing a host of benefits for many people in many regions.

David Pimentel et al.

Food Versus Biofuels: Environmental and Economic Costs

Introduction

With global shortages of fossil energy, especially oil and natural gas, and heavy biomass energy consumption occurring, a major focus has developed worldwide on biofuel production. Emphasis on biofuels as renewable energy sources has developed globally, including those made from crops such as corn, sugarcane, and soybean. Wood and crop residues also are being used as fuel. Though it may seem beneficial to use renewable plant materials for biofuel, the use of crop residues and other biomass for biofuels raises many environmental and ethical concerns.

Diverse conflicts exist in the use of land, water, energy and other environmental resources for food and biofuel production. Food and biofuels are dependent on the same resources for production: land, water, and energy. In the USA, about 19% of all fossil energy is utilized in the food system: about 7% for agricultural production, 7% for processing and packaging foods, and about 5% for distribution and preparation of food. In developing countries, about 50% of wood energy is used primarily for cooking in the food system.

The objective of this article is to analyze: (1) the uses and interdependencies among land, water, and fossil energy resources in food versus biofuel production and (2) the characteristics of the environmental impacts caused by food and biofuel production.

Food and Malnourishment

The Food and Agricultural Organization (FAO) of the United Nations confirms that worldwide food available per capita has been declining *continuously* based on availability of cereal grains during the past 23 years. Cereal grains make up an alarming 80% of the world's food supply. Although grain yields per hectare in both developed and developing countries are still gradually increasing, the rate of increase is slowing, while the world population and its food needs are rising. For example, from 1950 to 1980, US grain yields increased at about 3% per year. Since 1980, the annual rate of increase for corn and other grains is

From *Human Ecology*, February 2009, pp. 1–7, 9. Copyright © 2009 by Springer Science and Business Media. Reprinted by permission via Rightslink.

only approximately 1%. Worldwide the rate of increase in grain production is not keeping up with the rapid rate of world population growth of 1.1%.

The resulting decrease in food supply results in widespread malnutrition. There are more deaths from malnutrition than any other cause of death in the world today. The World Health Organization reports more than 3.7 billion people (56% of the global population) are currently malnourished, and that number is steadily increasing. Although much of the land worldwide is occupied by grains and other crops, malnutrition is still globally prevalent.

World Cropland and Water Resources

More than 99.7% of human food comes from the terrestrial environment, while less than 0.3% comes from the oceans and other aquatic ecosystems. Worldwide, of the total 13 billion hectares of land area, the percentages in use are: cropland, 11%; pasture land, 27%; forest land, 32%; urban, 9%; and other 21%. Most of the remaining land area (21%) is unsuitable for crops, pasture, and/or forests because the soil is too infertile or shallow to support plant growth or the climate and region are too harsh, too cold, dry, steep, stony, or wet. Most of the suitable cropland is already in use.

As the human population continues to increase rapidly, there has been an expansion of diverse human activities that have dramatically reduced cropland and pasture land. Much vital cropland and pastureland has been covered by transportation systems and urbanization. In the USA, about 0.4 ha (one acre) of land per person is covered with urbanization and highways. In 1960, when the world population numbered only three billion, approximately 0.5 ha was available per person for the production of a diverse, nutritious diet of plant and animal products. It is widely agreed that 0.5 ha is essential for a healthy diet. China's recent explosion in development provides an example of rapid declines in the availability of per capita cropland. The current available cropland in China is only 0.08 ha per capita. This relatively small amount of cropland provides the people in China with a predominantly vegetarian diet, which requires less energy, land, and biomass than the typical American diet.

In addition to land, water is a vital controlling factor in crop production. The production of 9 t/ha of corn requires about seven million liters of water (about 700,000 gallons of water per acre). Other crops also require large amounts of water. Irrigation provides much of the water for world food production. For example, 17% of the crops that are irrigated worldwide provide 40% of the world food supply. A major concern is that world-wide availability of irrigation water is projected to decline further because of global warming.

Energy Resources and Use

Since the industrial revolution of the 1850s, the rate of energy use from all sources has been growing even faster than the world population. For example, from 1970 to 1995, energy use increased at a rate of 2.5% per year (doubling every 30 years) compared with the worldwide population growth of 1.7% per year (doubling every 40 to 60 years). Developed countries annually consume about 70% of the fossil energy worldwide, while the developing nations, which have about 75% of the world population, use only 30% of world fossil energy.

Although about 50% of all the solar energy captured worldwide by photosynthesis is used by humans for food, forest products, and other systems, it is still inadequate to meet all human food production needs. To make up for this shortfall, about 473 quads (one quad = 1×10^{15} BTU) of fossil energy—mainly oil, gas, coal, and a small amount of nuclear—are utilized worldwide each year. Of these 473 quads, about 100 quads (or about 22%) of the world's total energy are utilized just in the United States, which has only 4.5% of the world's population.

Each year, the USA population uses three times more fossil energy than the total solar energy captured by all harvested US crops, forests, and grasses. Industry, transportation, home heating and cooling, and food production account for most of the fossil energy consumed in the United States. Per capita use of fossil energy in the United States per year amounts to about 9,500 l of oil equivalents—more than seven times per capita use in China. In China, most fossil energy is used by industry, although approximately 25% is now used for agriculture and in the food production system.

Worldwide, the earth's natural gas supply is considered adequate for about 40 years and that of coal for about 100 years. In the USA, natural gas is already in short supply: it is projected that the USA will deplete its natural gas resources in about 20 years. Many agree that the world reached peak oil and natural gas in 2007; from this point, these energy resources are declining slowly and continuously, until they run out altogether.

Youngquist reports that earlier estimates of the amount of oil and gas new exploration drilling would provide were very optimistic as to the amount of these resources to be found in the United States. Both the US oil production rate and existing reserves have continued to decline. Domestic oil and natural gas production has been decreasing for more than 30 years and are projected to continue to decline. Approximately 90% of US oil resources have already been exploited. At present, the United States is importing more than 63% of its oil, which puts its economy at risk due to fluctuating oil prices and difficult international political situations, as was seen previously during the 1973 oil crisis, the 1991 Gulf War, and the current Iraq War.

Biomass Resources

The total sustainable world biomass energy potential has been estimated to be about 92 quads (10^{15} BTU) per year, which represents 19% of total global energy use. The total forest biomass produced worldwide is 38 quads per year, which represents 8% of total energy use. In the USA, only 1% to 2% home heating is achieved with wood.

Global forest area removed each year totals 15 million ha. Global forest biomass harvested is just over 1,431 billion kg per year, of which 60% is industrial roundwood and 40% is fuelwood. About 90% of the fuelwood is utilized in developing countries. A significant portion (26%) of all forest wood is converted into charcoal. Production of charcoal causes between 30% and 50% of the wood energy to be lost and produces large quantities of smoke. On the other hand, charcoal is cleaner burning and thus produces less smoke than burning wood fuel directly; it is dirty to handle but lightweight.

Worldwide, most biomass is burned for cooking and heating. In developing countries, about 2 kcal of wood are utilized in cooking 1 kcal of food. Thus, more biomass and more land and water are needed to produce the biofuel for cooking than are needed to produce the food. However, biomass can also be converted into electricity. Assuming that optimal yield globally of three dry metric tons (t/ha) per year of woody biomass can be harvested sustainably, this would provide a gross energy yield of 13.5 million kcal/ha. Harvesting this wood biomass requires an energy expenditure of approximately 30 l of diesel fuel per hectare, plus the embodied energy for cutting and collecting wood for transport to an electric power plant. Thus, the energy input per output ratio for such a system is calculated to be 1:25.

Per capita consumption of woody biomass for heat in the USA amounts to 625 kg per year. The diverse biomass resources (wood, crop residues, and dung) used in developing nations averages about 630 kg per capita per year.

Woody biomass has the capacity to supply the USA with about five quads $(1.5 \times 10^{12}$ kWh thermal) of its total gross energy supply by the year 2050, provided that the amount of forest-land stays constant. A city of 100,000 people using the biomass from a sustainable forest (3 t/ha per year) for electricity requires approximately 200,000 ha of forest area, based on an average electrical demand of slightly more than one billion kilowatt-hours (electrical energy [e]) (860 kcal = 1 kWh.

Air quality impacts from burning biomass are less harmful than those associated with coal, but more harmful than those associated with natural gas. Biomass combustion releases more than 200 different chemical pollutants, including 14 carcinogens and four cocarcinogens, into the atmosphere. As a result, approximately four billion people globally suffer from continuous exposure to smoke. In the USA, wood smoke kills 30,000 people each year, although many of the pollutants from electric plants that use wood and other biomass can be mitigated. These controls include the same scrubbers that are frequently installed on coal-fired plants.

An estimated 2.0 billion tons of biomass is produced per year on US land area. This translates into about 32 quads of energy, which means that the solar energy captured by all the plants in the USA per year equates to only 32% of the energy currently consumed as fossil energy. There is insufficient US biomass for ethanol and biodiesel production to make the USA oil independent.

Of the total world land area in cropland, pasture, and forest, about 38% is cropland and pasture and about 30% is forests. Devoting a portion of this cropland and forest land to biofuels will stress both managed ecosystems and will not be sufficient to solve the world fuel problem.

Corn Ethanol

In the United States, ethanol constitutes 99% of all biofuels. For capital expenditures, new plant construction costs from $1.05 to $3.00 per gallon of ethanol. Fermenting and distilling corn ethanol requires large amounts of water. The corn is finely ground and approximately 15 l of water are added per 2.69 kg of ground corn. After fermentation, to obtain a liter of 95% pure ethanol from the 10% ethanol and 90% water mixture, 1 l of ethanol must be extracted from the approximately 10 l of the ethanol/water mixture. To be mixed with gasoline, the 95% ethanol must be further processed and more water must be removed, requiring additional fossil energy inputs to achieve 99.5% pure ethanol. Thus, a total of about 12 l of wastewater must be removed per liter of ethanol produced, and this relatively large amount of sewage effluent has to be disposed of at energy, economic, and environmental costs.

Manufacture of a liter of 99.5% ethanol uses 46% more fossil energy than it produces and costs $1.05 per liter ($3.97 per gallon). The corn feedstock alone requires more than 33% of the total energy input. The largest energy inputs in corn-ethanol production are for producing the corn feedstock plus the steam energy and electricity used in the fermentation/distillation process. The total energy input to produce a liter of ethanol is 7,474 kcal. However, a liter of ethanol has an energy value of only 5,130 kcal. Based on a net energy loss of 2,344 kcal of ethanol produced, 46% more fossil energy is expended than is produced as ethanol. The total cost, including the energy inputs for the fermentation/distillation process and the apportioned energy costs of the stainless steel tanks and other industrial materials, is $1,045 per 1,000 l of ethanol produced.

Subsidies for corn ethanol total more than $6 billion per year. This means that the subsidies per liter of ethanol are 60 times greater than the subsidies per liter of gasoline. In 2006, nearly 19 billion liters of ethanol were produced on 20% of US corn acreage. This 19 billion liters represents only 1% of total US petroleum use.

However, even if we completely ignore corn ethanol's negative energy balance and high economic cost, we still find that it is absolutely not feasible to use ethanol as a replacement for US oil. If all 341 billion kilograms of corn produced annually in the USA were converted into ethanol at the current rate of 2.69 kg per liter of ethanol, then 129 billion liters of ethanol could be produced. This would provide only 5% of total oil consumption in the USA. And of course, in this situation there would be no corn available for livestock and other needs.

In addition, the environmental impacts of corn ethanol are enormous:

1. Corn production causes more soil erosion than any other crop grown.
2. Corn production uses more nitrogen fertilizer than any other crop grown and is the prime cause of the dead zone in the Gulf of Mexico. In 2006, approximately 4.7 million tons of nitrogen was used in US corn production. Natural gas is required to produce nitrogen fertilizer. The USA now imports more than half of its nitrogen fertilizer. In addition, about 1.7 million tons of phosphorus was used in the USA.
3. Corn production uses more insecticides than any other crop grown.
4. Corn production uses more herbicides than any other crop grown.
5. More than 1,700 gallons of water are required to produce one gallon of ethanol.

6. Enormous quantities of carbon dioxide are produced. This is due to the large quantity of fossil energy used in production, and the immense amounts of carbon dioxide released during fermentation and soil tillage. All this speeds global warming.
7. Air pollution is a significant problem. Burning ethanol emits pollutants into air such as peroxyacetyl nitrate (PAN), acetaldehyde, alkylates, and nitrous oxide. These can have significant detrimental human health effects as well as impact other organisms and ecosystems.

In addition to corn ethanol's intensive environmental degradation and inefficient use of food-related resources, the production of corn ethanol also has a great effect on world food prices. For instance, the use of corn for ethanol production has increased the prices of US beef, chicken, pork, eggs, breads, cereals, and milk by 10% to 20%. Corn prices have more than doubled during the past year.

Grass and Cellulosic Ethanol

Tilman *et al.* suggest that all 235 million hectares of grassland available in the USA plus crop residues can be converted into cellulosic ethanol. This suggestion causes concerns among scientists. Tilman et al. recommend that crop residues, like corn stover, can be harvested and utilized as a fuel source. This would be a disaster for the agricultural ecosystem because crop residues are vital for protecting topsoil. Leaving the soil unprotected would intensify soil erosion by tenfold or more and may increase soil loss as much as 100-fold. Furthermore, even a partial removal of the stover can result in increased CO_2 emissions and intensify acidification and eutrophication due to increased runoff. Already, the US crop system is losing soil ten times faster than the sustainable rate. Soil formation rates at less than 1 t ha^{-1} $year^{-1}$, are extremely slow. Increased soil erosion caused by the removal of crop residues for use as biofuels facilitates soil–carbon oxidation and contributes to the greenhouse emissions problem.

Tilman *et al.* assume about 1,032 l of ethanol can be produced through the conversion of the 4 t ha^{-1} $year^{-1}$ of grasses harvested. However, Pimentel and Patzek report a negative 68% return in ethanol produced compared with the fossil energy inputs in switchgrass conversion. Furthermore, converting all 235 million hectares of US grassland into ethanol at the optimistic rate suggested by Tilman *et al.* would still provide only 12% of annual US consumption of oil. Verified data, however, confirm that the output in ethanol would require 1.5 l of oil equivalents to produce 1 l of ethanol.

To achieve the production of this much ethanol, US farmers would have to displace the 100 million cattle, seven million sheep, and four million horses that are now grazing on 324 million hectares of US grassland and rangeland. Already, overgrazing is a serious problem on US grassland and a similar problem exists worldwide. Thus, the assessment of the quantity of ethanol that can be produced on US and world grasslands by Tilman *et al.* appears to be unduly optimistic.

Converting switchgrass into ethanol results in a negative energy return. The negative energy return is 68% or a slightly more negative energy return than corn ethanol production. The cost of producing a liter of ethanol using switchgrass was 93¢.

Several problems exist [in] the conversion of cellulosic biomass into ethanol. First, it takes from two to five times more cellulosic biomass to achieve the same quantity of starches and sugars as are found in the same quantity of corn grain. Thus, two to five times more cellulosic material must be produced and handled compared with corn grain. In addition, the starches and sugars are tightly held in lignin in the cellulosic biomass. They can be released using a strong acid to dissolve the lignin. Once the lignin is dissolved, the acid action is stopped with an alkali. Now the solution of lignin, starches, and sugars can be fermented.

Some claim that the lignin can be used as a fuel. Clearly, it cannot when dissolved in water. Usually less than 25% of the lignin can be extracted from the water mixture using various energy intensive technologies.

Soybean Biodiesel

Processed vegetable oils from soybean, sunflower, rapeseed, oil palm, and other oil plants can be used as fuel in diesel engines. Unfortunately, producing vegetable oils for use in diesel engines is costly in terms of economics and energy. A slight net return on energy from soybean oil is possible only if the soybeans are grown without commercial nitrogen fertilizer. The soybean, since it is a legume, will under favorable conditions produce its own nitrogen. Still soy has a 63% net fossil energy loss.

The USA provides $500 million in subsidies for the production of 850 million liters of biodiesel, which is 74 times greater than the subsidies per liter of diesel fuel. The environmental impacts of producing soybean biodiesel are second only to that of corn ethanol:

1. Soybean production causes significant soil erosion, second only to corn production.
2. Soybean production uses large quantities of herbicides, second only to corn production. These herbicides cause major pollution problems with natural biota in the soybean production areas.
3. The USDA reports a soybean yield worldwide to be 2.2 tons per hectare.

With an average oil extraction efficiency of 18%, the average oil yield per year would be approximately 0.4 tons per hectare. This converts into 454 l of oil per hectare. Based on current US diesel consumption of 227 billion liters/year, this would require more than 500 million hectares of land in soybeans or more than half the total area of the USA planted just for soybeans! All 71 billion tons of soybeans produced in the USA could only supply 2.6% of total US oil consumption. . . .

Algae for Oil Production

Some cultures of algae consist of 30% to 50% oil. Thus, there is growing interest using algae to increase US oil supply based on the theoretical claims that 47,000 to 308,000 l ha^{-1} year^{-1} (5,000 to 33,000 gallons/acre) of oil could be produced using algae. The calculated cost per barrel would be $15. Currently, oil in the US market is selling for over $50 per barrel. If the above estimated production and price of oil produced from algae were exact, US annual oil needs could theoretically be met if 100% of all US land were in algal culture.

Despite all the algae-related research and claims dating back to 1970s, none of the projected algae and oil yields have been achieved. To the contrary, one calculated estimate based on all the included costs using algae would be $800 per barrel, not $15 per barrel, as quoted above. Algae, like all plants, require large quantities of nitrogen and water in addition to significant fossil energy inputs for the production system.

Conclusion

A rapidly growing world population and rising consumption of fossil fuels is increasing demand for both food and biofuels. That will exaggerate both food and fuel shortages. Producing biofuels requires huge amounts of both fossil energy and food resources, which will intensify conflicts among these resources.

Using food crops to produce ethanol raises major nutritional and ethical concerns. Nearly 60% of humans in the world are currently malnourished, so the need for grains and other basic foods is critical. Growing crops for fuel squanders land, water, and energy resources vital for the production of food for people. Using food and feed crops for ethanol production has brought increases in the prices of US beef, chicken, pork, eggs, breads, cereals, and milk of 10% to 20%. In addition, Jacques Diouf, Director General of the UN Food and Agriculture Organization reports that using food grains to produce biofuels already is causing food shortages for the poor of the world. Growing crops for biofuel not only ignores the need to reduce natural resource consumption, but exacerbates the problem of malnourishment worldwide by turning food grain into biofuel.

Recent policy decisions have mandated increased production of biofuels in the United States and worldwide. For instance, in the Energy Independence and Security Act of 2007, President Bush set "a mandatory renewable fuel standard (RFS) requiring fuel producers to use at least 36 billion gallons of biofuel in 2022." This would require 1.6 billion tons of biomass harvested per year and would require harvesting 80% of all biomass in the USA, including all agricultural crops, grasses, and forests. With nearly total biomass harvested, biodiversity and food supplies in the USA would be decimated.

Increased biofuel production also has the capability to impact the quality of food plants in crop systems. The release of large quantities of carbon dioxide associated with the planting and processing of plant materials for biofuels is reported to reduce the nutritional quality of major world foods, including wheat, rice, barley, potatoes, and soybeans. When crops are grown under high levels of carbon dioxide, protein levels may be reduced as much as 15%.

Many problems associated with biofuels have been ignored by some scientists and policy-makers. The production of biofuels that are being created in order to diminish dependence on fossil fuels actually depends on fossil fuels.

In most cases, more fossil energy is required to produce a unit of biofuel compared with the energy that it produces. Furthermore, the USA is importing oil and natural gas to produce biofuels, which is making the USA [less] oil independent. Publications promoting biofuels have used incomplete or insufficient data to support their claims. For instance, claims that cellulosic ethanol provides net energy have not been experimentally verified because most of their calculations are *theoretical*. Finally,environmental problems, including water pollution from fertilizers and pesticides, global warming, soil erosion and air pollution are intensifying with biofuel production. There is simply not enough land, water, and energy to produce biofuels.

Most conversions of biomass into ethanol and biodiesel result in a negative energy return based on careful up-to-date analysis of all the fossil energy inputs. Four of the negative energy returns are: corn ethanol at minus 46%; switchgrass at minus 68%; soybean biodiesel at minus 63%; and rapeseed at minus 58%. Even palm oil production in Thailand results in a minus 8% net energy return, when the methanol requirement for transesterification is considered in the equation.

Increased use of biofuels further damages the global environment and especially the world food system.

EXPLORING THE ISSUE

Are Biofuels a Reasonable Substitute for Fossil Fuels?

Critical Thinking and Reflection

1. Why is there so much interest in using ethanol as a portion of automobile fuel?
2. Why does the production of biofuels affect the food supply and food prices?
3. Is using biofuels to replace fossil fuels compatible with feeding the world's population?
4. What alternatives exist to using liquid fuels to power automobiles?
5. Which biofuels are likely to have the least undesirable impact on the environment and food supply?

Is There Common Ground?

There is agreement that using prime agricultural land and expensive fertilizers and processing to grow biofuel crops necessarily reduces resources needed to grow food and unsurprisingly has effects on food supply and prices. However, there are biofuel crops that are less dependent on agricultural resources. Investigate what is happening with "cellulosic" biofuels, algae-based biofuels, and bacteria-based biofuels and answer the following questions:

1. Can these alternatives be implemented in the near future?
2. Will their prices be competitive?
3. Can they meet demand?
4. Will their production compete with food production?

ISSUE 12

Is It Time to Revive Nuclear Power?

YES: **Allison MacFarlane**, from "Nuclear Power: A Panacea for Future Energy Needs?" *Environment* (March/April 2010)

NO: **Kristin Shrader-Frechette**, from "Five Myths About Nuclear Energy," *America* (June 23–30, 2008)

Learning Outcomes

After reading this issue, you should be able to:

- Compare the hazards of nuclear power and global warming.
- Explain how nuclear power avoids the release of greenhouse gases.
- Explain why it will be difficult to shift from fossil fuels to nuclear power rapidly.
- Describe the obstacles that must be overcome before nuclear power can be more widely used.

ISSUE SUMMARY

YES: Allison MacFarlane argues that although nuclear power poses serious problems to be overcome, it "offers a potential avenue to significantly mitigate carbon dioxide emissions while still providing baseload power required in today's world." However, it will take many years to build the necessary number of new nuclear power plants.

NO: Professor Kristin Shrader-Frechette argues that nuclear power is one of the most impractical and risky of energy sources. Renewable energy sources such as wind and solar are a sounder choice.

The technology of releasing for human use the energy that holds the atom together did not get off to an auspicious start. Its first significant application was military, and the deaths associated with the Hiroshima and Nagasaki explosions have ever since tainted the technology with negative associations. It did not help that for the ensuing half-century, millions of people grew up under the threat of nuclear Armageddon. But almost from the beginning, nuclear physicists and engineers wanted to put nuclear energy to more peaceful uses, largely in the form of power plants. Touted in the 1950s as an astoundingly cheap source of electricity, nuclear power soon proved to be more expensive than conventional sources, largely because safety concerns caused delays in the approval process and prompted elaborate built-in precautions. Safety measures have worked well when needed—Three Mile Island, often cited as a horrific example of what can go wrong, released very little radioactive material to the environment. The Chernobyl disaster occurred when safety measures were ignored. In both cases, human error was more to blame than the technology itself. The related issue of nuclear waste has also raised fears and proved to add expense to the technology.

It is clear that two factors—fear and expense—impede the wide adoption of nuclear power. If both could somehow be alleviated, it might become possible to gain the benefits of the technology. Among those benefits are that nuclear power does not burn oil, coal, or any other fuel, does not emit air pollution and thus contribute to smog and haze, does not depend on foreign sources of fuel and thus weaken national independence, and does not emit carbon. Avoiding the use of fossil fuels is an important benefit; see Robert L. Hirsch, Roger H. Bezdek, and Robert M. Wendling, "Peaking Oil Production: Sooner Rather than Later?" *Issues in Science and Technology* (Spring 2005). But avoiding carbon dioxide emissions may be more important at a time when society is concerned about global warming, and this is the benefit that prompted James Lovelock, creator of the Gaia Hypothesis and hero to environmentalists everywhere, to say, "If we had nuclear power we wouldn't be in this mess now, and whose fault was it? It was [the anti-nuclear environmentalists']." See his autobiography, *Homage to Gaia: The Life of an Independent Scientist* (Oxford University Press, 2001).

Others have also seen this point. The OECD's Nuclear Energy Agency ("Nuclear Power and Climate Change" (Paris, France, 1998); www.nea.fr/html/ndd/climate/climate.pdf) found that a greatly expanded deployment of nuclear power to combat global warming was both technically and economically feasible. Robert C. Morris published *The Environmental Case for Nuclear Power: Economic, Medical, and Political Considerations* (Paragon House) in 2000. "The time seems right to reconsider the future of nuclear power," say James A. Lake, Ralph G. Bennett, and John F. Kotek in "Next-Generation Nuclear Power," *Scientific American* (January 2002). Stewart Brand, for long a leading environmentalist, predicts in "Environmental Heresies," *Technology Review* (May 2005), that nuclear power will soon be seen as the "green" energy technology. David Talbot, "Nuclear Powers Up," *Technology Review* (September 2005), notes, "While the waste problem remains unsolved, current trends favor a nuclear renaissance. Energy needs are growing. Conventional energy sources will eventually dry up. The atmosphere is getting dirtier." Peter Schwartz and Spencer Reiss, "Nuclear Now!" *Wired* (February 2005), argue that nuclear power is the one practical answer to global warming and coming shortages of fossil fuels. Iain Murray, "Nuclear Power? Yes, Please," *National Review* (June 16, 2008), argues that the world's experience with nuclear power has shown it to be both safe and reliable.

Costs can be contained, and if one is concerned about global warming, the case for nuclear power is unassailable.

Robert Evans, "Nuclear Power: Back in the Game," *Power Engineering* (October 2005), reports that a number of power companies are considering new nuclear power plants. See also Eliot Marshall, "Is the Friendly Atom Poised for a Comeback?" and Daniel Clery, "Nuclear Industry Dares to Dream of a New Dawn," *Science* (August 19, 2005). Nuclear momentum is growing, says Charles Petit, "Nuclear Power: Risking a Comeback," *National Geographic* (April 2006), thanks in part to new technologies. Karen Charman, "Brave Nuclear World? (Part I)" *World Watch* (May/June 2006), objects that producing nuclear fuel uses huge amounts of electricity derived from fossil fuels, so going nuclear can hardly prevent all releases of carbon dioxide (although using electricity derived from nuclear power would reduce the problem). She also notes that "Although no comprehensive and integrated study comparing the collateral and external costs of energy sources globally has been done, all currently available energy sources have them. . . . Burning coal—the single largest source of air pollution in the United States—causes global warming, acid rain, soot, smog, and other toxic air emissions and generates waste ash, sludge, and toxic chemicals. Landscapes and ecosystems are completely destroyed by mountain-top removal mining, while underground mining imposes high fatality, injury, and sickness rates. Even wind energy kills birds, can be noisy, and, some people complain, blights landscapes."

Stephen Ansolabehere et al., "The Future of Nuclear Power," *An Interdisciplinary MIT Study* (MIT, 2003), note that in 2000 there were 352 nuclear power plants in the developed world as a whole, and a mere 15 in developing nations, and that even a very large increase in the number of nuclear power plants—from 1000 to 1500—will not stop all releases of carbon dioxide. In fact, if carbon emissions double by 2050 as expected, from 6500 to 13,000 million metric tons per year, the 1800 million metric tons not emitted because of nuclear power will seem relatively insignificant. Nevertheless, say John M. Deutch and Ernest J. Moniz, "The Nuclear Option," *Scientific American* (September 2006), such a cut in carbon emissions would be "significant." Christine Laurent, in "Beating Global Warming with Nuclear Power?" *UNESCO Courier* (February 2001), notes, "For several years, the nuclear energy industry has attempted to cloak itself in different ecological robes. Its credo: nuclear energy is a formidable asset in the battle against global warming because it emits very small amounts of greenhouse gases. This stance, first presented in the late 1980s when the extent of the phenomenon was still the subject of controversy, is now at the heart of policy debates over how to avoid droughts, downpours and floods." Laurent adds that it makes more sense to focus on reducing carbon emissions by reducing energy consumption.

Even though President Obama declared support for "a new generation of safe, clean nuclear power plants" in his January 27, 2010, State of the Union speech and the Department of Energy soon proposed massive loan guarantees for the industry (see Pam Russell and Pam Hunter, "Nuclear Resurgence Poised for Liftoff," *ENR: Engineering News-Record* (March 1, 2010)), the debate over the future of nuclear power is likely to remain vigorous for some time to come. But as Richard A. Meserve says in a *Science* editorial ("Global Warming and Nuclear Power," *Science* (January 23, 2004)), "For those who are serious about confronting global warming, nuclear power should be seen as part of the solution. Although it is unlikely that many environmental groups will become enthusiastic proponents of nuclear power, the harsh reality is that any serious program to address global warming cannot afford to jettison any technology prematurely. . . . The stakes are large, and the scientific and educational community should seek to ensure that the public understands the critical link between nuclear power and climate change." Paul Lorenzini, "A Second Look at Nuclear Power," *Issues in Science and Technology* (Spring 2005), argues that the goal must be energy "sufficiency for the foreseeable future with minimal environmental impact." Nuclear power can be part of the answer, but making it happen requires that we shed ideological biases. "It means ceasing to deceive ourselves about what might be possible." Charles Forsberg, "The Real Path to Green Energy: Hybrid Nuclear-Renewable Power," *Bulletin of the Atomic Scientists* (November/December 2009), suggests that the best use of nuclear power will be to provide energy for biofuel refineries and as backup for solar and wind power.

Alvin M. Weinberg, former director of the Oak Ridge National Laboratory, notes in "New Life for Nuclear Power," *Issues in Science and Technology* (Summer 2003), that to make a serious dent in carbon emissions would require perhaps four times as many reactors as suggested in the MIT study. The accompanying safety and security problems would be challenging. If the challenges can be met, says John J. Taylor, retired vice president for nuclear power at the Electric Power Research Institute, in "The Nuclear Power Bargain," *Issues in Science and Technology* (Spring 2004), there are a great many potential benefits. Are new reactor technologies needed? Richard K. Lester, "New Nukes," *Issues in Science and Technology* (Summer 2006), says that better centralized waste storage is what is needed, at least in the short term, despite the Obama Administration's declaration that Yucca Mountain, the only U.S. storage site under development, will no longer be supported. On the other hand, some new technologies are available already; see Carol Matlack, "A High-End Bet on Nuclear Power," *BusinessWeek* (March 15, 2010).

Environmental groups such as Friends of the Earth are adamantly opposed, but there are signs that some environmentalists do not agree; see William M. Welch, "Some Rethinking Nuke Opposition," *USA Today* (March 23, 2007). Judith Lewis, "The Nuclear Option," *Mother Jones* (May/June 2008), concludes, "When rising seas flood our coasts, the idea of producing electricity from the most terrifying force ever harnessed may not seem so frightening—or expensive—at all."

In the YES selection, Allison MacFarlane argues that although nuclear power poses serious problems to be overcome, it "offers a potential avenue to significantly mitigate carbon dioxide emissions while still providing baseload power required in today's world." However, it will take many years to build the necessary number of new nuclear power plants. Professor Kristin Shrader-Frechette argues that nuclear power is one of the most impractical and risky of energy sources. Renewable energy sources such as wind and solar are a sounder choice.

YES

Allison MacFarlane

Nuclear Power: A Panacea for Future Energy Needs?

Each week seems to bring further evidence that the Earth is warming at a faster rate than previously estimated. Pressure is building to replace power sources that emit carbon dioxide with those that do not. It is in this "climate" that nuclear energy is getting a second look. Once relegated to the junk heap after the Three Mile Island and Chernobyl disasters brought the dangers associated with nuclear power to everyone's attention, nuclear power may now be undergoing a "renaissance," as the nuclear industry likes to say. Environmentalists such as Stewart Brand, originator of the Whole Earth Catalog, and Patrick Moore of Greenpeace have started pushing nuclear power as a ready solution to the problem of electricity production without carbon emissions.

Over the past few years, more than 40 countries that currently do not have nuclear power have expressed interest in acquiring it to address their future energy needs, and a few are making significant progress. "Every country has the right to make use of nuclear power, as well as the responsibility to do it in accordance with the highest standards of safety, security, and non-proliferation," stated Mohamed ElBaradei, then–Director General of the International Atomic Energy Agency (IAEA).

All this enthusiasm for nuclear power must be tempered by plain reality. Construction of new nuclear power reactors is still one of the most capital-intensive ventures compared with other energy sources such as coal and natural gas. High-level nuclear waste from the nuclear industry waits in above-ground temporary storage until a final solution is implemented. And the connection between nuclear power and nuclear weapons continues to be perceived as a threat by the international community. Given the pushes and pulls for nuclear power, what is the likelihood that the future will entail a vast expansion of nuclear power on a global scale, and if so, what are its implications, and can this expansion be considered "sustainable" in a development sense?

How Does Nuclear Perform Now?

Globally, there are 436 reactors with a generating capacity of 372 GWe (gigawatts electric) located in 31 countries. [In] 2006, nuclear power generated 15 percent of the world's electricity. Currently, the IAEA estimates that

From *Environment* by Allison MacFarlane, March/April 2010, pp. 36–44. Copyright © 2010. Reprinted by permission of Taylor & Francis Group, LLC. www.taylorandfrancis.com

52 reactors are under construction, whereas the World Nuclear Association lists 71 reactors that will come online between 2009 and 2015. Of the 52 listed by the IAEA, . . . over 85 percent of them are located in Asia and Eastern Europe, and they would add 46 GWe to the existing generating capacity. Of the 52 under construction, 12 of them began construction before 1990, and it is not clear that these projects will ever be completed. At the same time, existing reactors are aging. The average age of the operating fleet in 2007 was 23 years. Most reactors were originally designed for a 40-year life span, but some countries, such as the United States, have recently been granting 20-year life extensions to their fleet.

Carbon Dioxide Emissions Reductions

Nuclear reactors do not emit carbon dioxide to produce electricity because their fuel is uranium-based. This is not to say that nuclear power is emission free—it is not. Carbon dioxide is emitted during the lifecycle of nuclear power production, particularly during the uranium mining, milling, and fuel fabrication processes, and during the construction of new nuclear plants. Nonetheless, nuclear power displaces large volumes of carbon dioxide in comparison with fossil fuel plants.

For instance, in 2006, global nuclear power provided 2594 TWh of electricity. If that electricity was produced by coal and natural gas plants combined, it would have added 3904 million tons of carbon-dioxide equivalent to the atmosphere. Compared to annual emissions of 29,195 million tons of CO_2 equivalent from fossil fuel burning, nuclear power saved about 13 percent.

Certainly if nuclear power grew its capacity, it could contribute significantly to carbon dioxide emissions reductions. Scholars have suggested that 1000 GWe nuclear capacity (in addition to the existing capacity) could reduce potential carbon dioxide emissions growth by 15–25 percent by 2050. Of course, adding 1000 GWe is equivalent to building 1000 new large-scale 1000-MWe plants over the next 40 years, which requires that 25 plants be built per year, an ambitious schedule, but doable. Such plans are not on anyone's drawing board at the moment, though.

Past and Present Safety Issues

Since the Chernobyl accident in 1986, there have been no catastrophic failures at nuclear power plants. That's not to say that there have not been problems, but the overall global safety record since 1986 has been good. This is, in large part, due to vigilant oversight of the nuclear industry and redundant design in safety features.

Issues may arise from aging plants as equipment fails. A case in point is the Davis-Besse reactor in Ohio, which developed an undetected hole in the reactor pressure vessel head, leaving only 0.95 cm of stainless steel to protect against a loss of coolant accident. The hole, which was almost 13 cm in diameter and 17 cm deep, was caused by boric acid corrosion from the borated cooling water in the reactor. The U.S. Nuclear Regulatory Commission's (NRC)

Inspector General found that the NRC itself did not demand more rapid action to address the potentially dangerous issue because it wanted to "lessen the financial impact" on the utility plant operator.

The best way to continue to ensure reactor safety is to follow the advice of the Kemeney Commission, which investigated the 1979 Three Mile Island accident in the United States. They recommended that constant vigilance was required to ensure reactor safety; complacency about the safety of nuclear power would lead to accidents.

Current Waste Management Strategies

With over 50 years of power reactor operation, by 2007 the world had accumulated almost $365,000\,m^3$ of high-level nuclear waste, including spent nuclear fuel and reprocessing wastes; over 3 million m^3 of long-lived low- and intermediate-level waste; and over 2 million m^3 of short-lived low- and intermediate-level waste. Though many countries have low-level waste disposal facilities and a few have intermediate-level facilities, none have opened a high-level nuclear waste disposal facility. Most spent nuclear fuel remains in cooling pools and dry storage casks at power reactors; some has been reprocessed, and the resulting high-level waste has been vitrified and sits in storage facilities.

Most countries with significant nuclear power generation have decided that high-level nuclear waste should be disposed of in mined geologic repositories. Four countries have been focusing on a single site within their borders. In the United States, the Yucca Mountain, Nevada, site is currently undergoing licensing review for repository construction by the Nuclear Regulatory Commission, although the current Obama Administration has said it will not support the Yucca Mountain site. . . .

Currently there are two management pathways for handling the spent nuclear fuel generated by power reactors. One is the "open cycle," in which spent fuel is considered high-level waste and will be directly disposed of in a repository. The United States, Sweden, Finland, Canada, and others follow this pathway. The second is the "closed cycle," as practiced by France, the United Kingdom, Japan, Russia, and India, in which spent nuclear fuel is reprocessed, the plutonium and uranium extracted, and the remaining waste turned into glass logs destined for a repository. The separated plutonium is then made into new nuclear fuel known as MOX (mixed oxide), but as of yet, it is not reprocessed a second time. The extracted uranium tends to be considered a waste product, as it is too expensive to clean up for reuse.

Reprocessing does reduce the volume of high-level waste by a factor of 4 to 10 times. But this volume reduction does not imply a corresponding capacity reduction in a geologic repository, because volume is not the relevant unit of measure for capacity; heat production and radionuclide composition of the waste are. In addition, reprocessing generates large volumes of low- and intermediate-level wastes, some of which require their own repository. As a result, reprocessing as it is currently practiced has a limited impact on repository size. And no matter which cycle is used, open or closed, a repository will be required in the end.

Issues for the Future

For developed countries, nuclear power can significantly offset carbon dioxide emissions, whereas for developing countries, nuclear power has the potential to electrify the country as well as provide desalination services to address dwindling water supplies. An expansion of nuclear power will not occur without overcoming significant hurdles, though. Whether nuclear power will be sustainable is another issue. The UK Sustainable Development Commission recently concluded that nuclear power was not necessary to achieve a low-carbon future. Some issues associated with nuclear power are unique when compared with other electricity sources, especially in terms of safety and security.

Safety

To operate safely, nuclear power requires a large, active, and well-established infrastructure. The industry needs a trained workforce for plant construction and operation, and also for the regulatory system. In places like the United States, the existing nuclear engineering workforce is aging, and questions remain as to how rapidly it could be replaced. The construction workforce for nuclear power plants requires specialized skills, and the current number of these workers is small. Moreover, there are supply chain issues globally for the nuclear industry. Most significant is the limited capacity for heavy forgings for the reactor pressure vessels. At the moment, a single Japanese plant is capable of making these, and its availability is limited because of commitments to other industries.

For countries with no nuclear plants, an indigenous workforce would have to obtain training abroad, they would need to hire foreign workers, or they would have to rely on the reactor suppliers for a workforce. The same will be true for a regulatory infrastructure. Countries with no existing reactors will need to establish a nuclear infrastructure, again by developing it from the bottom up or by hiring/adapting one from abroad. For example, the United Arab Emirates has hired a former U.S. Nuclear Regulatory Commission official to be the Director-General of its Federal Authority for Nuclear Regulation.

Another key issue for emerging nuclear energy countries is that of liability in the case of nuclear accident for adequate protection of the victims and predictability for nuclear suppliers and insurers. A catastrophic nuclear accident could result in compensation costs that would run into the hundreds of billions of dollars. There is no single global liability regime, and currently 236 of the 436 operating reactors are not covered by liability conventions. Those countries that are party to liability conventions must assure that the legal liability is the responsibility of the reactor operator, and that there are monetary and temporal limits on liability claims.

Cost

Costs associated with the construction of new nuclear reactors may pose the greatest roadblock to the global expansion of nuclear power. . . .

Historically, reactor construction experiences delays and cost escalation. Costs of constructing U.S. nuclear power plants exceed original estimates by 200 percent to 400 percent on average. Experience in the U.S. also shows that increased construction times lead to greatly increased costs from accumulating interest costs. And construction time has not improved that much. Average construction time globally, in most recent experience between 2001–2005, was 6.8 years; at best it was five years, on average.

In the United States, nuclear reactors may cost even more per unit. Florida Power and Light, the utility poised to build two new reactors at its Turkey Point site, recently released figures of $5,780/kW to $8,071/kW, for a total of $12.1 billion to $24.3 billion, depending on the type of plants built. Costs such as these will be prohibitive for many developing countries and even for most utility companies in developed countries. Indeed, the U.S. nuclear industry is seeking further loan guarantees from the federal government over the original $18 billion laid out in the Energy Policy Act of 2005.

All of the reactor designs currently available for construction ("generation III") are large reactors, ranging in size between 600 MWe and 1700 MWe, with most averaging around 1000 MWe. The reason for this is the economies of scale associated with large reactors. Costs associated with reactor design, licensing, regulation, and operation are independent of reactor size, so larger reactors tend to cost less per kilowatt than smaller reactors. These large reactors require large, sophisticated electrical grid networks, something that many developing countries lack. Developing countries seeking nuclear power for electricity production or desalination would be better served by smaller reactors, but none is currently available, though new designs are being put forth.

Capital costs can be decreased by shortening construction time, building standardized designs (in the United States, all existing plants have unique designs), using factory construction of plant components, relying on local resources for labor and construction materials, and building more than one reactor at a site, which allows for cost sharing in licensing. For example, Westinghouse's new AP-1000, under construction in Sanmen, China, uses a modular design in which factory prefabricated components can be assembled at the reactor site and put in place using a large crane. Smaller reactors might be helpful in reducing capital costs, because the industry might be able get the economic advantages of producing many units, thus overcoming the problems associated with site-built reactors. Moreover, for small, modular reactors, it may be possible to do the licensing at the manufacturing plant instead of the individual site, greatly reducing costs.

Security

One of the main security concerns about nuclear power programs is their connection to nuclear weapons development. This concern is the reason for international agreements, such as the Nuclear Nonproliferation Treaty (NPT), which guarantees that countries that do not have nuclear weapons are allowed nuclear energy technology, and for institutions such as the IAEA, whose main charge is to monitor nuclear energy programs. The general problem is that the fuel for nuclear power plants is also the same fuel for nuclear weapons: highly enriched uranium or plutonium.

Running a nuclear power reactor poses little proliferation threat. The problem comes either at the "front-end" of the nuclear fuel cycle, associated with fuel production, or the "back-end," associated with the reprocessing of spent nuclear fuel. . . .

One additional security issue requires comment. Nuclear power plants could become terrorist targets in the future. Terrorists might desire to cause a loss of coolant accident of some sort. Particularly vulnerable in this situation are older plants with near-full or full cooling pools in which the fuel is placed in high density arrangements. These pools are located in different areas depending on the reactor design, though some are located several stories above ground. Were the pool to lose coolant, a fire could ensue, causing the release of huge amounts of radioactivity (more than Chernobyl). Fortunately, unlike dealing with proliferation issues associated with the fuel cycle, there is a relatively straightforward solution to this problem. Older fuel can be moved from the cooling pools into dry cask storage on site. Dry casks are passively cooled and therefore much safer if attacked. One of the only reasons reactors have not gone to low-density fuel arrangement in pools is cost.

Environmental/Health Impacts

Life cycle wastes from nuclear power are relatively small in volume in comparison to other energy sources such as fossil fuels. Nuclear power has low emissions from reactor operation, and overall waste production is low, especially in the open fuel cycle. More wastes are generated by reprocessing spent nuclear fuel, though again, in comparison with fossil fuels, they are small. . . .

The bulk of the volume of wastes associated with nuclear power originates from mining and processing the uranium ore for fabrication into fuel. The largest and most problematic wastes associated with mining are uranium mill tailings, which are the rock residue from the extraction of uranium from the ore. The problem is that all the uranium daughter products, in particular radium (Ra-226) and its daughter radon (Rn-222), from millions to billions of years of decay, remain with the residual rock. These pose risks from windblown dust exposure, radon emissions, and leaching into groundwater. . . .

High-level nuclear wastes will increase if nuclear power expands. At current rates of production, an additional 1000 GWe will produce an extra 20,000–30,000 metric tons of spent fuel annually. This amount is approximately one-third to one-half the current statutory capacity for the proposed Yucca Mountain repository in the United States. Thus, an expanded nuclear world would require more or larger repositories than are envisaged today.

Prospects for the Future

Nuclear power offers a potential avenue to significantly mitigate carbon dioxide emissions while still providing baseload power required in today's world. But nuclear power cannot provide significant baseload power over the next 15 years or so, because it will take many years before a considerable number of new plants are licensed and built. The high capital costs of the plants and the risks of investing in this technology will likely enforce a slow start to large-scale nuclear build. New policies that institute a price on carbon dioxide emissions will certainly improve nuclear power's prospects. In some ways, though, nuclear power suffers from a chicken-and-egg situation: more manufacturing experience may reduce prices, but to achieve the necessary experience, many new plants must first be paid for and built.

New reactors in emerging nuclear energy countries will take even longer because of the large infrastructure required and the need for the international community to respond in an organized fashion to the potential proliferation threats posed by nuclear energy. As a result, nuclear power may begin to provide significant carbon emissions relief in the longer term, starting 20 or 30 years from now.

Unlike most other energy sources, nuclear power does have "doomsday" scenarios. History may repeat itself. Both the accidents at Three Mile Island and Chernobyl had a chilling effect on the global nuclear industry. In the United States, no new nuclear plants have been ordered since the Three Mile Island accident. In Sweden in the 1980s, the country voted to end its use of nuclear power (though they have more recently reversed this decision). If a similar accident occurs, nuclear power's future will grow dim. Similarly, if a country breaks out and develops nuclear weapons using its nuclear energy technology, expansion of nuclear power to non-nuclear energy countries will stop. Countries with nuclear energy technology will be loathe to provide it to emerging nuclear energy nations. Unlike fossil fuels and renewable resources, nuclear energy's success rests on the performance and behavior of all of those involved in producing nuclear power, wherever they are. And this is a somewhat tenuous situation on which to base an industry.

Kristin Shrader-Frechette  **NO**

Five Myths About Nuclear Energy

Atomic energy is among the most impractical and risky of available fuel sources. Private financiers are reluctant to invest in it, and both experts and the public have questions about the likelihood of safely storing lethal radioactive wastes for the required million years. Reactors also provide irresistible targets for terrorists seeking to inflict deep and lasting damage on the United States. The government's own data show that U.S. nuclear reactors have more than a one-in-five lifetime probability of core melt, and a nuclear accident could kill 140,000 people, contaminate an area the size of Pennsylvania, and destroy our homes and health.

In addition to being risky, nuclear power is unable to meet our current or future energy needs. Because of safety requirements and the length of time it takes to construct a nuclear-power facility, the government says that by the year 2050 atomic energy could supply, at best, 20 percent of U.S. electricity needs; yet by 2020, wind and solar panels could supply at least 32 percent of U.S. electricity, at about half the cost of nuclear power. Nevertheless, in the last two years, the current U.S. administration has given the bulk of taxpayer energy subsidies—a total of $20 billion—to atomic power. Why? Some officials say nuclear energy is clean, inexpensive, needed to address global climate change, unlikely to increase the risk of nuclear proliferation and safe.

On all five counts they are wrong. Renewable energy sources are cleaner, cheaper, better able to address climate change and proliferation risks, and safer. The government's own data show that wind energy now costs less than half of nuclear power; that wind can supply far more energy, more quickly, than nuclear power; and that by 2015, solar panels will be economically competitive with all other conventional energy technologies. The administration's case for nuclear power rests on at least five myths. Debunking these myths is necessary if the United States is to abandon its current dangerous energy course.

Myth 1. Nuclear Energy Is Clean

The myth of clean atomic power arises partly because some sources, like a pro-nuclear energy analysis published in 2003 by several professors at the Massachusetts Institute of Technology, call atomic power a "carbon-free source" of energy. On its Web site, the U.S. Department of Energy, which is also a proponent of nuclear energy, calls atomic power "emissions free." At best, these claims are half-truths because they "trim the data" on emissions.

While nuclear reactors themselves do not release greenhouse gases, reactors are only part of the nine-stage nuclear fuel cycle. This cycle includes mining uranium ore, milling it to extract uranium, converting the uranium to gas, enriching it, fabricating fuel pellets, generating power, reprocessing spent fuel, storing spent fuel at the reactor and transporting the waste to a permanent storage facility. Because most of these nine stages are heavily dependent on fossil fuels, nuclear power thus generates at least 33 grams of carbon-equivalent emissions for each kilowatt-hour of electricity that is produced. (To provide uniform calculations of greenhouse emissions, the various effects of the different greenhouse gases typically are converted to carbon-equivalent emissions.) Per kilowatt-hour, atomic energy produces only one-seventh the greenhouse emissions of coal, but twice as much as wind and slightly more than solar panels.

Nuclear power is even less clean when compared with energy-efficiency measures, such as using compact-fluorescent bulbs and increasing home insulation. Whether in medicine or energy policy, preventing a problem is usually cheaper than curing or solving it, and energy efficiency is the most cost-effective way to solve the problem of reducing greenhouse gases. Department of Energy data show that one dollar invested in energy-efficiency programs displaces about six times more carbon emissions than the same amount invested in nuclear power. Government figures also show that energy-efficiency programs save $40 for every dollar invested in them. This is why the government says it could immediately and cost-effectively cut U.S. electricity consumption by 20 percent to 45 percent, using only existing strategies, like time-of-use electricity pricing. (Higher prices for electricity used during daily peak-consumption times—roughly between 8 a.m. and 8 p.m.—encourage consumers to shift their time of energy use. New power plants are typically needed to handle only peak electricity demand.)

Myth 2. Nuclear Energy Is Inexpensive

Achieving greater energy efficiency, however, also requires ending the lop-sided system of taxpayer nuclear subsidies that encourage the myth of inexpensive electricity from atomic power. Since 1949, the U.S. government has provided about $165 billion in subsidies to nuclear energy, about $5 billion to solar and wind together, and even less to energy-efficiency programs. All government efficiency programs—to encourage use of fuel-efficient cars, for example, or to provide financial assistance so that low-income citizens can insulate their homes—currently receive only a small percentage of federal energy monies.

After energy-efficiency programs, wind is the most cost-effective way both to generate electricity and to reduce greenhouse emissions. It costs about half as much as atomic power. The only nearly finished nuclear plant in the West, now being built in Finland by the French company Areva, will gener-

WHAT DOES THE CHURCH SAY?

Though neither the Vatican nor the U.S. bishops have made a statement on nuclear power, the church has outlined the ethical case for renewable energy. In *Centesimus Annus* Pope John Paul II wrote that just as Pope Leo XIII in 1891 had to confront "primitive capitalism" in order to defend workers' rights, he himself had to confront the "new capitalism" in order to defend collective goods like the environment. Pope Benedict XVI warned that pollutants "make the lives of the poor especially unbearable." In their 2001 statement *Global Climate Change,* the U.S. Catholic bishops repeated his point: climate change will "disproportionately affect the poor, the vulnerable, and generations yet unborn."

The bishops also warn that "misguided responses to climate change will likely place even greater burdens on already desperately poor peoples." Instead they urge "energy conservation and the development of alternate renewable and clean-energy resources." They argue that renewable energy promotes care for creation and the common good, lessens pollution that disproportionately harms the poor and vulnerable, avoids threats to future generations and reduces nuclear-proliferation risks.

---

ate electricity costing 11 cents per kilowatt-hour. Yet the U.S. government's Lawrence Berkeley National Laboratory calculated actual costs of new wind plants, over the last seven years, at 3.4 cents per kilowatt-hour. Although some groups say nuclear energy is inexpensive, their misleading claims rely on trimming the data on cost. The 2003 M.I.T. study, for instance, included neither the costs of reprocessing nuclear material, nor the full interest costs on nuclear-facility construction capital, nor the total costs of waste storage. Once these omissions—from the entire nine-stage nuclear fuel cycle—are included, nuclear costs are about 11 cents per kilowatt-hour.

The cost-effectiveness of wind power explains why in 2006 utility companies worldwide added 10 times more wind-generated, than nuclear, electricity capacity. It also explains why small-scale sources of renewable energy, like wind and solar, received $56 billion in global private investments in 2006, while nuclear energy received nothing. It explains why wind supplies 20 percent of Denmark's electricity. It explains why, each year for the last several years, Germany, Spain and India have each, alone, added more wind capacity than all countries in the world, taken together, have added in nuclear capacity.

In the United States, wind supplies up to 8 percent of electricity in some Midwestern states. The case of Louis Brooks is instructive. Utilities pay him $500 a month for allowing 78 wind turbines on his Texas ranch, and he can still use virtually all the land for farming and grazing. Wind's cost-effectiveness also explains why in 2007 wind received $9 billion in U.S. private investments, while nuclear energy received zero. U.S. wind energy has been growing by nearly 3,000 megawatts each year, annually producing new electricity equivalent to what three new nuclear reactors could generate. Meanwhile, no new U.S. atomic-power reactors have been ordered since 1974.

Should the United States continue to heavily subsidize nuclear technology? Or, as the distinguished physicist Amory Lovins put it, is the nuclear industry dying of an "incurable attack of market forces"? Standard and Poor's, the credit- and investment-rating company, downgrades the rating of any utility that wants a nuclear plant. It claims that even subsidies are unlikely to make nuclear investment wise. *Forbes* magazine recently called nuclear investment "the largest managerial disaster in business history," something pursued only by the "blind" or the "biased."

Myth 3. Nuclear Energy Is Necessary to Address Climate Change

Government, industry and university studies, like those recently from Princeton, agree that wind turbines and solar panels already exist at an industrial scale and could supply one-third of U.S. electricity needs by 2020, and the vast majority of U.S. electricity by 2050—not just the 20 percent of electricity possible from nuclear energy by 2050. The D.O.E. says wind from only three states (Kansas, North Dakota and Texas) could supply all U.S. electricity needs, and 20 states could supply nearly triple those needs. By 2015, according to the D.O.E., solar panels will be competitive with all conventional energy technologies and will cost 5 to 10 cents per kilowatt hour. Shell Oil and other fossil-fuel companies agree. They are investing heavily in wind and solar.

From an economic perspective, atomic power is inefficient at addressing climate change because dollars used for more expensive, higher-emissions nuclear energy cannot be used for cheaper, lower-emissions renewable energy. Atomic power is also not sustainable. Because of dwindling uranium supplies, by the year 2050 reactors would be forced to use low-grade uranium ore whose greenhouse emissions would roughly equal those of natural gas. Besides, because the United States imports nearly all its uranium, pursuing nuclear power continues the dangerous pattern of dependency on foreign sources to meet domestic energy needs.

Myth 4. Nuclear Energy Will Not Increase Weapons Proliferation

Pursuing nuclear power also perpetuates the myth that increasing atomic energy, and thus increasing uranium enrichment and spent-fuel reprocessing, will increase neither terrorism nor proliferation of nuclear weapons. This myth has been rejected by both the International Atomic Energy Agency and the U.S. Office of Technology Assessment. More nuclear plants means more weapons materials, which means more targets, which means a higher risk of terrorism and proliferation. The government admits that Al Qaeda already has targeted U.S. reactors, none of which can withstand attack by a large airplane. Such an attack, warns the U.S. National Academy of Sciences, could cause fatalities as far away as 500 miles and destruction 10 times worse than that caused by the nuclear accident at Chernobyl in 1986.

Nuclear energy actually increases the risks of weapons proliferation because the same technology used for civilian atomic power can be used for weapons, as the cases of India, Iran, Iraq, North Korea and Pakistan illustrate. As the Swedish Nobel Prize winner Hannes Alven put it, "The military atom and the civilian atom are Siamese twins." Yet if the world stopped building nuclear-power plants, bomb ingredients would be harder to acquire, more conspicuous and more costly politically, if nations were caught trying to obtain them. Their motives for seeking nuclear materials would be unmasked as military, not civilian.

Myth 5. Nuclear Energy Is Safe

Proponents of nuclear energy, like Patrick Moore, cofounder of Greenpeace, and the former Argonne National Laboratory adviser Steve Berry, say that new reactors will be safer than current ones—"meltdown proof." Such safety claims also are myths. Even the 2003 M.I.T. energy study predicted that tripling civilian nuclear reactors would lead to about four core-melt accidents. The government's Sandia National Laboratory calculates that a nuclear accident could cause casualties similar to those at Hiroshima or Nagasaki: 140,000 deaths. If nuclear plants are as safe as their proponents claim, why do utilities need the U.S. Price-Anderson Act, which guarantees utilities protection against 98 percent of nuclear-accident liability and transfers these risks to the public? All U.S. utilities refused to generate atomic power until the government established this liability limit. Why do utilities, but not taxpayers, need this nuclear-liability protection?

Another problem is that high-level radioactive waste must be secured "in perpetuity," as the U.S. National Academy of Sciences puts it. Yet the D.O.E. has already admitted that if nuclear waste is stored at Nevada's Yucca Mountain, as has been proposed, future generations could not meet existing radiation standards. As a result, the current U.S. administration's proposal is to allow future releases of radioactive wastes, stored at Yucca Mountain, provided they annually cause no more than one person—out of every 70 persons exposed to them—to contract fatal cancer. These cancer risks are high partly because Yucca Mountain is so geologically unstable. Nuclear waste facilities could be breached by volcanic or seismic activity. Within 50 miles of Yucca Mountain, more than 600 seismic events, of magnitude greater than two on the Richter scale, have occurred since 1976. In 1992, only 12 miles from the site, an earthquake (5.6 on the Richter scale) damaged D.O.E. buildings. Within 31 miles of the site, eight volcanic eruptions have occurred in the last million years. These facts suggest that Alvin Weinberg was right. Four decades ago, the then-director of the government's Oak Ridge National Laboratory warned that nuclear waste required society to make a Faustian bargain with the devil. In exchange for current military and energy benefits from atomic power, this generation must sell the safety of future generations.

Yet the D.O.E. predicts harm even in this generation. The department says that if 70,000 tons of the existing U.S. waste were shipped to Yucca

Mountain, the transfer would require 24 years of dozens of daily rail or truck shipments. Assuming low accident rates and discounting the possibility of terrorist attacks on these lethal shipments, the D.O.E. says this radioactive-waste transport likely would lead to 50 to 310 shipment accidents. According to the D.O.E., each of these accidents could contaminate 42 square miles, and each could require a 462-day cleanup that would cost $620 million, not counting medical expenses. Can hundreds of thousands of mostly unguarded shipments of lethal materials be kept safe? The states do not think so, and they have banned Yucca Mountain transport within their borders. A better alternative is onsite storage at reactors, where the material can be secured from terrorist attack in "hardened" bunkers.

Where Do We Go From Here?

If atomic energy is really so risky and expensive, why did the United States begin it and heavily subsidize it? As U.S. Atomic Energy Agency documents reveal, the United States began to develop nuclear power for the same reason many other nations have done so. It wanted weapons-grade nuclear materials for its military program. But the United States now has more than enough weapons materials. What explains the continuing subsidies? Certainly not the market. *The Economist* (7/7/05) recently noted that for decades, bankers in New York and London have refused loans to nuclear industries. Warning that nuclear costs, dangers and waste storage make atomic power "extremely risky," *The Economist* claimed that the industry is now asking taxpayers to do what the market will not do: invest in nuclear energy. How did *The Economist* explain the uneconomical $20 billion U.S. nuclear subsidies for 2005–7? It pointed to campaign contributions from the nuclear industry.

Despite the problems with atomic power, society needs around-the-clock electricity. Can we rely on intermittent wind until solar power is cost-effective in 2015? Even the Department of Energy says yes. Wind now can supply up to 20 percent of electricity, using the current electricity grid as backup, just as nuclear plants do when they are shut down for refueling, maintenance and leaks. Wind can supply up to 100 percent of electricity needs by using "distributed" turbines spread over a wide geographic region—because the wind always blows somewhere, especially offshore.

Many renewable energy sources are safe and inexpensive, and they inflict almost no damage on people or the environment. Why is the current U.S. administration instead giving virtually all of its support to a riskier, more costly nuclear alternative?

EXPLORING THE ISSUE

Is It Time to Revive Nuclear Power?

Critical Thinking and Reflection

1. Which is more dangerous? Nuclear power or global warming? (See *The Anatomy of a Silent Crisis,* Global Humanitarian Forum Geneva (2009).
2. How are nuclear wastes handled today?
3. Pretend that your campus is a nuclear power plant. Now devise an evacuation plan to protect the residents of neighboring communities in case of a serious accident.
4. What are the advantages of a "closed cycle" approach to handling spent nuclear fuel?

Is There Common Ground?

There is no argument that today's society needs vast amounts of energy to function. The debate largely centers on how to meet the need (which is projected to grow greatly in the future). The greatest agreement may well be in the solutions people refuse to consider. As an exercise, consider the following major uses of energy:

Global trade

Commuting

Air-conditioning

Central heating

1. In what ways could we change our lifestyles to reduce energy use in each of these areas?
2. Would reducing population help?

Internet References . . .

The Agriculture Network Information Center

The Agriculture Network Information Center is a guide to quality agricultural (including biotechnology) information on the Internet as selected by the National Agricultural Library, land-grant universities, and other institutions.

www.agnic.org/

American Fisheries Society

The mission of the American Fisheries Society is to improve the conservation and sustainability of fishery resources and aquatic ecosystems by advancing fisheries and aquatic science and promoting the development of fisheries professionals.

www.fisheries.org/afs/index.html

EarthTrends

The World Resources Institute offers data on biodiversity, fisheries, agriculture, population, and a great deal more.

http://earthtrends.wri.org

Organic Farming Research Foundation

The Organic Farming Research Foundation sponsors research related to organic farming practices, disseminates research results, and educates the public and decision makers about organic farming issues.

http://ofrf.org/index.html

The Population Council

Established in 1952, the Population Council "is an international, nonprofit institution that conducts research on three fronts: biomedical, social science, and public health. This research—and the information it produces—helps change the way people think about problems related to reproductive health and population growth." Many of the council's publications are available online.

www.popcouncil.org/

United Nations Population Division

The United Nations Population Division is responsible for monitoring and appraising a broad range of areas in the field of population. This site offers a wealth of recent data and links.

www.un.org/esa/population/unpop.htm

Food and Population

*T*o many, *"sustainability" means arranging things so that the natural world—plants and animals, forests and coral reefs, and fresh water and landscapes—can continue to exist more or less (mostly less) as it did before human beings multiplied, developed technology, and began to cause extinctions, air and water pollution, soil erosion, desertification, climate change, and so on. To others, "sustainability" means arranging things so that humankind can continue to survive and thrive, even keeping up its history of growth, technological development, and energy use—as if the environment and its resources were infinite.*

The two visions of "sustainability" are logically incompatible. Many fear that we are on a collision course with limited resources, with the National Geographic *warning of "The End of Plenty" (June 2009). Can we avoid disaster? Must we reduce the numbers of people on the planet, their use of technology, and their standard of living? If we do, will human well-being be lessened? If we do not, how can we continue to feed everyone? Can ocean fisheries be maintained? Is genetic engineering the answer? Is organic farming the answer? All of these issues provoke considerable debate.*

- Do We Have a Population Problem?
- Does Commercial Fishing Have a Future?
- Can Organic Farming Feed the World?

ISSUE 13

Do We Have a Population Problem?

YES: **David Attenborough**, from "This Heaving Planet," *New Statesman* (April 25, 2011)

NO: **Tom Bethell**, from "Population, Economy, and God," *The American Spectator* (May 2009)

Learning Outcomes

After reading this issue, you should be able to:

- Explain the nature of the : "population problem."
- Define the concept of "carrying capacity."
- List the potential benefits of stabilizing or reducing population.
- List the potential drawbacks of stabilizing or reducing population.

ISSUE SUMMARY

YES: Sir David Attenborough argues that the environmental problems faced by the world are exacerbated by human numbers. Without population reduction, the problems will become ever more difficult—and ultimately impossible—to solve.

NO: Tom Bethell argues that population alarmists project their fears onto popular concerns, currently the environment, and every time their scare-mongering turns out to be based on faulty premises. Blaming environmental problems will be no different. Societies are sustained not by population control but by belief in God.

In 1798, the British economist Thomas Malthus published his *Essay on the Principle of Population*. In it, he pointed with alarm at the way the human population grew geometrically (a hockey-stick-shaped curve of increase) and at how agricultural productivity grew only arithmetically (a straight-line increase). It was obvious, he said, that the population must inevitably outstrip its food supply and experience famine. Contrary to the conventional wisdom of the time, population growth was not necessarily a good thing. Indeed, it led inexorably to catastrophe. For many years, Malthus was something of a laughing stock. The doom he forecast kept receding into the future as new lands were opened to agriculture, new agricultural technologies appeared, new ways of preserving food limited the waste of spoilage, and the birth rate dropped in the industrialized nations (the "demographic transition"). The food supply kept ahead of population growth and seemed likely—to most observers—to continue to do so. Malthus's ideas were dismissed as irrelevant fantasies.

Yet overall population kept growing. In Malthus's time, there were about 1 billion human beings on Earth. By 1950—when Warren S. Thompson worried that civilization would be endangered by the rapid growth of Asian and Latin American populations during the next five decades (see "Population," *Scientific American* [February 1950])—there were a little over 2.5 billion. In 1999, the tally passed 6 billion. It passed 7 billion in 2011. By 2025 it will be over 8 billion. Statistics like these, which are presented in *World Resources 2008: Roots of Resilience—Growing the Wealth of the Poor* (World Resources Institute, 2008) (www.wri.org/publication/world-resources-2008-roots-of-resilience), published in collaboration with the United Nations Development Programme, the United Nations Environment Programme, and the World Bank, are positively frightening. The Worldwatch Institute's yearly *State of the World* reports (W.W. Norton) are no less so. By 2050 the UN expects the world population to be about 9 billion (see *World Population Prospects: The 2010 Revision Population Database*; http://esa.un.org/unpd/wpp/index.htm; United Nations, 2010). By 2100, it will be 10.1 billion; see Jocelyn Keiser, "10 Billion Plus: Why World Population Projections Were Too Low," *ScienceInsider* (May 4, 2011) (http://scim.ag/_worldpop). While global agricultural production has also increased, it has not kept up with rising demand, and—because of the loss of topsoil to erosion, the exhaustion of aquifers for irrigation water, and the high price of energy for making fertilizer (among other things)—the prospect of improvement seems exceedingly slim to many observers.

Two centuries never saw Malthus's forecasts of doom come to pass. Population continued to grow, and environmentalists pointed with alarm at a great many problems that resulted from human use of the world's resources (air and water pollution, erosion, loss of soil fertility and groundwater, loss of species, and a great deal more). "Cornucopian" economists such as the late Julian Simon insisted that the more people there are on Earth, the more people there are to solve problems and that humans can find ways around all possible resource shortages. See Simon's essay, "Life on Earth Is Getting Better, Not Worse," *The Futurist* (August 1983). See also David Malakoff, "Are More People Necessarily a Problem?" *Science* (July 29, 2011) (a special issue on population).

Was Malthus wrong? Both environmental scientists and many economists now say that if population continues to grow, problems are inevitable. But earlier predictions of a world population of 10 or 12 billion by 2050 are no longer looking very likely. The UN's population statistics show a slowing of growth, to be followed by an actual decline in population size.

Fred Pearce, *The Coming Population Crash: and Our Planet's Surprising Future* (Beacon, 2010), is optimistic about the effects on human well-being of the coming decline in population. Do we still need to work on controlling population? Historian Matthew Connolly, *Fatal Misconception: The Struggle to Control World Population* (Belknap Press, 2010), argues that the twentieth-century movement to control population was an oppressive movement that failed to deliver on its promises. Now that population growth is slowing, the age of population control is over. Yet there remains the issue of "carrying capacity," defined very simply as the size of the population that the environment can support, or "carry," indefinitely, through both good years and bad. It is not the size of the population that can prosper in good times alone, for such a large population must suffer catastrophically when droughts, floods, or blights arrive or the climate warms or cools. It is a long-term concept, where "long-term" means not decades or generations, nor even centuries, but millennia or more. See Mark Nathan Cohen, "Carrying Capacity," *Free Inquiry* (August/September 2004), and T. C. R. White, "The Role of Food, Weather, and Climate in Limiting the Abundance of Animals," *Biological Reviews* (August 2008).

Andrew R. B. Ferguson, in "Perceiving the Population Bomb," *World Watch* (July/August 2001), sets the maximum sustainable human population at about 2 billion. Sandra Postel, in the Worldwatch Institute's *State of the World 1994* (W. W. Norton, 1994), says, "As a result of our population size, consumption patterns, and technology choices, we have surpassed the planet's carrying capacity. This is plainly evident by the extent to which we are damaging and depleting natural capital (including land and water). The point is reiterated by Robert Kunzig, "By 2045 Global Population Is Projected to Reach Nine Billion. Can the Planet Take the Strain?" *National Geographic* (January 2011) (*National Geographic* ran numerous articles on population-related issues during 2011). Thomas L. Friedman, "The Earth Is Full," *New York Times* (June 7, 2011), thinks a crisis is imminent, but we will learn and move on; see also Paul Gilding, *The Great Disruption: Why the Climate Crisis Will Bring on the End of Shopping and the Birth of a New World* (Bloomsbury Press, 2011).

If population growth is now declining and world population will actually begin to decline during this century, there is clearly hope. But most estimates of carrying capacity put it at well below the current world population size, and it will take a long time for global population to fall far enough to reach such levels. We seem to be moving in the right direction, but it remains an open question whether our numbers will decline far enough soon enough (i.e., before environmental problems become critical). On the other hand, Jeroen Van den Bergh and Piet Rietveld, "Reconsidering the Limits to World Population: Meta-Analysis and Meta-Prediction," *Bioscience* (March 2004), set their best estimate of human global carrying capacity at 7.7 billion, which is distinctly reassuring. However, there is still concern that global population will not stop at that point; see David R. Francis, "'Birth Dearth' Worries Pale in Comparison to Overpopulation," *Christian Science Monitor* (July 14, 2008).

How high a level will population actually reach? Fertility levels are definitely declining in many developed nations; see Alan Booth and

Ann C. Crouter, eds., *The New Population Problem: Why Families in Developed Countries Are Shrinking and What It Means* (Lawrence Erlbaum Associates, 2005). The visibility of this fertility decline is among the reasons mentioned by Martha Campbell, "Why the Silence on Population?" *Population and Environment* (May 2007). Yet Doug Moss, "What Birth Dearth?" *E—The Environmental Magazine* (November–December 2006), reminds us that there is still a large surplus of births—and therefore a growing population—in the less developed world. If we think globally, there is no shortage of people. And David E. Bloom, "7 Billion and Counting," *Science* (July 29, 2011), notes, "Despite alarmist predictions, historical increases in population have not been economically catastrophic. Moreover, changes in population age structure [providing for more workers] have opened the door to increased prosperity."

Some people worry that a decline in population will not be good for human welfare. Michael Meyer, "Birth Dearth," *Newsweek* (September 27, 2004), argues that a shrinking population will mean that the economic growth that has meant constantly increasing standards of living must come to an end, government programs (from war to benefits for the poor and elderly) will no longer be affordable, a shrinking number of young people will have to support a growing elderly population, and despite some environmental benefits, quality of life will suffer. China is already feeling some of these effects; see Wang Feng, "China's Population Destiny: The Looming Crisis," *Current History* (September 2010), and Mara Hvistendahl, "Has China Outgrown the One-Child Policy?" *Science* (September 17, 2010). Julia Whitty, "The Last Taboo," *Mother Jones* (May–June 2010), argues that even though the topic of overpopulation has become unpopular, it is clear that we are already using the Earth's resources faster than they can be replenished and the only answer is to slow and eventually reverse population growth. Scott Victor Valentine, "Disarming the Population Bomb," *International Journal of Sustainable Development and World Ecology* (April 2010), calls for "a renewed international focus on managed population reduction as a key enabler of sustainable development." As things stand, the current size and continued growth of population threaten the United Nations' Millennium Development Goals (including alleviating global poverty, improving health, and protecting the environment; see www.un.org/millenniumgoals/); see Willard Cates, Jr., et al., "Family Planning and the Millennium Development Goals," *Science* (September 24, 2010).

In the following selections, Sir David Attenborough argues that the environmental problems faced by the world are exacerbated by human numbers. Without population reduction, the problems will become ever more difficult—and ultimately impossible—to solve. Tom Bethell argues that population alarmists project their fears onto popular concerns, currently the environment, and every time their scare-mongering turns out to be based on faulty premises. Blaming environmental problems will be no different. Societies are sustained not by population control but by belief in God.

YES

David Attenborough

This Heaving Planet

Fifty years ago, on 29 April 1961, a group of far-sighted people in this country got together to warn the world of an impending disaster. Among them were a distinguished scientist, Sir Julian Huxley; a bird-loving painter, Peter Scott; an advertising executive, Guy Mountford; a powerful and astonishingly effective civil servant, Max Nicholson—and several others.

They were all, in addition to their individual professions, dedicated naturalists, fascinated by the natural world not just in this country but internationally. And they noticed what few others had done—that all over the world, charismatic animals that were once numerous were beginning to disappear.

The Arabian oryx, which once had been widespread all over the Arabian Peninsula, had been reduced to a few hundred. In Spain, there were only about 90 imperial eagles left. The Californian condor was down to about 60. In Hawaii, a goose that once lived in flocks on the lava fields around the great volcanoes had been reduced to 50. And the strange rhinoceros that lived in the dwindling forests of Java—to about 40. These were the most extreme examples. Wherever naturalists looked they found species of animals whose populations were falling rapidly. This planet was in danger of losing a significant number of its inhabitants, both animals and plants.

Something had to be done. And that group determined to do it. They would need scientific advice to discover the causes of these impending disasters and to devise ways of slowing them and, they hoped, of stopping them. They would have to raise awareness and understanding of people everywhere; and, like all such enterprises, they would need money to enable them to take practical action.

They set about raising all three. Since the problem was an international one, they based themselves not in Britain but in the heart of Europe, in Switzerland. They called the organisation that they created the World Wildlife Fund (WWF).

As well as the international committee, separate action groups would be needed in individual countries. A few months after that inaugural meeting in Switzerland, Britain established one—and was the first country to do so.

The methods the WWF used to save these endangered species were several. Some, such as the Hawaiian goose and the oryx, were taken into captivity in zoos, bred up into a significant population and then taken back to their original home and released. Elsewhere—in Africa, for example—great areas of unspoiled country were set aside as national parks, where the animals

From *New Statesman*, April 25, 2011, pp. 29–32. Copyright © 2011 by New Statesman, Ltd. Reprinted by permission.

could be protected from poachers and encroaching human settlement. In the Galápagos Islands and in the home of the mountain gorillas in Rwanda, ways were found of ensuring that local people who also had claims on the land where such animals lived were able to benefit financially by attracting visitors.

Ecotourism was born. The movement as a whole went from strength to strength. Twenty four countries established their own WWF national appeals. Existing conservation bodies, of which there were a number in many parts of the world but which had been working largely in isolation, acquired new zest and international links. New ones were founded focusing on particular areas or particular species. The world awoke to conservation. Millions—billions—of dollars were raised. And now, 50 years on, conservationists who have worked so hard and with such foresight can justifiably congratulate themselves on having responded magnificently to the challenge.

Yet now, in spite of a great number of individual successes, the problem seems bigger than ever. True, thanks to the vigour and wisdom of conservationists, no major charismatic species has yet disappeared. Many are still trembling on the brink, but they are still hanging on. Today, however, overall there are more problems not fewer, more species at risk of extinction than ever before. Why?

Fifty years ago, when the WWF was founded, there were about three billion people on earth. Now there are almost seven billion—over twice as many—every one of them needing space. Space for their homes, space to grow their food (or to get others to grow it for them), space to build schools, roads and airfields. Where could that come from? A little might be taken from land occupied by other people but most of it could only come from the land which, for millions of years, animals and plants had had to themselves—the natural world.

But the impact of these extra billions of people has spread even beyond the space they physically claimed. The spread of industrialisation has changed the chemical constituents of the atmosphere. The oceans that cover most of the surface of the planet have been polluted and increasingly acidified. The earth is warming. We now realise that the disasters that continue increasingly to afflict the natural world have one element that connects them all—the unprecedented increase in the number of human beings on the planet.

There have been prophets who have warned us of this impending disaster. One of the first was Thomas Malthus. His surname—Malthus—leads some to suppose that he was some continental European philosopher, a German perhaps. But he was not. He was an Englishman, born in Guildford, Surrey, in the middle of the 18th century. His most important book, *An Essay on the Principle of Population,* was published in 1798. In it, he argued that the human population would increase inexorably until it was halted by what he termed "misery and vice". Today, for some reason, that prophecy seems to be largely ignored—or, at any rate, disregarded. It is true that he did not foresee the so-called Green Revolution (from the 1940s to the late 1970s), which greatly increased the amount of food that can be produced in any given area of arable land. And there may be other advances in our food producing skills that we ourselves still cannot foresee. But such advances only delay things. The fundamental truth that Malthus proclaimed remains the truth: there cannot be more people on this earth than can be fed.

Many people would like to deny that this is so. They would like to believe in that oxymoron "sustainable growth."

Kenneth Boulding, President Kennedy's environmental adviser 45 years ago, said something about this: "Anyone who believes in indefinite growth in anything physical, on a physically finite planet, is either mad—or an economist."

The population of the world is now growing by nearly 80 million a year. One and a half million a week. A quarter of a million a day. Ten thousand an hour. In this country [UK] it is projected to grow by 10 million in the next 22 years. That is equivalent to ten more Birminghams.

All these people, in this country and worldwide, rich or poor, need and deserve food, water, energy and space. Will they be able to get it? I don't know. I hope so. But the government's chief scientist and the last president of the Royal Society have both referred to the approaching "perfect storm" of population growth, climate change and peak oil production, leading inexorably to more and more insecurity in the supply of food, water and energy.

Consider food. For animals, hunger is a regular feature of their lives. The stoical desperation of the cheetah cubs whose mother failed in her last few attempts to kill prey for them, and who consequently face starvation, is very touching. But that happens to human beings, too. All of us who have travelled in poor countries have met people for whom hunger is a daily background ache in their lives. There are about a billion such people today—that is four times as many as the entire human population of this planet a mere 2,000 years ago, at the time of Christ.

You may be aware of the government's Foresight project, Global Food and Farming Futures. It shows how hard it is to feed the seven billion of us alive today. It lists the many obstacles that are already making this harder to achieve—soil erosion, salinisation, the depletion of aquifers, over-grazing, the spread of plant diseases as a result of globalisation, the absurd growing of food crops to turn into biofuels to feed motor cars instead of people—and so on. So it underlines how desperately difficult it is going to be to feed a population that is projected to stabilise "in the range of eight to ten billion people by the year 2050." It recommends the widest possible range of measures across all disciplines to tackle this. And it makes a number of eminently sensible recommendations, including a second green revolution.

But, surprisingly, there are some things that the project report does not say. It doesn't state the obvious fact that it would be much easier to feed eight billion people than ten billion. Nor does it suggest that the measures to achieve such a number—such as family planning and the education and empowerment of women—should be a central part of any programme that aims to secure an adequate food supply for humanity. It doesn't refer to the prescient statement 40 years ago by Norman Borlaug, the Nobel laureate and father of the first green revolution.

Borlaug produced new strains of high-yielding, short-strawed and disease-resistant wheat and in doing so saved many thousands of people in India, Pakistan, Africa and Mexico from starvation. But he warned us that all he had done was to give us a "breathing space" in which to stabilise our numbers. The government's report anticipates that food prices may rise with oil prices, and makes it clear that this will affect poorest people worst and discusses various way to help them. But it doesn't mention what every mother subsisting on the equivalent of a dollar a day already knows—that her children would be better fed if there were four of them around the table instead of ten. These are strange omissions.

How can we ignore the chilling statistics on arable land? In 1960 there was more than one acre of good cropland per person in the world—enough to sustain a reasonable European diet. Today, there is only half an acre each. In China, it is only a quarter of an acre, because of their dramatic problems of soil degradation.

Another impressive government report on biodiversity published this year, *Making Space for Nature in a Changing World,* is rather similar. It discusses all the rising pressures on wildlife in the UK—but it doesn't mention our growing population as being one of them—which is particularly odd when you consider that England is already the most densely populated country in Europe.

Most bizarre of all was a recent report by a royal commission on the environmental impact of demographic change in this country which denied that population size was a problem at all—as though 10 million extra people, more or less, would have no real impact. Of course it is not our only or even our main environmental problem but it is absurd to deny that, as a multiplier of all the others, it is a problem.

I suspect that you could read a score of reports by bodies concerned with global problems—and see that population is one of the drivers that underlies all of them—and yet find no reference to this obvious fact in any of them. Climate change tops the environmental agenda at present. We all know that every additional person will need to use some carbon energy, if only firewood for cooking, and will therefore create more carbon dioxide—though a rich person will produce vastly more than a poor one. Similarly, we can all see that every extra person is—or will be—an extra victim of climate change—though the poor will undoubtedly suffer more than the rich. Yet not a word of it appeared in the voluminous documents emerging from the Copenhagen and Cancún climate summits.

Why this strange silence? I meet no one who privately disagrees that population growth is a problem. No one—except flat-earthers—can deny that the planet is finite. We can all see it—in that beautiful picture of our earth taken by the Apollo mission. So why does hardly anyone say so publicly? There seems to be some bizarre taboo around the subject.

This taboo doesn't just inhibit politicians and civil servants who attend the big conferences. It even affects the environmental and developmental non-governmental organisations, the people who claim to care most passionately about a sustainable and prosperous future for our children.

Yet their silence implies that their admirable goals can be achieved regardless of how many people there are in the world or the UK, even though they all know that they can't.

I simply don't understand it. It is all getting too serious for such fastidious niceties. It remains an obvious and brutal fact that on a finite planet human population will quite definitely stop at some point. And that can only happen in one of two ways. It can happen sooner, by fewer human births—in a word, by contraception. That is the humane way, the powerful option that allows all of us to deal with the problem, if we collectively choose to do so. The alternative is an increased death rate—the way that all other creatures must suffer, through famine or disease or predation. That, translated into human terms, means famine or disease or war—over oil or water or food or minerals or grazing rights or just living space. There is, alas, no third alternative of indefinite growth.

The sooner we stabilise our numbers, the sooner we stop running up the "down" escalator. Stop population increase—stop the escalator—and we have some chance of reaching the top; that is to say, a decent life for all.

To do that requires several things. First and foremost, it needs a much wider understanding of the problem, and that will not happen while the taboo on discussing it retains such a powerful grip on the minds of so many worthy and intelligent people. Then it needs a change in our culture so that while everyone retains the right to have as many children as they like, they understand that having large families means compounding the problems their children and everyone else's children will face in the future.

It needs action by governments. In my view, all countries should develop a population policy—as many as 70 countries already have them in one form or another—and give it priority. The essential common factor is to make family planning and other reproductive health services freely available to everyone, and empower and encourage them to use it—though without any kind of coercion.

According to the Global Footprint Network, there are already more than a hundred countries whose combination of numbers and affluence have already pushed them past the sustainable level. They include almost all developed countries. The UK is one of the worst. There the aim should be to reduce over time both the consumption of natural resources per person and the number of people—while, needless to say, using the best technology to help maintain living standards. It is tragic that the only current population policies in developed countries are, perversely, attempting to increase their birth rates in order to look after the growing number of old people. The notion of ever more old people needing ever more young people, who will in turn grow old and need even more young people, and so on ad infinitum, is an obvious ecological Ponzi scheme.

I am not an economist, nor a sociologist, nor a politician, and it is from their disciplines that answers must come. But I am a naturalist. Being one means that I know something of the factors that keep populations of different species of animals within bounds and what happens when they aren't.

I am aware that every pair of blue tits nesting in my garden is able to lay over 20 eggs a year but, as a result of predation or lack of food, only one or two will, at best, survive. I have watched lions ravage the hundreds of wildebeest fawns that are born each year on the plains of Africa. I have seen how increasing numbers of elephants can devastate their environment until, one year when the rains fail on the already over-grazed land, they die in hundreds.

But we are human beings. Because of our intelligence, and our ever-increasing skills and sophisticated technologies, we can avoid such brutalities. We have medicines that prevent our children from dying of disease. We have developed ways of growing increasing amounts of food. But we have removed the limiters that keep animal populations in check. So now our destiny is in our hands.

There is one glimmer of hope. Wherever women have the vote, wherever they are literate and have the medical facilities to control the number of children they bear, the birth rate falls. All those civilised conditions exist in the southern Indian state of Kerala. In India as a whole, the total fertility rate is 2.8 births per woman. In Kerala, it is 1.7 births per woman. In Thailand last year, it was 1.8 per woman, similar to that in Kerala. But compare that with the mainly Catholic Philippines, where it is 3.3.

Here and there, at last, there are signs of a recognition of the problem. Save the Children mentioned it in its last report. The Royal Society has assembled a working party of scientists across a wide range of disciplines who are examining the problem.

But what can each of us do? Well, there is just one thing that I would ask. Break the taboo, in private and in public—as best you can, as you judge right. Until it is broken there is no hope of the action we need. Wherever and whenever we speak of the environment, we should add a few words to ensure that the population element is not ignored. If you are a member of a relevant NGO, invite them to acknowledge it.

If you belong to a church—and especially if you are a Catholic, because its doctrine on contraception is a major factor in this problem—suggest they consider the ethical issues involved. I see the Anglican bishops in Australia have dared to do so. If you have contacts in government, ask why the growth of our population, which affects every department, is as yet no one's responsibility. Big empty Australia has appointed a sustainable population minister, so why can't small crowded Britain?

The Hawaiian goose, the oryx, and the imperial eagle that sounded the environmental alarm 50 years ago were, you might say, the equivalent of canaries in coal mines—warnings of impending and even wider catastrophe.

Make a list of all the other environmental problems that now afflict us and our poor battered planet—the increase of greenhouse gases and consequential global warming, the acidification of the oceans and the collapse of fish stocks, the loss of rainforest, the spread of deserts, the shortage of arable land, the increase in violent weather, the growth of mega-cities, famine, migration patterns. The list goes on and on. But they all share one underlying cause. Every one of these global problems, social as well as environmental, becomes more difficult—and ultimately impossible—to solve with ever more people.

Population, Economy, and God

World population, once "exploding," is still increasing, and "momentum" ensures that it will do so for decades to come. But fertility rates have tumbled. In Europe every country has fallen below replacement level. Some governments, especially France's, are beginning to use financial incentives to restore fertility rates but the effort, if generous enough to work—by paying women to have a third child—could bankrupt the welfare state.

In rich countries, a total fertility rate of 2.1 babies per woman is needed if population is to remain stable. But in the European Union as a whole the rate is down to 1.5. Germany is at 1.4, and Italy, Spain, and Greece are at 1.3. The fertility rate in France is now 2.0, or close to replacement. But the uneasy question is whether this is due to subsidies or to the growing Muslim population.

All over the world, with a few anomalies, there is a strong inverse correlation between GDP per capita and babies per family. It's a paradox, because wealthier people can obviously afford lots of children. But very predictably they have fewer. Hong Kong (1.02), Singapore, and Taiwan are three of the richest countries in the world, and three of the four lowest in total fertility. The countries with the highest fertility rates are Mali (7.4), Niger, and Uganda. Guess how low they are on the wealth chart.

Here's a news item. Carl Djerassi, one of the inventors of the birth control pill, recently deplored the sharp decline of total fertility in Austria (1.4), the country of his birth. A Catholic news story seized on that and reported that one of the pill's inventors had said the pill had caused a "demographic catastrophe." Austria's leading Catholic, Cardinal Schönborn, said the Vatican had predicted 40 years ago that the pill would promote a dramatic fall in birth rates.

Djerassi, 85, an emeritus professor of chemistry at Stanford, did warn of a catastrophe and he said that Austria should admit more immigrants. But he denied that people have smaller families "because of the availability of birth control." They do so "for personal, economic, cultural, and other reasons," of which "changes in the status of women" was the most important. Japan has an even worse demographic problem, he said, "yet the pill was only legalized there in 1999 and is still not used widely." (Japan's fertility rate is 1.22.) (In fact, if the pill and abortion really were illegal more children surely would be born, if only because unintentional pregnancies would come to term.)

Austrian families who had decided against children wanted "to enjoy their schnitzels while leaving the rest of the world to get on with it," Djerassi also said. That may have rankled because the country had just put his face on a postage stamp.

So what is causing these dramatic declines? It's under way in many countries outside Europe too. In Mexico, fertility has moved down close to replacement level—having been as high as six babies per woman in the 1970s.

Obviously economic growth has been the dominant factor but there are other considerations.

Young couples hardly read Paul Ehrlich before deciding whether to have children, but scaremongering authors have played a key role in creating our anti-natalist mood. Books warning of a (then) newfangled emergency, the "population explosion," began appearing soon after World War II. Consider *Road to Survival* (1948), by William Vogt, or *People! Challenge to Survival*, by the same author. An anti-people fanatic before his time, Vogt was hypnotized by the Malthusian doctrine that population growth would overtake the food supply. That would lead to a war of all against all. Paul Ehrlich projected that the 1980s would see massive die-offs from starvation. (Obesity turned out to be the greater health threat.)

In that earlier period, the population controllers didn't feel they had to mince words. Vogt wrote in 1960 that "tens of thousands of children born every year in the United States should, solely for their own sakes, never have seen the light of day. . . . There are hundreds of thousands of others, technically legitimate since their parents have engaged in some sort of marriage ritual, but whose birth is as much of a crime against them as it is against the bastards."

At a time when the world population still had not reached 3 billion—today it is 6.7 billion—Vogt thought "drastic measures are inescapable." He warned of "mounting population pressures in the Soviet Union," where, by the century's end, "there may be 300 million Russians." It was time for them "to begin control of one of the most powerful causes of war—overpopulation."

Note: the population of Russia by 2000 was 145 million; today it is 141 million. (Fertility rate: 1.4.)

Population alarmists have long enjoyed the freedom to project their fears onto whatever cause is uppermost in the progressive mind. Then it was war. Today it is the environment, which, we are told, human beings are ruining. This will be shown to have been as false as the earlier warnings, but not before our environmental scares have done much harm to a fragile economy (at the rate things are going with Obama). All previous scares were based on faulty premises, and the latest one, based on "science," will be no different.

I believe that two interacting factors shape population growth or decline: economic prosperity and belief in God. As to the first, there is no doubt that rising material prosperity discourages additional children. Fewer infants die; large families are no longer needed to support older parents. The welfare state—which only rich countries can afford—has greatly compounded this effect. When people believe that the government will take care of them, pay their pensions and treat their maladies, children do seem less essential.

A rise in prosperity also encourages people to think that they can dispense with God. Religion diminishes when wealth increases—that's my theory. But with a twist that I shall come to. Wealth generates independence, including independence from God, or (if you will) Providence. God is gradually forgotten, then assumed not to exist. This will tend to drive childbearing down even further. Hedonism will become predominant. Remember, Jesus warned that it's the rich, not the poor, who are at spiritual hazard.

The legalization of abortion reflected the decline of religious faith in America, but it must also have led others to conclude that God was no longer to be feared. That's why I don't quite believe Djerassi when he tries to disassociate the pill from fertility. The ready availability of the pill told society at large that sex without consequences was perfectly acceptable. Then, by degrees, that self-indulgent view became an anti-natalist worldview.

It became so ingrained that many people now think it obvious. Sex became a "free" pastime as long as it was restricted to consenting adults. Furthermore, anyone who questioned that premise risked denunciation as a bigot.

The U.S. has been seen as the great stumbling block to any theory linking prosperity, lack of faith, and low fertility. Prosperity here has been high, and overall fertility is at replacement. But I am wary of this version of American exceptionalism. How much lower would U.S. fertility fall without the influx of Latino immigrants and their many offspring? Nicholas Eberstadt, a demographer at AEI, tells me that Mexican immigrants now actually have a higher fertility rate in the U.S. than they do in Mexico. (Maybe because they come to American hospitals for free medical care?)

I wonder also if religious vitality here is what it's cracked up to be. Surely it has weakened considerably. A recent survey by Trinity College in Hartford, funded by the Lilly Endowment, showed that the percentage of Americans identifying themselves as Christian dropped to 76 percent from 86 percent in 1990; those with "no" religion, 8.2 percent of the population in 1990, are now 15 percent.

As a social force, the U.S. Catholic bishops have withered away to a shocking extent. Hollywood once respected and feared their opinion. Today, the most highly placed of these bishops are unwilling to publicly rebuke pro-abortion politicians who call themselves Catholic, even when they give scandal by receiving Communion in public. How the mitered have fallen. They daren't challenge the rich and powerful.

But there is another factor. Calling yourself a Christian when the pollster phones imposes no cost and self-reported piety may well be inflated. We have to distinguish between mere self-labelers and actual churchgoers. And beyond that there are groups with intense religious belief who retain the morale to ignore the surrounding materialism and keep on having children.

The ultra-Orthodox in Israel are the best example. Other Jewish congregations may go to synagogue, but they have children at perhaps one-third the ultra-Orthodox rate. At about seven or eight children per family, theirs is one of the highest fertility rates in the world. And they don't permit birth control—Carl Djerassi, please note. In the U.S. Orthodox Jews again far out-breed their more secular sisters.

The Mormons are also distinctive. Utah, about two-thirds Mormon, has the highest fertility rate (2.63 in 2006) among the 50 states; Vermont has the lowest (1.69). In the recent Trinity Survey, Northern New England is now "the least religious section of the country." Vermont is the least religious state; 34 percent of residents say they have "no religion." So minimal faith and low fertility are demonstrably linked. Mormon fertility is declining, to be sure, and I recognize that I am flirting with a circular argument: deciding which groups are the most fervent by looking at their birth rates.

Then there's the Muslim concern. It's hard to avoid concluding that the lost Christian zeal has been appropriated by Islam. In the U.S., Muslims have doubled since 1990 (from a low base, to 0.6% of the population). The rise of Islam suggests that the meager European fertility rates would be even lower if Muslims had not contributed disproportionately to European childbearing.

It's hard to pin down the numbers, though. Fertility in France has risen, but Nick Eberstadt tells me that the French government won't reveal how many of these babies are born to Muslim parents. "They treat it as a state secret," he said. In other countries such as Switzerland, where lots of guest workers are employed, the fertility rate would be much lower than it already is (1.44) were it not for the numerous offspring of those guest workers.

When a population is not replacing itself, the welfare state creates its own hazard. Lots of new workers are needed to support the retirees. Germany's low fertility will require an annual immigration of 200,000 just to maintain the current population. Where will they come from? Many arrive from Turkey, where the fertility rate has also declined (to a bout 2 .0). But not as far as it has declined among native Germans. So the concern is that in the welfare states of Europe, believing Muslims are slowly replacing the low-morale, low-fertility, materialistic non-believers who once formed a Christian majority.

I could summarize the argument with this overstatement: The intelligentsia stopped believing in God in the 19th century. In the 20th it tried to build a new society, man without God. It failed. Then came a new twist. Man stopped believing in himself. He saw himself as a mere polluter—a blot on the landscape. Theologians tell us that creatures cannot exist without the support of God. A corollary may be that societies cannot long endure without being sustained by a *belief* in God.

EXPLORING THE ISSUE

Do We Have a Population Problem?

Critical Thinking and Reflection

1. Is it possible to have too many people on Earth?
2. What is wrong with the statement that there is no population problem because all of Earth's human population could fit inside the state of Texas?
3. What does population have to do with sustainability?
4. What is more important for long-term survival of the human species—population control or belief in God?

Is There Common Ground?

The essayists for this issue agree that human population continues to grow and that long-term human survival (or sustainability) matters. They disagree on the best way to achieve long-term human survival.

1. Does quality of life seem likely to suffer more with a declining population or a growing population?
2. What are the key features of "quality of life"? (One good place to start your research is www.foe.co.uk/community/tools/isew/.)
3. How might we determine what the Earth's carrying capacity for human beings really is?
4. What is the influence (if any) of religious faith on carrying capacity?

ISSUE 14

Does Commercial Fishing Have a Future?

YES: Carl Safina, from "A Future for U.S. Fisheries," *Issues in Science and Technology* (Summer 2009)

NO: Food and Agriculture Organization of the United Nations, from "World Review of Fisheries and Aquaculture," *The State of World Fisheries and Aquaculture, 2010,* (FAO, 2010)

Learning Outcomes

After reading this issue, you should be able to:

- Explain what overfishing is.
- Explain the factors that drive people to overexploit ecosystem resources such as marine fish.
- Describe possible ways to prevent destruction of oceanic fisheries, as well as possible replacements for them.

ISSUE SUMMARY

YES: Carl Safina argues that despite an abundance of bad news about the state of the oceans and commercial fisheries, there are some signs that conservation and even restoration of fish stocks to a sustainable state are possible.

NO: The Food and Agriculture Organization of the United Nations argues that the proportion of marine fish stocks that are overexploited has increased tremendously since the 1970s. Despite some progress, there remains "cause for concern." The continuing need for fish as food means there will be continued growth in aquaculture.

Carl Safina called attention to the poor state of the world's fisheries in "Where Have All the Fishes Gone?" *Issues in Science and Technology* (Spring 1994), and "The World's Imperiled Fish," *Scientific American* (November 1995).

Expanding population, improved fishing technology, and growing demand had combined to drive down fish stocks around the world. Fishers going further from shore and deploying larger nets kept the catch growing, but the the Food and Agriculture Organization of the United Nations had noted that the fisheries situation was already "globally non-sustainable, and major ecological and economic damage [was] already visible."

In 1998, the UN declared the International Year of the Ocean. Kieran Mulvaney, "A Sea of Troubles," E—The Environmental Magazine (January–February 1998), reported, "According to the United Nations Food and Agriculture Organization (FAO), an estimated 70 percent of global fish stocks are 'over-exploited,' 'fully exploited,' 'depleted' or recovering from prior over-exploitation. By 1992, FAO had recorded 16 major fishery species whose global catch had declined by more than 50 percent over the previous three decades—and in half of these, the collapse had begun after 1974." Ocean fishing did not seem sustainable. Christian Mullon, Pierre Freon, and Philippe Cury, "The Dynamics of Collapse in World Fisheries," Fish and Fisheries (June 2005), find that current trends are likely to cause a "global collapse of many more fisheries" than hitherto.

As a species becomes depleted, fishers catch more younger fish, and fewer survive to reproduce; see Allen W. L. To and Yvonne Sadovy de Mitcheson, "Shrinking Baseline: The Growth in Juvenile Fisheries, with the Hong Kong Grouper Fishery as a Case Study," Fish and Fisheries (December 2009). Daniel Pauly and Reg Watson, "Counting the Last Fish," Scientific American (July 2003), note that desirable fish tend to be top predators, such as tuna and cod. When the numbers of these fish decline due to overfishing, fishers shift their attention to fish lower on the food chain and consumers see a change in what is available at the market. The cod are smaller, and monkfish and other once less desirable fish join them on the crushed ice at the market. And the change is an indicator of trouble in the marine ecosystem.

Karin E. Limburg and John R. Waldman, "Dramatic Declines in North Atlantic Diadromous Fishes," Bioscience (December 2009), report that for 24 fish species studied, "populations have declined dramatically from original baselines." Oran R. Young, "Taking Stock: Management Pitfalls in Fisheries Science," Environment (April 2003), notes that despite putting great effort into assessing fish stocks and managing fisheries for sustainable yield, marine fish stocks have continued to decline. This is partly the "result of the inability of managers to resist pressures from interest groups to set total allowable catches too high, even in the face of warnings from scientists about the dangers of triggering stock depletions. The problem also arises, however, from repeated failures on the part of analysts and policy makers to anticipate the collapse of major stocks or to grasp either the current condition or the reproductive dynamics of important stocks." He cautions against putting "blind faith in the validity of scientific assessments" and suggests more use of the precautionary principle despite the risk that this would set allowed catch levels lower than many people would like. According to Marta Coll et al., "Ecosystem Overfishing in the Ocean," PLoS One (www.plosone.org) (2008), "At present, total catch per capita from Large Marine Ecosystems is at least twice the value estimated to ensure fishing at moderate sustainable levels." Lynda D. Rodwell and Callum M. Roberts, "Fishing and the Impact of Marine Reserves in a Variable Environment," *Canadian Journal of Fisheries & Aquatic Sciences* (November 2004), find that in variable marine environments, reserves can increase catches, reduce variability of catches, and make planning more efficient. However, it is worth noting that Martin D. Smith, Junjie Zhang, and Felicia C. Coleman, "Effectiveness of Marine Reserves for Large-Scale Fisheries Management," *Canadian Journal of Fisheries & Aquatic Sciences* (January 2006), find that the effect of at least some marine reserves is not beneficial. Results may vary according to size of reserve and the fish species of interest; see Robert E. Blyth-Skyrme et al. "Conservation Benefits of Temperate Marine Protected Areas: Variation Among Fish Species," *Conservation Biology* (June 2006). Christopher Pala, "Victory at Sea," *Smithsonian* (September 2008), describes the world's largest marine reserve surrounding the Phoenix Islands in the South Pacific.

In September 2004, the U.S. Commission on Ocean Policy issued its report, "An Ocean Blueprint for the 21st Century" (http://www.oceancommission .gov/documents/full_color_rpt/welcome.html), calling for improved management systems "to handle mounting pollution, declining fish populations and coral reefs, and promising new industries such as aquaculture." In many ways it agreed with the Pew report. Carl Safina and Sarah Chasis, "Saving the Oceans," *Issues in Science and Technology* (Fall 2004), discuss the two reports and say that it is time for Congress to craft a new approach to ocean policy, including fisheries protection, and put scientists in charge of policy. James N. Sanchirico and Susan S. Hanna, "Sink or Swim Time for U.S. Fishery Policy," *Issues in Science and Technology* (Fall 2004), add that an ecosystem approach is essential. According to Rainer Froese and Alexander Proelß, "Rebuilding Fish Stocks No Later than 2015: Will Europe Meet the Deadline?" *Fish & Fisheries* (June 2010), "Maintaining or restoring fish stocks at levels that are capable of producing maximum sustainable yield is a legal obligation under the United Nations Convention on the Law of the Sea (UNCLOS) and has been given the deadline of no later than 2015 in the Johannesburg Plan of Implementation of 2002. . . . But even if fishing were halted in 2010, 22% of the stocks are so depleted that they cannot be rebuilt by 2015. If current trends continue, Europe will miss the 2015 deadline by more than 30 years." They note that just passing laws is not enough.

David Helvarg, in "The Last Fish," *Earth Island Journal* (Spring 2003), concludes that about half of America's commercial seafood species are now overfished. Globally, the figure is still over 70 percent. And the North Atlantic contains only a third as much biomass of commercially valuable fish as it did in the 1950s. He recommends more buying out of excess fishing capacity, limiting the number of people allowed to enter the fishing industry, creating marine reserves, and perhaps most importantly, taking fisheries management out of the hands of people with a vested interest in the status quo.

In June 2000, the independent Pew Oceans Commission undertook the first national review of ocean policies in more than 30 years. Its report, "America's Living Oceans: Charting a Course for Sea Change" (Pew Oceans Commission, 2003) (http://www.pewoceans.org/), noted that many commercially fished species are in decline (North Atlantic cod, haddock, and yellowtail flounder reached historic lows in 1989). The reasons include intense fishing pressure to feed demand for seafood, as well as pollution, coastal development, fishing practices such as bottom dragging that destroy habitat, and fragmented ocean policy that makes it difficult to prevent or control the damage. The answers, the Commission suggests, must include such things as the no-fishing zones known as marine protected areas or reserves, which have been shown to restore habitat and fish populations, with clear benefits for commercial fishing. See also Julia Whitty, "The Fate of the Ocean," *Mother Jones* (March/April 2006); Michelle Allsopp et al., *Oceans in Peril: Protecting Marine Biodiversity* (Worldwatch, 2007); and "The Global Fish Crisis," a special report in *National Geographic* (April 2007). There is also a pressing need to rein in illegal fishing; see Stefan Flothmann et al., "Closing Loopholes: Getting Illegal Fishing Under Control," *Science* (June 4, 2010).

When Boris Worm et al., "Impacts of Biodiversity Loss on Ocean Ecosystem Services," *Science* (November 3, 2006), argued that human activities, including overfishing, so threaten marine biodiversity that before the mid-twenty-first century populations of all those ocean fish currently sought will be so reduced that commercial fishing will have ended, critics were vocal. After some debate, according to Eric Stokstad, "Détente in the Fisheries War," *Science* (April 10, 2009), Worm and one major critic Ray Hilborn of the University of Washington in Seattle teamed up with other researchers and graduate students to study larger data sets in search of a better vision of the future. One recent report from Worm's group (Camilo Mora et al., "Management Effectiveness of the World's Marine Fisheries," *Public Library of Science Biology* (June 2009) (http://www.plosbiology.org/article/info%3Adoi%2F10.1371%2Fjournal.pbio.1000131), finds that demand for seafood is rising far beyond what can be met sustainably and only a handful of countries have "a robust scientific basis for management recommendations." Boris Worm, Ray Hilborn et al., "Rebuilding Global Fisheries," *Science* (July 31, 2009), examine efforts to restore damaged marine ecosystems and fisheries and find that although there is some success, "63% of assessed fish stocks worldwide still require rebuilding, and even lower exploitation rates are needed to reverse the collapse of vulnerable species." Global fisheries remain in danger. Christopher L. Delgado et al., "Outlook for Fish to 2020" (International Food Policy Research Institute, 2003), note the prospect that the demand for fish will outpace the available supply, forcing the continuing expansion of aquaculture. Kjellrun Hiis Hauge, Belinda Cleeland, and Douglas Clyde Wilson, "Fisheries Depletion and Collapse" (International Risk Governance Council, 2009), note that overfishing is only one cause of fisheries depletion and collapse.

In the following selections, Carl Safina argues that despite an abundance of bad news about the state of the oceans and commercial fisheries, there are some signs that conservation and even restoration of fish stocks to a sustainable state are possible. The Food and Agriculture Organization of the United Nations argues that the proportion of marine fish stocks that are overexploited has increased tremendously since the 1970s. Despite some progress, there remains "cause for concern." The continuing need for fish as food means there will be continued growth in aquaculture.

YES

Carl Safina

A Future for U.S. Fisheries

For the fishing industry in the United States, and for the fishery resources on which the industry depends, there is good news and bad news. Bad news still predominates, as many commercial fishers and their communities have suffered severe financial distress and many fish stocks have declined considerably in numbers. Poor management by the National Marine Fisheries Service (NMFS), which regulates the fishing industry, and some poor choices by many fishers have contributed to the problems. But there are some bright spots, small and scattered, that suggest that improvements are possible.

Starting with the bad news, the federal government's fisheries management remains primitive, simplistic, and, in important cases, ineffectual, despite a fund of knowledge and conceptual tools that could be applied. In many regions—New England and the Pacific Northwest, among others—failed management costs more than the receipts from fisheries. This does not suggest that management should be given up as a lost cause, leaving the industry in a free-for-all, although this strategy might, in fact, be cheaper and not much less effective.

As a key problem, most management efforts today are based primarily on catch quotas that regulate how much fishers can harvest of a particular species in some set period, perhaps a season or a year. The problem is that quotas are set according to estimates of how much of the resource can be taken out of the ocean, rather than on how much should be left in. This may sound like two sides of the same coin, but in practice the emphasis on extraction creates a continual bias on the part of fisheries agencies and unrealistic short-term expectations among fishers. For example, a basic tenet of these approaches is that a virgin fish population should be reduced by about two-thirds to make it more "productive." But this notion is belied in the real world, where it has been proven that larger breeding populations are more productive.

The failure of this approach is readily apparent. The Sustainable Fisheries Act of 1996, reaffirmed by Congress in 2006, states that fish populations may not be fished down below about one-third of their estimated virgin biomass. It also states that in cases where fish stocks already have been pushed below that level, they must be restored (in most cases) to that level within a decade. On paper, this act looked good. (Full disclosure: I drafted the quantitative overfishing and recovery goals and triggers mandated by the act.) Unfortunately, the NMFS wrote implementing regulations interpreting the mandates as meaning that overfishing could continue for some time before rebuilding was required.

Reprinted with permission from *Issues in Science and Technology,* by Carl Safina, Summer 2009, pp. 43–46. Copyright © 2009 by the University of Texas at Dallas, Richardson, TX.

This too-liberal interpretation blurred the concept and delayed benefits. In its worst cases, it acknowledged that fish populations must be rebuilt in a decade but said that overfishing could continue in the meantime.

Clearly, the nation needs to take a different approach, based solidly on science. As a foundation, regulatory and management agencies must move from basing their actions on "how much can we take?" to concentrating on "how much must we leave?" The goal must be keeping target fish populations and associated living communities functioning, with all components being highly productive and resilient.

The nation must confront another reality as well. So many fisheries are so depleted that the only way to restore them will be to change the basic posture of regulations and management programs to one of recovery. Most fish populations could recover within a decade, even with some commercial fishing. But continuing to bump along at today's depleted levels robs fishing families and communities of income and risks resource collapse.

Ingredients for Success

Moving to a new era of fisheries management will require revising some conventional tools that are functioning below par and adopting an array of new "smart tools." Regulations that set time frames for overfishing and recovery can play a valuable role, if properly interpreted. For example, traditional catch quotas must be based firmly on scientific knowledge about fish stocks, and they must be enforced with an eye toward protecting the resource. Newer tools, adapted to specific environments and needs, would include:

Tradable Catch Shares

In this approach, now being used in some regions in varying degrees, fishery managers allot to fishers specific shares of the total allowable catch and give them the flexibility and the accountability for reaching their shares. Thus, fishers do not own the fish; rather, they own a percentage of the total allowed catch, which may fluctuate from year to year if management agencies adjust it up or down.

In expanding the use of such programs, managers must establish the shares based on the advice of independent scientists who are insulated from industry lobbying. Managers also should allot shares only to working fishers, not to corporations or processors. Of course, finding equitable ways of determining which fishers get catch shares will be critical. Methods of allocating shares may vary from location to location, but the key is ensuring an open process that accounts for fishers' legitimate interests and maintains conservation incentives. In many cases, fewer fishers will be eligible to keep fishing. But those not selected would likely have been forced out of business anyway by the combination of pressure from more successful fishers and reduced fish stocks.

By significantly reducing competition that breeds a race for fish, this approach offers several benefits. For one, it makes for safer fishing. Fishers who own shares know that they have the whole season to fill their quota regardless of what other boats are catching, so they are less likely to feel forced to head out in dangerous weather. In addition, owning a share helps ensure (other factors permitting) that a fisher can earn a decent living, so local, state, or regional politicians will feel less pressure to protect their fishing constituents and push for higher catch quotas. At the same time, marginal operators granted shares would no longer feel trapped, because they would have something to sell if they wished to exit the fishery. By promoting longer-term thinking among fishers and politicians alike, catch-share programs help foster a sense of future investment in which quota holders will benefit from high or recovered fish populations.

The impact of tradable catch shares can be seen in experiences in several regions. In Alaska, where fisheries managers once kept a tight cap on the halibut catch, the fishing season shrank to two days annually because there were so many competing boats. After managers introduced tradable catch shares, the number of boats fell precipitously and the season effectively expanded to whenever the fishers wanted to work toward filling their shares. Safety improved markedly, and the halibut population remained robust. In New England, where the industry resisted tradable shares, the story ended differently. Managers allotted individual fishers a shrinking number of days at sea, which progressively crippled their economic viability, gave them no option to exit the fishery short of foreclosure, and kept fishing pressure so high that the fish stocks never recovered.

Area-Based Fisheries

Although this concept may be relatively new in Western fisheries management, it has underpinned the management of fishing in Pacific islands for millennia. In practice, this approach is most applicable where fish populations spawn in localized areas and do not migrate far from their spawning area. For example, consider the case of clams, which spawn in limited areas and never move far away. In many regions, clamming is regulated on a township-by-township basis. Thus, conserving clams off one port will benefit that port, even if (especially if) the next port eliminates its own clam beds. This model holds promise for greater use with various fish species as well. In New England waters, cod once spawned in many local populations, many of which are now extinct. Overall regional quotas and regional mobility of boats contributed to their extinction. Had managers established local area-based restrictions, these populations might well have been saved, to the benefit of local communities.

In implementing area-based fisheries, managers will need to move deliberately, being mindful of what is scientifically supported and careful not to unduly raise people's expectations. If managers move too hastily, the restrictions may meet a lot of social skepticism and may not work as well as advertised, setting back not only the health of the fish stocks but also the credibility of the managers and scientists who support such actions.

Closed Areas

In recent years, fisheries managers have decided that some stocks are so threatened that the only choice is to close all or part of their habitat to fishing. Such efforts are to be applauded, although they have been too few and too limited in scale to achieve major success. Still, the lessons are instructive, as closures have been found to result in increases in fish populations, in the size of individual fish, and in greater diversity of species.

On Georges Bank in the north Atlantic, for example, success has been mixed, but tantalizing. Managers closed some of the grounds in an effort to protect northern cod, in particular, whose stocks had become severely depleted. So far, cod stocks have not rebounded, for a suite of reasons. But populations of several other important species, notably haddock and sea scallops, have mushroomed. These recovered populations have yielded significant financial benefits to the region, although in the case of sea scallops, fishing interests successfully lobbied to be allowed back into the closed areas, hampering full recovery of the resource.

Mixed Zoning

In many resource-based industries, even competing interests often agree on one thing: They do not want an area closed to them. Yet regarding fishing, conservationists too often have insisted that protected areas be closed to all extraction, and their single-minded pursuit of all-or-nothing solutions has made it easy for commercial interests to unite in demanding that the answer be nothing. A more nuanced approach is needed.

A comprehensive zoning program should designate a mix of areas, including areas that are entirely open to any kind of fishing at any time, areas that are closed to fishers using mobile gear, areas that are closed to fishers using gear that drags along the seafloor, areas that are closed in some seasons, and areas that are fully protected no-take zones. Such integrated zoning would better protect sensitive seafloor habitats and aquatic nursery areas from the kinds of activities that hurt those areas, while allowing harmless activities to proceed. For instance, tuna fishing could be banned in tuna breeding or nursery areas, yet allowed in ocean canyons, even those with deep coral and other important sedentary bottom communities. This type of zoning would also be most likely to gain the support of competing interests, as each party would get something it wants.

Reduction of Incidental Catch

Almost all methods of commercial fishing catch undersized or unmarketable individuals of the target species. Few of these can be returned alive. Fortunately, a number of simple changes in fishing methods and gear, such as the use of nets with larger mesh size, have been developed that can reduce incidental kill by more than 90%, and the government should adopt regulations that require use of these cleaner techniques. In some cases, however, it may be appropriate to require fishers to keep all fish caught—no matter their size, appearance, or even species—in order to reduce the waste that otherwise would result.

Commercial fishers also often catch creatures other than fish, with fatal results. For some creatures, such as sea turtles, capture may endanger their species' very survival. Here, too, advances in fishing technology are helping, but regulators must pay increased attention to finding ways to reduce this problem.

Protection Based on Size

Managers may be able to protect some fish stocks by setting regulations based on graduated fish sizes. This approach, taken almost by default, has led to a spectacular recovery of striped bass along the Atlantic coast. At one time, this population had become deeply depleted, and reproduction rates had fallen precipitously. But one year, environmental conditions arose that favored the survival of eggs and larvae and led to a slight bump in the number of young fish. After much rancor and debate, federal fisheries managers forced states to cooperate in shepherding this class of juveniles to adulthood. They did this primarily by placing a continually increasing limit on the minimum size of fish that fishers could keep. Over the course of more than a decade, the limits protected the fish as they grew and, ultimately, began reproducing. The limits also protected fish hatched in subsequent years, and they, too, grew into adulthood. This simple approach—protecting fish until they have had a chance to reproduce—did more to recover a highly valued, highly sought species than all of the complex calculations, models, and confused politics of previous management efforts.

Subsidy Reform

The federal government provides various segments of the fishing industry with major subsidies that have resulted in a number of adverse consequences. Improperly designed and sized subsidies have propped up bloated and over-capitalized fisheries that have systematically removed too many fish from the seas. Of course, some subsidies will remain necessary. But in most cases, subsidy amounts should be reduced. Also, many subsidies should be redirected to support efforts to develop cleaner technologies and to ease the social pain that fishers and their communities might face in adopting the improved technologies.

Ecologically Integrated Management

Perhaps the worst mistake of traditional fisheries management is that it consider each species in isolation. For example, simply focusing on how much herring fishers can take from the ocean without crashing herring stocks does not address the question of how much herring must be left to avoid crashing the tuna, striped bass, and humpback whales that feed on herring. Management regulations must be revised to reflect such broader food-web considerations.

Sustainable Aquaculture

During the past quarter-century, many nations have turned increasingly to aquaculture to supplement or even replace conventional commercial fishing. Although not at the head of this effort, the United States offers various forms of assistance and incentives to aid the development of the industry. But fish farming is not a panacea. Some operations raise unsustainable monocultures of fish, shrimp, and other aquatic species. Some destroy natural habitats such as marshes that are vital to wild fish. Some transfer pathogens to wild populations. Some pollute natural waters with food, feces, or pesticides necessary to control disease in overcrowded ponds and pens.

As the nation expands fish farming, doing it right should trump doing it fast. Generally, aquaculture will be most successful if it concentrates on raising smaller species and those lower on the food chain. Fish are not cabbages; they do not grow on sunlight. They have to be fed something, and what most fish eat is other fish. Just as the nation's ranchers raise cows and not lions, fish farmers should raise species such as clams, oysters, herring, tilapia, and other vegetarian fish, but not tuna. Farming large carnivores would take more food out of the ocean to feed them than the farming operation would produce. The result would be a loss of food for people, a loss of fish to other fisheries, and a loss to the ocean. Done poorly, aquaculture is as much of a ticking time bomb as were overcapitalized fisheries.

Working Together

Given the magnitude of the problems facing the nation's commercial fishers and fisheries, the various stakeholders must draw together. Although some recent experiences may suggest otherwise, fishers and scientists need each other in order to succeed. Fishers might lack the training to understand the scientific techniques, especially data analysis, that underpin improved management tools, and scientists might lack the experience required to understand the valid concerns and observations of fishers. But without more trust and understanding, adversarial postures that undermine wise management will continue to waste precious time as resources continue to deteriorate and communities and economies suffer. This need not be the case.

Similarly, fishers, fishery managers, and scientists should work together to better inform the public about the conditions and needs of the nation's fishing industry and fish stocks. Consider the example of marine zoning. The less people understand about fishing, the more they insist that closed, no-take marine reserves are the answer. Similarly, the less people understand about conservation, the more they insist that traditional methods of fisheries management, which typically ignore the need for reserves, are adequate tools for protecting fish stocks. As in many other areas, knowledge breeds understanding—and very often solutions.

World Review of Fisheries and Aquaculture

Fisheries Resources: Trends in Production, Utilization and Trade

Overview

Capture fisheries and aquaculture supplied the world with about 142 million tonnes of fish in 2008. . . . Of this, 115 million tonnes was used as human food, providing an estimated apparent per capita supply of about 17 kg (live weight equivalent), which is an all-time high. Aquaculture accounted for 46 percent of total food fish supply, a slightly lower proportion than reported in *The State of World Fisheries and Aquaculture 2008* owing to a major downward revision of aquaculture and capture fishery production statistics by China (see below), but representing a continuing increase from 43 percent in 2006. Outside China, per capita supply has remained fairly static in recent years as growth in supply from aquaculture has offset a small decline in capture fishery production and a rising population. In 2008, per capita food fish supply was estimated at 13.7 kg if data for China are excluded. In 2007, fish accounted for 15.7 percent of the global population's intake of animal protein and 6.1 percent of all protein consumed. Globally, fish provides more than 1.5 billion people with almost 20 percent of their average per capita intake of animal protein, and 3.0 billion people with at least 15 percent of such protein. In 2007, the average annual per capita apparent fish supply in developing countries was 15.1 kg, and 14.4 kg in low-income food-deficit countries (LIFDCs). In LIFDCs, which have a relatively low consumption of animal protein, the contribution of fish to total animal protein intake was significant—at 20.1 percent—and is probably higher than that indicated by official statistics in view of the under-recorded contribution of small-scale and subsistence fisheries.

China remains by far the largest fish-producing country, with production of 47.5 million tonnes in 2008 (32.7 and 14.8 million tonnes from aquaculture and capture fisheries, respectively). These figures were derived using a revised statistical methodology adopted by China in 2008 for all aquaculture and capture fishery production statistics and applied to statistics for 2006 onwards. The revision was based on the outcome of China's 2006 National Agricultural Census, which contained questions on fish production for the first time, as well as on results from various pilot sample surveys, most of which were conducted in collaboration with FAO. While revisions varied according to species, area and sector, the overall result was a downward correction of fishery and aquaculture production statistics for 2006 of about 13.5 percent. FAO subsequently estimated revisions for its historical statistics for China for 1997–2005. Notice of the impending revision by China had been given in *The State of World Fisheries and Aquaculture 2008*. Because of the major importance of China in the global context, China is in some cases discussed separately from the rest of the world in this publication.

Global capture fisheries production in 2008 was about 90 million tonnes, with an estimated first-sale value of US$93.9 billion, comprising about 80 million tonnes from marine waters and a record 10 million tonnes from inland waters. World capture fisheries production has been relatively stable in the past decade, with the exception of marked fluctuations driven by catches of anchoveta—a species extremely susceptible to oceanographic conditions determined by the El Niño Southern Oscillation—in the Southeast Pacific. Fluctuations in other species and regions tend to compensate for each other to a large extent. In 2008, China, Peru and Indonesia were the top producing countries. China remained by far the global leader with production of about 15 million tonnes.

Although the revision of China's fishery statistics reduced reported catches by about 2 million tonnes per year in the Northwest Pacific, this area still leads by far the ranking of marine fishing areas, followed by the Southeast Pacific, the Western Central Pacific and the Northeast Atlantic. The same species have dominated marine catches since 2003, with the top ten species accounting for about 30 percent of all marine catches. Catches from inland waters, two-thirds of which were reported as being taken in Asia in 2008, have shown a slowly but steadily rising trend since 1950, owing in part to stock enhancement practices and possibly also to some improvements in reporting, which still remains poor for inland water fisheries (with small-scale and subsistence fisheries substantially underrepresented in the statistics).

Aquaculture continues to be the fastest-growing animal-food-producing sector and to outpace population growth, with per capita supply from aquaculture increasing from 0.7 kg in 1970 to 7.8 kg in 2008, an average annual growth rate of 6.6 percent. It is set to overtake capture fisheries as a source of food fish. While aquaculture production (excluding aquatic plants) was less than 1 million tonnes per year in the early 1950s, production in 2008 was 52.5 million tonnes, with a value of US$98.4 billion. Aquatic plant production by aquaculture in 2008 was 15.8 million tonnes (live weight equivalent), with a value of US$7.4 billion, representing an average annual growth rate in terms of weight of almost 8 percent since 1970. Thus, if aquatic plants are included, total global aquaculture production in 2008 amounted to 68.3 million tonnes with a first-sale value of US$106 billion. World aquaculture is heavily dominated by the Asia–Pacific region, which accounts for 89 percent of production in terms of quantity and 79 percent in terms of value. This dominance is mainly because of China's enormous production, which accounts for 62 percent of global production in terms of quantity and 51 percent of global value.

Growth rates for aquaculture production are slowing, reflecting the impacts of a wide range of factors, and vary greatly among regions. Latin America and the Caribbean showed the highest average annual growth in the period 1970–2008 (21.1 percent), followed by the Near East (14.1 percent) and Africa (12.6 percent). China's aquaculture production increased at an average annual growth rate of 10.4 percent in the period 1970–2008, but in the new millennium it has declined to 5.4 percent, which is significantly lower than in the 1980s (17.3 percent) and 1990s (12.7 percent). The average annual growth in aquaculture production in Europe and North America since 2000 has also slowed substantially to 1.7 percent and 1.2 percent, respectively. The once-leading countries in aquaculture development such as France, Japan and Spain have shown falling production in the past decade. It is expected that, while world aquaculture production will continue to grow in the coming decade, the rate of increase in most regions will slow. . . .

The proportion of marine fish stocks estimated to be underexploited or moderately exploited declined from 40 percent in the mid-1970s to 15 percent in 2008, whereas the proportion of overexploited, depleted or recovering stocks increased from 10 percent in 1974 to 32 percent in 2008. The proportion of fully exploited stocks has remained relatively stable at about 50 percent since the 1970s. In 2008, 15 percent of the stock groups monitored by FAO were estimated to be underexploited (3 percent) or moderately exploited (12 percent) and able to produce more than their current catches. This is the lowest percentage recorded since the mid-1970s. Slightly more than half of the stocks (53 percent) were estimated to be fully exploited and, therefore, their current catches are at or close to their maximum sustainable productions, with no room for further expansion. The remaining 32 percent were estimated to be either overexploited (28 percent), depleted (3 percent) or recovering from depletion (1 percent) and, thus, yielding less than their maximum potential production owing to excess fishing pressure, with a need for rebuilding plans. This combined percentage is the highest in the time series. The increasing trend in the percentage of overexploited, depleted and recovering stocks and the decreasing trend in underexploited and moderately exploited stocks give cause for concern.

Most of the stocks of the top ten species, which account in total for about 30 percent of the world marine capture fisheries production in terms of quantity, are fully exploited. The two main stocks of anchoveta (*Engraulis ringens*) in the Southeast Pacific and those of Alaska pollock (*Theragra chalcogramma*) in the North Pacific and blue whiting (*Micromesistius poutassou*) in the Atlantic are fully exploited. Several Atlantic herring (*Clupea harengus*) stocks are fully exploited, but some are depleted. Japanese anchovy (*Engraulis japonicus*) in the Northwest Pacific and Chilean jack mackerel (*Trachurus murphyi*) in the Southeast Pacific are considered to be fully exploited. Some limited possibilities for expansion may exist for a few stocks of chub mackerel (*Scomber japonicus*), which are moderately exploited in the Eastern Pacific, while the stock in the Northwest Pacific was estimated to be recovering. In 2008, the largehead hairtail (*Trichiurus lepturus*) was estimated to be overexploited in the main fishing area in the Northwest Pacific. Of the 23 tuna stocks, most are more or less fully exploited (possibly up to 60 percent), some are overexploited or depleted (possibly up to 35 percent) and only a few appear to be underexploited (mainly skipjack). In the long term, because of the substantial demand for tuna and the significant overcapacity of tuna fishing fleets, the status of tuna stocks may deteriorate further if there is no improvement in their management. Concern about the poor status of some bluefin stocks and the difficulties in managing them led to a proposal to the Convention on International Trade in Endangered Species of Wild Fauna and Flora (CITES) in 2010 to ban the international trade of Atlantic bluefin. Although it was hardly in dispute that the stock status of this high-value food fish met the biological criteria for listing on CITES Appendix I, the proposal was ultimately rejected. Many parties that opposed the listing stated that in their view the International Commission for the Conservation of Atlantic Tunas (ICCAT) was the appropriate body for the management of such an important commercially exploited aquatic species. Despite continued reasons for concern in the overall situation, it is encouraging to note that good progress is being made in reducing exploitation rates and restoring overfished fish stocks and marine ecosystems through effective management actions in some areas such as off Australia, on the Newfoundland–Labrador Shelf, the Northeast United States Shelf, the Southern Australian Shelf, and in the California Current ecosystems.

Inland fisheries are a vital component in the livelihoods of people in many parts of the world, in both developing and developed countries. However, irresponsible fishing practices, habitat loss and degradation, water abstraction, drainage of wetlands, dam construction and pollution (including eutrophication) often act together, thus compounding one another's effects. They have caused substantial declines and other changes in inland fishery resources. Although these impacts are not always reflected by a discernable decrease in fishery production (especially when stocking is practised), the fishery may change in composition and value. The poor state of knowledge on inland fishery resources and their ecosystems has led to differing views on the actual status of many resources. One view maintains that the sector is in serious trouble because of the multiple uses of and threats to inland water ecosystems. The other view holds that the sector is in fact growing, that much of the production and growth has gone unreported and that stock enhancement through stocking and other means has played a significant role. Irrespective of these views, the role of inland fisheries in poverty alleviation and food security needs to be better reflected in development and fisheries policies and strategies. The tendency to undervalue inland fisheries in the past has resulted in inadequate representation in national and international agendas. . . .

The Outlook

In spite of the trend of gradually increasing inland catches, it is reported that the abundance of inland water species populations declined by 28 percent between 1970 and 2003. Action is required to secure conservation of aquatic ecosystems and safeguard the resources that form the basis for inland fisheries. A range of factors will directly or indirectly drive the development of the sector. However, there is the possibility to mitigate some negative impacts through technological advances, wealth creation and better management.

Drivers of Inland Fisheries

A General Scenario   For inland fisheries to have a future, there must be fish resources that can be exploited to satisfy people's needs for food, income and/ or recreation.

Those now engaged in inland fisheries have fundamentally different reasons to be involved. Commercial, full-time and part-time fishers pursue fisheries because they see the activity as one of their best possibilities to secure a livelihood for themselves and their families. Occasional and subsistence fishers go fishing for additional income or to add fish to their meals, and recreational fishers do so because it is for most of them a leisure-time occupation. However, the sector is highly dynamic with possibilities for people to enter or leave it or increase or decrease their participation in response to developments and available opportunities inside and outside fisheries.

The status of the fisheries resources depends to some extent on the number of fishers and how they are regulated. However, the threats coming from outside the fisheries sector are often more important and can lead to fishers being deprived of their resources and their livelihoods. General social and economic development is a major force influencing the drivers within and outside the fisheries sector, in both a positive and negative manner.

Need for More Food   According to the projections by the United Nations Population Division, the world population will increase from 6.8 billion today to 9 billion by 2050. As stated above, 65–90 percent of the inland capture fish production takes place in the developing and low-income food-deficit countries. The World Bank's forecast for 2020 suggests that 826 million people, or 12.8 percent, of developing country citizens will be living on US$1.25 a day or less and that there will be almost 2 billion poor people living at or below the US$2 a day poverty line. The growing population will need significant increases in food production at affordable prices.

More land (including wetlands) will be used, and some will be used more intensively, as agricultural food production expands during coming decades. This will result in increased use of agrochemicals with serious negative consequences for inland fisheries.

The demand for water for both irrigation and domestic purposes will continue to increase, leading to reduced water availability for fisheries, especially during the dry season. There will be attempts to transfer water between separate basins, with unpredictable consequences for biodiversity. There are also already plans to connect large rivers and transform them into shipping lanes linking distant cities, provinces and countries in areas with poorly developed rail and road infrastructure. There is expected to be increased demand for energy, including hydropower—leading to further damming of rivers.

The need for animal protein, including fish, will increase. Most marine fish stocks are already fully exploited. Notwithstanding increases in aquaculture production, fishing pressure will increase on inland fish stocks, and there will probably be a rise in unsustainable fishing methods, such as the use of explosives and poison, electrofishing and dry pumping of small natural waterbodies. These methods are all capable of killing large amounts of fish indiscriminately.

Aquaculture will continue to grow, and high-value species and products will increasingly come from farms rather than wild stocks. This may reduce capture fishing pressure. In developing countries, improvements in aquaculture technology will allow more fish to be sold more cheaply but, in some markets, cultured species will have problems competing with wild fish because of the need for feed based on fishmeal and fish oil. However, progress is being made on developing feed alternatives derived from locally available animal-waste products or using plant-based proteins instead of animal protein. Where water is available, culture-based and enhanced fisheries will become increasingly important in poor countries with rapidly growing populations because of the lower levels of investment and running costs, but they will require hatcheries to provide the seed. This development will tend to concentrate access to fishing among fewer groups, and the role of fishing as a safety net for the poorest of the poor is likely to be threatened. . . .

EXPLORING THE ISSUE

Does Commercial Fishing Have a Future?

Critical Thinking and Reflection

1. Oceanic fisheries offer an example of a "commons"—a resource "owned" by everyone. Describe at least one problem that follows from this situation.
2. Who should regulate oceanic fisheries?
3. How can one nation's regulators control what happens beyond that nation's own marine boundary (the 200-mile limit)?

Is There Common Ground?

There is a great deal of agreement that global fisheries are in trouble. There is also considerable agreement about the steps that must be taken to make those fisheries sustainable. The problem is the conflict between sustainability and current human needs. If human needs come before the needs of the environment, there may not be much change.

1. Would reducing human needs help achieve sustainability?
2. Is it possible to reduce human needs?
3. How could one possibly reduce human needs without increasing human suffering?

Online Resource

The Mid-Atlantic Fisheries Management Council

In the United States, eight regional Fisheries Management Councils manage fisheries in their region. Members of the councils are appointed by the Secretary of Commerce.

www.mafmc.org/

ISSUE 15

Can Organic Farming Feed the World?

YES: Ed Hamer and Mark Anslow, from "10 Reasons Why Organic Can Feed the World," *Ecologist* (March 2008)

NO: D. J. Connor, from "Organic Agriculture Cannot Feed the World," *Field Crops Research* (March 2008)

Learning Outcomes

After reading this issue, you should be able to:

- Explain how organic agriculture differs from conventional agriculture.
- Explain how organic agriculture benefits the environment.
- Discuss whether organic agriculture can supply future world populations with adequate food.
- Describe the problems that must be solved before agriculture of any kind can feed future world populations.

ISSUE SUMMARY

YES: Ed Hamer and Mark Anslow argue that organic agriculture can feed the world if people are willing to eat less meat. It would also use less energy and water, emit fewer greenhouse gases, provide better nutrition, protect ecosystems, and increase employment.

NO: D. J. Connor argues that a major report claiming that organic methods could produce enough food to sustain a global human population even larger than that of today has serious faults. At best organic methods could support a population less than half as large as today's (over 7 billion).

There was a time when all farming was organic. Fertilizer was compost and manure. Fields were periodically left fallow (unfarmed) to recover soil moisture and nutrients. Crops were rotated to prevent nutrient exhaustion. Pesticides were nonexistent. Farmers were at the mercy of periodic droughts (despite irrigation) and insect infestations.

As population grew, so did the demand for food. In Europe and America, the concomitant demand for fertilizer led in the nineteenth century to a booming trade in guano mined from Caribbean and Pacific islands where deposits of seabird dung could be a hundred and fifty foot thick. When the guano deposits were exhausted, there was an agricultural crisis that was relieved only by the invention of synthetic nitrogen-containing fertilizers early in the twentieth century. See Jimmy Skaggs, *The Great Guano Rush: Entrepreneurs and American Overseas Expansion* (St. Martin's, 1994), and G. J. Leigh, *The World's Greatest Fix: A History of Nitrogen and Agriculture* (Oxford University Press, 2004). Unfortunately, synthetic fertilizers do not maintain the soil's content of organic matter (humus). This deficit can be amended by tilling in sewage sludge, but the public is not usually very receptive to the idea, partly because of the "yuck factor," but also because sewage sludge may contain human pathogens and chemical contaminants.

Synthetic pesticides, beginning with DDT, came into use in the 1940s. Their history is nicely outlined in Keith S. Delaplane, *Pesticide Usage in the United States: History, Benefits, Risks, and Trends* (University of Georgia Extension Bulletin 1121, 1996) (http://entweb.clemson.edu/pesticid/program/SRPIAP/pestuse.pdf). When they turned out to have problems—target species quickly became resistant, and when the chemicals reached human beings and wildlife on food and in water, they proved to be toxic—some people sought alternatives. These alternatives to synthetic fertilizers and pesticides (among other things) comprise what is usually meant by "organic farming." Proponents of organic farming have called it holistic, biodynamic, ecological, and natural and claimed a number of advantages for its practice. They say it both preserves the health of the soil and provides healthier food for people. They also argue that it should be used more, even to the point of replacing chemical-based "industrial" agriculture. Because proponents of chemicals hold that fertilizers and pesticides are essential to produce food in the quantities that a world population of over 7 billion people requires, and to hold food prices down to affordable levels, one strand of debate has been over whether organic agriculture can do the job. Jules Pretty, "Can Ecological Agriculture Feed Nine Billion People?" *Monthly Review* (November 2009), puts the emphasis more on achieving sustainability and minimizing damage to the environment and notes that future agricultural systems are likely to include features of both organic and conventional agriculture. No technologies—such as genetic engineering—or practices can be ruled out by fiat.

Catherine Badgley et al., "Organic Agriculture and the Global Food Supply," *Renewable Agriculture & Food Systems* (June 2007), is among the latest to insist that organic methods could produce enough food to sustain a global human population even larger than that of today, and without needing more farmland. Organic agriculture would also decrease the undesirable environmental effects of conventional farming. It is worth noting that Alex Avery, author of *The Truth About Organic Foods* (Henderson, 2006) and director of research and education for the Hudson Institute's Center for Global Food

Issues, argues in "'Organic Abundance' Report: Fatally Flawed," Center for Global Food Issues, Hudson Institute (September 2007), that Badgley et al., are guilty of misreporting data, inflating averages by counting high organic yields multiple times, and counting as organic farming clearly nonorganic methods. John J. Miller, "The Organic Myth," *National Review* (February 9, 2004), argues that organic farming is not productive enough to feed today's population, much less larger future populations, it is prone to dangerous biological contamination, and it is not sustainable. "Wishful thinking is at the heart of the organic-food movement."

Quantitative yield comparisons are not the only organic-related topics that prompt debate. Is organic food better or safer for the consumer than nonorganic food? Faidon Magkos et al., "Organic Food: Buying More Safety or Just Peace of Mind? A Critical Review of the Literature," *Critical Reviews in Food Science & Nutrition* (January 2006), report that the quality and safety of organic food are largely a matter of perception. "Relevant scientific evidence . . . is scarce, while anecdotal reports abound." Pesticide and herbicide residues may be lower, but even on nonorganic food they are low. Environmental contaminants are likely to affect both organic and nonorganic foods. "'Organic' does not automatically equal 'safe.'"

Is organic farming better for the environment? Soil fertility is in decline in many parts of the world; see Alfred E. Herteminck, "Assessing Soil Fertility Decline in the Tropics Using Soil Chemical Data," *Advances in Agronomy* (2006). In Africa, the situation is extraordinarily serious. According to the International Center for Soil Fertility and Agricultural Development, "About 75 percent of the farmland in sub-Saharan Africa is plagued by severe degradation, losing basic soil nutrients needed to grow the crops that feed Africa, according to a new report . . . on the precipitous decline in African soil health from 1980 to 2004. Africa's crisis in food production and battle with hunger are largely rooted in this 'soil health crisis.'" Proponents of organic farming argue that organic methods are essential to relieving the crisis, but one study of changes in soil fertility, as indicated by crop yield, earthworm numbers, and soil properties, after converting from conventional to organic practices, found that different soils responded differently, with some improving and some not; see Anne Kjersti Bakken et al., "Soil Fertility in Three Cropping Systems after Conversion from Conventional to Organic Farming," *Acta Agriculturae Scandinavica: Section B, Soil & Plant Science* (June 2006). Richard Wood et al., "A Comparative Study of Some Environmental Impacts of Conventional and Organic Farming in Australia," *Agricultural Systems* (September 2006), find in a comparison of organic and conventional farms "that direct energy use, energy related emissions, and greenhouse gas emissions are higher for the former."

Paul Collier, "The Politics of Hunger," *Foreign Affairs* (November–December 2008), says that part of the problem is "the middle- and upper-class love affair with peasant [organic] agriculture." What the world needs is more commercial agriculture and more science, as well as an end to subsidies for biofuels production. According to Catherine M. Cooney, "Sustainable Agriculture Delivers the Crops," *Environmental Science & Technology* (February 15, 2006), "Sustainable agriculture, such as crop rotation, organic farming, and genetically modified seeds, increased crop yields by an average of 79 percent" while also improving the lives of farmers in developing countries. Cong Tu et al., "Responses of Soil Microbial Biomass and N Availability to Transition Strategies from Conventional to Organic Farming Systems," *Agriculture, Ecosystems & Environment* (April, 2006), note that a serious barrier to changing from conventional to organic farming, despite soil improvements, is an initial reduction in yield and increase in pests. David Pimentel et al., "Environmental, Energetic, and Economic Comparisons of Organic and Conventional Farming Systems," *Bioscience* (July 2005), find that organic farming uses less energy, improves soil, and has yields comparable to those of conventional farming, but crops probably cannot be grown as often (because of fallowing), which thus reduces long-term yields. However, "because organic foods frequently bring higher prices in the marketplace, the net economic return per [hectare] is often equal to or higher than that of conventionally produced crops." This is one of the factors that prompts Craig J. Pearson to argue in favor of shifting "from conventional open or leaky systems to more closed, regenerative systems" in "Regenerative, Semiclosed Systems: A Priority for Twenty-First-Century Agriculture," *Bioscience* (May 2007).

Better economic return means that organic farming is currently good for the organic farmer. However, the advantage would disappear if the world converted to organic farming. Initial declines in yield mean the conversion would be difficult, but the longer we wait and the more population grows, the more difficult it will be. Whether the conversion is essential depends on the availability of alternative solutions to the problem, and it is worth noting that high energy prices make chemical fertilizers increasingly expensive. Sadly, Stacey Irwin, "Battle High Fertilizer Costs," *Farm Industry News* (January 2006), does not mention the possibility of using organic methods.

In the following selections, Ed Hamer and Mark Anslow argue that organic agriculture can feed the world if people are willing to eat less meat. It would also use less energy and water, emit fewer greenhouse gases, provide better nutrition, protect ecosystems, and increase employment. D. J. Connor argues that a major report (Badgely, et al.) claiming that organic methods could produce enough food to sustain a global human population even larger than that of today has serious faults. At best, organic methods could support a population less than half as large as today's (over 7 billion).

YES

Ed Hamer and Mark Anslow

10 Reasons Why Organic Can Feed the World

1. Yield

Switching to organic farming would have different effects according to where in the world you live and how you currently farm.

Studies show that the less-industrialised world stands to benefit the most. In southern Brazil, maize and wheat yields doubled on farms that changed to green manures and nitrogen-fixing leguminous vegetables instead of chemical fertilisers. In Mexico, coffee-growers who chose to move to fully organic production methods saw increases of 50 percent in the weight of beans they harvested. In fact, in an analysis of more than 286 organic conversions in 57 countries, the average yield increase was found to be an impressive 64 percent.

The situation is more complex in the industrialised world, where farms are large, intensive facilities, and opinions are divided on how organic yields would compare.

Research by the University of Essex in 1999 found that, although yields on US farms that converted to organic initially dropped by between 10 and 15 percent, they soon recovered, and the farms became more productive than their all-chemical counterparts. In the UK, however, a study by the Elm Farm Research Centre predicted that a national transition to all-organic farming would see cereal, rapeseed and sugar beet yields fall by between 30 and 60 percent. Even the Soil Association admits that, on average in the UK, organic yields are 30 percent lower than non-organic.

So can we hope to feed ourselves organically in the British Isles and Northern Europe? An analysis by former *Ecologist* editor Simon Fairlie in *The Land journal* suggests that we can, but only if we are prepared to rethink our diet and farming practices. In Fairlie's scenario, each of the UK's 60 million citizens could have organic cereals, potatoes, sugar, vegetables and fruit, fish, pork, chicken and beef, as well as wool and flax for clothes and biomass crops for heating. To achieve this we'd each have to cut down to around 230 g of beef (½ lb), compared to an average of 630 g (1½ lb) today, 252 g of pork/bacon, 210 g of chicken and just under 4 kg (9 lb) of dairy produce each week—considerably more than the country enjoyed in 1945. We would probably need to supplement our diet with homegrown vegetables, save our food scraps as livestock feed and reform the sewage system to use our waste as an organic fertiliser.

2. Energy

Currently, we use around 10 calories of fossil energy to produce one calorie of food energy. In a fuel-scarce future, which experts think could arrive as early as 2012, such numbers simply won't stack up.

Studies by the Department for Environment, Food and Rural affairs over the past three years have shown that, on average, organically grown crops use 25 percent less energy than their chemical cousins. Certain crops achieve even better reductions, including organic leeks (58 percent less energy) and broccoli (49 percent less energy).

When these savings are combined with stringent energy conservation and local distribution and consumption (such as organic box schemes), energy-use dwindles to a fraction of that needed for an intensive, centralised food system. A study by the University of Surrey shows that food from Tolhurst Organic Produce, a smallholding in Berkshire, which supplies 400 households with vegetable boxes, uses 90 percent less energy than if non-organic produce had been delivered and bought in a supermarket.

Far from being simply 'energy-lite', however, organic farms have the potential to become self-sufficient in energy—or even to become energy exporters. The 'Dream Farm' model, first proposed by Mauritius-born agro-scientist George Chan, sees farms feeding manure and waste from livestock and crops into biodigesters, which convert it into a methane-rich gas to be used for creating heat and electricity. The residue from these biodigesters is a crumbly, nutrient-rich fertiliser, which can be spread on soil to increase crop yields or further digested by algae and used as a fish or animal feed.

3. Greenhouse Gas Emissions and Climate Change

Despite organic farming's low-energy methods, it is not in reducing demand for power that the techniques stand to make the biggest savings in greenhouse gas emissions.

The production of ammonium nitrate fertiliser, which is indispensable to conventional farming, produces vast quantities of nitrous oxide—a greenhouse gas with a global warming potential some 320 times greater than that of CO_2. In fact, the production of one tonne of ammonium nitrate creates 6.7 tonnes of greenhouse gases (CO_2e), and was responsible for around 10 percent of all industrial greenhouse gas emissions in Europe in 2003.

The techniques used in organic agriculture to enhance soil fertility in turn encourage crops to develop deeper roots, which increase the amount of organic matter in the soil, locking up carbon underground and keeping it out of the atmosphere. The opposite happens in conventional farming: high quantities of artificially supplied nutrients encourage quick growth and shallow roots. A study published in 1995 in the journal *Ecological Applications* found that levels of carbon in the soils of organic farms in California were as much as 28 percent higher as a result. And research by the Rodale Institute shows that if the US were to convert all its corn and soybean fields to organic methods, the amount of carbon that could be stored in the soil would equal 73 percent of the country's Kyoto targets for CO_2 reduction.

Organic farming might also go some way towards salvaging the reputation of the cow, demonised in 2007 as a major source of methane at both ends of its digestive tract. There's no doubt that this is a problem: estimates put global methane emissions from ruminant livestock at around 80 million tonnes a year, equivalent to around two billion tonnes of CO_2, or close to the annual CO_2 output of Russia and the UK combined. But by changing the pasturage on which animals graze to legumes such as clover or birdsfoot trefoil (often grown anyway by organic farmers to improve soil nitrogen content), scientists at the Institute of Grassland and Environmental Research believe that methane emissions could be cut dramatically. Because the leguminous foliage is more digestible, bacteria in the cow's gut are less able to turn the fodder into methane. Cows also seem naturally to prefer eating birdsfoot trefoil to ordinary grass.

4. Water Use

Agriculture is officially the most thirsty industry on the planet, consuming a staggering 72 percent of all global freshwater at a time when the UN says 80 percent of our water supplies are being overexploited.

This hasn't always been the case. Traditionally, agricultural crops were restricted to those areas best suited to their physiology, with drought-tolerant species grown in the tropics and water-demanding crops in temperate regions. Global trade throughout the second half of the last century led to a worldwide production of grains dominated by a handful of high- yielding cereal crops, notably wheat, maize and rice. These thirsty cereals—the 'big three'—now account for more than half of the world's plant-based calories and 85 percent of total grain production.

Organic agriculture is different. Due to its emphasis on healthy soil structure, organic farming avoids many of the problems associated with compaction, erosion, salinisation and soil degradation, which are prevalent in intensive systems. Organic manures and green mulches are applied even before the crop is sown, leading to a process known as 'mineralisation'—literally the fixing of minerals in the soil. Mineralised organic matter, conspicuously absent from synthetic fertilisers, is one of the essential ingredients required physically and chemically to hold water on the land.

Organic management also uses crop rotations, undersowing and mixed cropping to provide the soil with near-continuous cover. By contrast, conventional farm soils may be left uncovered for extended periods prior to sowing, and again following the harvest, leaving essential organic matter fully exposed to erosion by rain, wind and sunlight.

In the US, a 25-year Rodale Institute experiment on climatic extremes found that, due to improved soil structure, organic systems consistently achieve higher yields during periods both of drought and flooding.

5. Localisation

The globalisation of our food supply, which gives us Peruvian apples in June and Spanish lettuces in February, has seen our food reduced to a commodity in an increasingly volatile global marketplace. Although year-round availability makes for good marketing in the eyes of the biggest retailers, the costs to the environment are immense.

Friends of the Earth estimates that the average meal in the UK travels 1,000 miles from plot to plate. In 2005, Defra released a comprehensive report on food miles in the UK, which valued the direct environmental, social and economic costs of food transport in Britain at £9 billion each year. In addition, food transport accounted for more than 30 billion vehicle kilometres, 25 percent of all HGV journeys and 19 million tonnes of carbon dioxide emissions in 2002 alone.

The organic movement was born out of a commitment to provide local food for local people, and so it is logical that organic marketing encourages localisation through veg boxes, farm shops and stalls. Between 2005 and 2006, organic sales made through direct marketing outlets such as these increased by 53 percent, from £95 to £146 million, more than double the sales growth experienced by the major supermarkets. As we enter an age of unprecedented food insecurity, it is essential that our consumption reflects not only what is desirable, but also what is ultimately sustainable. While the 'organic' label itself may inevitably be hijacked, 'organic and local' represents a solution with which the global players can simply never compete.

6. Pesticides

It is a shocking testimony to the power of the agrochemical industry that in the 45 years since Rachel Carson published her pesticide warning *Silent Spring,* the number of commercially available synthetic pesticides has risen from 22 to more than 450.

According to the World Health Organization there are an estimated 20,000 accidental deaths worldwide each year from pesticide exposure and poisoning. More than 31 million kilograms of pesticide were applied to UK crops alone in 2005, 0.5 kilograms for every person in the country. A spiralling dependence on pesticides throughout recent decades has resulted in a catalogue of repercussions, including pest resistance, disease susceptibility, loss of natural biological controls and reduced nutrient-cycling.

Organic farmers, on the other hand, believe that a healthy plant grown in a healthy soil will ultimately be more resistant to pest damage. Organic systems encourage a variety of natural methods to enhance soil and plant health, in turn reducing incidences of pests, weeds and disease.

First and foremost, because organic plants grow comparatively slower than conventional varieties they have thicker cell walls, which provide a tougher natural barrier to pests. Rotations or 'break-crops', which are central to organic production, also provide a physical obstacle to pest and disease lifecycles by removing crops from a given plot for extended periods. Organic systems also rely heavily on a rich agro-ecosystem in which many agricultural pests can be controlled by their natural predators.

Inevitably, however, there are times when pestilence attacks are especially prolonged or virulent, and here permitted pesticides may be used. The use of organic pesticides is heavily regulated and the International Federation of Organic Agriculture Movements (IFOAM) requires specific criteria to be met before pesticide applications can be justified.

There are in fact only four active ingredients permitted for use on organic crops: copper fungicides, restricted largely to potatoes and occasionally orchards; sulphur, used to control additional elements of fungal diseases; Retenone, a naturally occurring plant extract, and soft soap, derived from potassium soap and used to control aphids. Herbicides are entirely prohibited.

7. Ecosystem Impact

Farmland accounts for 70 percent of UK land mass, making it the single most influential enterprise affecting our wildlife. Incentives offered for intensification under the Common Agricultural Policy are largely responsible for negative ecosystem impacts over recent years. Since 1962, farmland bird numbers have declined by an average of 30 percent. During the same period more than 192,000 kilometres of hedgerows have been removed, while 45 percent of our ancient woodland has been converted to cropland.

By contrast, organic farms actively encourage biodiversity in order to maintain soil fertility and aid natural pest control. Mixed farming systems ensure that a diversity of food and nesting sites are available throughout the year, compared with conventional farms where autumn sow crops leave little winter vegetation available.

Organic production systems are designed to respect the balance observed in our natural ecosystems. It is widely accepted that controlling or suppressing one element of wildlife, even if it is a pest, will have unpredictable impacts on the rest of the food chain. Instead, organic producers regard a healthy ecosystem as essential to a healthy farm, rather than a barrier to production.

In 2005, a report by English Nature and the RSPB on the impacts of organic farming on biodiversity reviewed more than 70 independent studies of flora, invertebrates, birds and mammals within organic and conventional farming systems. It concluded that biodiversity is enhanced at every level of the food chain under organic management practices, from soil microbiota right through to farmland birds and the largest mammals.

8. Nutritional Benefits

While an all-organic farming system might mean we'd have to make do with slightly less food than we're used to, research shows that we can rest assured it would be better for us.

In 2001, a study in the *Journal of Complementary Medicine* found that organic crops contained higher levels of 21 essential nutrients than their conventionally grown counterparts, including iron, magnesium, phosphorus and vitamin C. The organic crops also contained lower levels of nitrates, which can be toxic to the body.

Other studies have found significantly higher levels of vitamins—as well as polyphenols and antioxidants—in organic fruit and veg, all of which are thought to play a role in cancer-prevention within the body.

Scientists have also been able to work out why organic farming produces more nutritious food. Avoiding chemical fertiliser reduces nitrates levels in the food; better-quality soil increases the availability of trace minerals, and reduced levels of pesticides mean that the plants' own immune systems grow stronger, producing higher levels of antioxidants. Slower rates of growth also mean that organic food frequently contains higher levels of dry mass, meaning that fruit and vegetables are less pumped up with water and so contain more nutrients by weight than intensively grown crops do.

Milk from organically fed cows has been found to contain higher levels of nutrients in six separate studies, including omega-3 fatty acids, vitamin E, and beta-carotene, all of which can help prevent cancer. One experiment discovered that levels of omega-3 in organic milk were on average 68 percent higher than in non-organic alternatives.

But as well as giving us more of what we do need, organic food can help to give us less of what we don't. In 2000, the UN Food and Agriculture Organization (FAO) found that organically produced food had 'lower levels of pesticide and veterinary drug residues' than non-organic did. Although organic farmers are allowed to use antibiotics when absolutely necessary to treat disease, the routine use of the drugs in animal feed—common on intensive livestock farms—is forbidden. This means a shift to organic livestock farming could help tackle problems such as the emergence of antibiotic-resistant bacteria.

9. Seed-Saving

Seeds are not simply a source of food; they are living testimony to more than 10,000 years of agricultural domestication. Tragically, however, they are a resource that has suffered unprecedented neglect. The UN FAO estimates that 75 percent of the genetic diversity of agricultural crops has been lost over the past 100 years.

Traditionally, farming communities have saved seeds year-on-year, both in order to save costs and to trade with their neighbours. As a result, seed varieties evolved in response to local climatic and seasonal conditions, leading to a wide variety of fruiting times, seed size, appearance and flavour. More importantly, this meant a constant updating process for the seed's genetic resistance to changing climatic conditions, new pests and diseases.

By contrast, modern intensive agriculture depends on relatively few crops—only about 150 species are cultivated on any significant scale worldwide. This is the inheritance of the Green Revolution, which in the late 1950s perfected varieties Filial 1, or F1 seed technology, which produced hybrid seeds with specifically desirable genetic qualities. These new high-yield seeds were widely adopted, but because the genetic makeup of hybrid F1 seeds becomes diluted following the first harvest, the manufacturers ensured that farmers return for more seed year-on-year.

With its emphasis on diversity, organic farming is somewhat cushioned from exploitation on this scale, but even Syngenta, the world's third-largest biotech company, now offers organic seed lines. Although seed-saving is not a prerequisite for organic production, the holistic nature of organics lends itself well to conserving seed.

In support of this, the Heritage Seed Library, in Warwickshire, is a collection of more than 800 open-pollinated organic varieties, which have been carefully preserved by gardeners across the country. Although their seeds are not yet commercially available, the Library is at the forefront of addressing the alarming erosion of our agricultural diversity.

Seed-saving and the development of local varieties must become a key component of organic farming, giving crops the potential to evolve in response to what could be rapidly changing climatic conditions. This will help agriculture keep pace with climate change in the field, rather than in the laboratory.

10. Job Creation

There is no doubt British farming is currently in crisis. With an average of 37 farmers leaving the land every day, there are now more prisoners behind bars in the UK than there are farmers in the fields.

Although it has been slow, the decline in the rural labour force is a predictable consequence of the industrialisation of agriculture. A mere one percent of the UK workforce is now employed in land-related enterprises, compared with 35 percent at the turn of the last century.

The implications of this decline are serious. A skilled agricultural workforce will be essential in order to maintain food security in the coming transition towards a new model of post-fossil fuel farming. Many of these skills have already been eroded through mechanisation and a move towards more specialised and intensive production systems.

Organic farming is an exception to these trends. By its nature, organic production relies on labour-intensive management practices. Smaller, more diverse farming systems require a level of husbandry that is simply uneconomical at any other scale. Organic crops and livestock also demand specialist knowledge and regular monitoring in the absence of agrochemical controls.

According to a 2006 report by the University of Essex, organic farming in the UK provides 32 percent more jobs per farm than comparable non-organic farms. Interestingly, the report also concluded that the higher employment observed could not be replicated in non-organic farming through initiatives such as local marketing. Instead, the majority (81 percent) of total employment on organic farms was created by the organic production system itself. The report estimates that 93,000 new jobs would be created if all farming in the UK were to convert to organic.

Organic farming also accounts for more younger employees than any other sector in the industry. The average age of conventional UK farmers is now 56, yet organic farms increasingly attract a younger more enthusiastic workforce, people who view organics as the future of food production. It is for this next generation of farmers that Organic Futures, a campaign group set up by the Soil Association in 2007, is striving to provide a platform.

D. J. Connor

 **NO**

Organic Agriculture Cannot
Feed the World

Introduction

Organic agriculture (OA) currently occupies 0.3% of agricultural land, mostly in developed countries. This land is farmed according to rules administered by various OA-regulating associations that, in the case of crops, disallow the use of most inorganic compounds for crop nutrition, synthetic compounds for pest, disease and weed control, and more recently, genetically modified cultivars. The rules also encourage rotations and intercrops to build soil fertility, improve crop nutrition, and to control or limit production problems associated with pests, diseases, and weeds. These latter aspects of OA are practiced much more widely outside OA, but those systems, here referred to as conventional agriculture (AG), vary enormously in range and amount of "OA-prohibited" inputs. They include, for example, many farms in developing countries where agrochemicals are not used because they are either not available or are too expensive.

The acceptance of OA in developed countries is increasing, albeit more slowly than in the previous decade, driven by consumer concern for food and environmental safety and supported by premium prices that consumers will pay for OA-labelled products. The biggest markets are in the USA, where 0.3% of agricultural land is devoted to OA, and Europe with 3.4%, where additional subsidies are available. At the same time, AG is also responding to the same food and environmental safety concerns with alternative labels developed by food retailers based on "good farming practices" for products from production systems that integrate the use of agrochemicals with biological processes. The same trends are evident in developing countries too, but the markets for labelled products are smaller and price premiums depend on the access they give to export markets in developed countries.

An important issue to the acceptance of OA is found in the question of its productivity. Existing analyses [such as by V. Smil] have put the carrying capacity of OA at 3–4 billion, well below the present world population (6.2 billion) and that projected for 2050 (9 billion). Those analyses are based on the performance of OA systems as practiced in [the nineteenth century] before the widespread use of inorganic fertilizers and when the world population was around 1 billion. They remain relevant to the present discussion, because advances in crop yield since those times have not changed the essential

From *Field Crops Research*, March 2008, pp. 187–190. Copyright © 2008 by Elsevier Science Ltd. Reprinted by permission via Rightslink.

metabolism of plant growth and nutrient requirement to support it. A recent paper by Badgley et al. ["Organic Agriculture and the Global Food Supply," *Renewable Agriculture & Food Systems* (June 2007)] has reopened this debate by presenting an analysis purporting to show that OA cannot only greatly increase productivity in developing countries but could feed the entire world also. The paper, initially presented in May 2007 at the FAO Conference on Organic Agriculture, has received much attention in the popular press and science magazines. OA does not need the ability to feed the world to contribute within agricultural production but this report provides some with justification to solve perceived and real production and environmental challenges in agriculture and food supply in a single step by large-scale transformation to OA. In this, the alternative development paths that resource-poor farmers of the developing world might take are a particular focus of attention. The conclusions of the study are, however, invalid because data are misinterpreted and calculations accordingly are erroneous.

A Major Overestimation of the Productivity of OA

Badgley et al. estimate OA productivity by applying OA/AG yield ratios calculated from the literature to national food production statistics from the FAO database. For developed countries the ratios are dominantly between high-yielding crops well supplied with either organic nutrients (OA) or fertilizer (AG). The mean ratio is 0.91. For developing countries, in contrast, the ratios (mean 1.74) are obtained largely from comparisons of low-yielding crops that received either organic nutrients (OA) or little or no fertilizer (AG). In most cases, organic nutrients were from external sources. It is the inequality of inputs to crops grown below full nutrition that explains the different and high ratio (>1) in the developing country data. The authors, however, avoid that explanation, defend the comparison as valid with the unproven suggestion that OA is able to out yield AG "by eliciting different pathways of gene expression," and rather focus on the low value of the ratio from developed countries. That outcome they explain results because many agricultural soils there "have been degraded by years of tillage, synthetic fertilizers, and pesticide residues." Neither of those explanations can be identified, however, in the comparisons from developed countries. The authors have seriously misinterpreted the ratios for developing countries, but more importantly have misunderstood the applicability of those ratios to their subsequent calculations of the productivity of OA. Both contribute to major overestimation.

The calculation of productivity following large-scale transformation to OA is made in two ways. First, OA/AG ratios for developed countries are applied to entire world food production, and second, individual ratios are applied to developed and developing country food production, respectively. The first alternative, it is proposed, assumes AG in the developing world is equally well supported by fertilizer as the developed world, the second that it is not supported by fertilizer at all. The first alternative, unsurprisingly, estimates a reduction (5%) in total food supply (7% in crop production) and the second a 57% increase (43% in crops). The latter, it is concluded, is sufficient to feed a greater population than the current 6.2 billion. The conclusion is wrong because, quite apart from the error in the yield ratio itself, food production in developing countries is not achieved by OA. Agriculture there is dominated by conventional methods that currently consume 70% of total world fertilizer use. In other words the authors have applied an overestimate of the relative yield of OA in developing countries to a gross overestimate of its productivity there. If crop production by organic methods in the developing world were even 70% of current recorded value, then that combined with a yield ratio (1.74) that is substantially greater than a defensible maximum (=1), would lead to an overestimation of 250% in that calculation of productivity of OA.

So much for arithmetic, but there is another serious agronomic misunderstanding. The authors have failed to realize that any significant increase in OA from its current small base of world agricultural area (0.3%) will increase competition for limited organic nutrients. That in turn will reduce the beneficial impact of OA on the low-input component of agriculture in developing countries and increase the current disadvantage of OA in developed countries. Crop yields and/or cropped areas will fall as an increasing proportion of land is devoted to biological regeneration of fertility. The ratios will fall to values below the current estimate for developed countries, further explaining why the method of calculation has seriously overestimated productivity of large-scale OA.

On the Limited Availability of Organic Nutrients

If large-scale conversion to OA were possible, where would the organic nutrients to support the high productivity required to feed a large and expanding world population come from? The second part of Badgley et al. seeks to confirm the conclusion of adequate productivity of OA by a parallel estimate of the potential of leguminous cover crops to provide the required organic N. A survey of published N-fixation rates and crop performance in legume rotations is used to establish equivalent N fertilizer contributions from such crops as 95 and 108 kg N/ha for temperate (developed) and tropical (developing) conditions, respectively. There is no space here to analyse the relevance of all the data that contribute to these estimates save noting that the mean values may be overestimates because of some obvious outliers and because the entries are not restricted to short-term cover crops as required by the analysis. At those high rates, however, the authors calculate that an additional legume cover crop added to 1362 Mha of cropped area (total world crop area less 170 Mha already in leguminous forage production) would provide 140 Mt N/year. Since this is substantially greater than the current annual worldwide N fertilizer use of 91 Mt, the authors used 82 Mt for their 2001 calculation, the ability of OA to feed the world is confirmed. Or is it? What are the limitations of this calculation?

In practice all existing cropland cannot be provided annually with N by an additional leguminous cover crop without significant disruption to crop area and production. Here, the authors demonstrate limited appreciation of crop ecology and agronomy. First, because much productive land already carries multiple crops. In needy places such as Bangladesh, for example, average cropping intensity is already 2.5/year. Second, because most land that now carries one crop or less each year does so for limitations of temperature or water supply that exclude the possibility of a second crop. In those places, legume cover crops could only be introduced in a 2-year cropping sequence. In others, for example southern Australia, cover crops could replace legume-based pastures now in rotations with crops but with an uncertain advantage to N gain. Farmers there, in a region where a significant proportion (70%) of N is provided by biological fixation, have shown no enthusiasm for that option. There are no current statistics to estimate the proportion of world current cropland that could accept an additional legume cover crop, but it is certainly much less than 100% as used by Badgley et al., and could only be increased significantly, in areas with suitable thermal regimes, by massive expansion of irrigation for which water is not available. If the authors had sought to explain why Smil was mistaken in his assessment of the small legume-based productivity of OA (maximum support for 3–4 billion) they would have been exposed to more realistic cropping scenarios than that they proposed.

An Evaluation of the Productivity of Legume-Based Agriculture

Fortunately, there is a way to evaluate the legume-based cropping strategy proposed by Badgley et al. The provision of 100 kg N/ha fertilizer equivalent to all cropland would support, with inevitable in-crop losses to drainage and volatilisation of say 25%, the production of 4170 kg crop dry matter/ha at 1.8% N. That amount of crop growth would, with a creditable harvest index of 0.4, establish an N-limited cereal yield of 1.7 t grain/ha, sufficient over 1362 Mha and allowing 10% for storage losses and seed for the following crop, to provide an adequate diet for 4.2 billion using the Standard Nutritional Unit (SNU) of 500 kg grain equivalent/person/year. If instead, 50% of cropland could only support cover crops in 2-year rotations, then production would support 3.1 billion, comparable with the estimate of 3.2 billion made by Smil.

The issue of legume rotations for the biological regeneration of fertility emphasizes, as stressed earlier, why OA/AG yield ratios used in the previous calculations must fall substantially under large-scale transformation to OA. The cropping strategy proposed by Badgley et al., although unrealistic in detail, properly presents crops as components of closed systems. Leguminous crops and pastures build up N fertility, non-legume crops extract it either directly or as applied green or animal manure, and the cycle continues. That is not, however, the sort of system from which the yield ratios used in the productivity assessment were calculated. Rather they were dominated by yields of individual crops for which organic manure was just another external input. In cropping systems, the entire area involved in production of grain crop, cover crop, pasture, and animals must be included in the calculations of productivity. It is this cost of biological N-fixation (land, labour, water) that most disadvantages OA relative to AG and suppresses the system production ratios well below unity. If a grain crop can be grown after a legume crop in 1 year then fertilizer may allow the growth of two grain crops. If legume and grain crops can only be grown in successive years, then successive grain crops are possible with fertilizer. This is exactly the issue that faces millions of resource-poor farmers in Asia where the yields required for survival can only be achieved by multiple food crops. How, for example in the rice–wheat system of south-east Asia and China could OA provide the quantity of N (300–400 kg/ha) to produce the required annual combined grain yields of 10+ t/ha.

Conclusion

A critical analysis of the nature and use of OA/AG yield ratios does not support the proposition that large-scale OA productivity would be sufficient to feed the world or that legume cover crops could replace N fertilizer use without disrupting current food production. There is, therefore, no newly established production frontier for OA so that those who use the conclusions of the study by Badgley et al. to promote or support OA will have been misled and limited resources for research and development would be misallocated. The biggest losers are likely to be resource-poor farmers in developing countries. That organic nutrients can increase the now low yields of nutrient-limited crops is not in dispute. What is in dispute is the promotion of a transient OA solution as the sustainable solution when fertilizers, that can provide a complementary route to increasing yields now, will be essential for the high productivity that will be required in the future.

A Way Forward

Transformation to OA is, of course, not only about N supply but rather the interaction of a number of social, environmental and economic concerns and outcomes. Important issues include maintenance of soil condition, provision of nutrients other than N, human labour, control of pests, diseases and weeds, product quality and safety, and minimizing off-site environmental effects. Surely now is the time, for unbiased analyses of alternative agricultural production systems in the search for optimal management solutions. The world needs a highly productive agriculture that can save as much land as possible for nature. There is much emphasis in AG to improve fertilizer formulations and use site-specific application methods, timing and amounts to optimise fertilizer use for production and environment. At the same time, crop nutrition, and not just N, is a critical issue in OA because there are few locations where sufficient quantities of organic nutrients are available. There is recognition within the broad reach of the OA movement that revised practices would be advantageous. There are proposals, on the one hand, to more closely align practice with low-input approaches to crop production and, on the other, a move to "replacement" organics that allows fertilizer when essential. Such adjustments would remove much heat from debates as both ends of the farming spectrum seek to optimise inputs for productivity and environmental sustainability. This seems especially pertinent at a time when the enthusiasm for biofuel as a solution to energy security and climate change is set to revolutionize agricultural practice by seeking an additional productivity that, even for a small replacement of current liquid transport fuel, could easily outstrip that required now for food production.

EXPLORING THE ISSUE

Can Organic Farming Feed the World?

Critical Thinking and Reflection

1. What aspects of the food we consume are essential to a modern standard of living?
2. What aspects of the food we consume are essential to human health and well-being?
3. The answers to the previous two questions should be different. Does the difference have implications for how much food an agricultural system must be able to produce?
4. Why are people willing to pay more for organic food?

Is There Common Ground?

Both sides in this debate agree that agricultural systems can be improved. In fact, Connor notes that conventional agriculture can profit by using organic techniques, and organic agriculture can benefit by using conventional techniques.

1. How does this agreement show up in the context of fertilizer use?
2. How does this agreement show up in the context of soil quality?
3. To what extent do both organic and conventional agriculture seek to protect the environment?
4. Discuss how changes in population size affect the answer to the issue question.

Internet References . . .

The Center for Health, Environment & Justice (CHEJ)

The Center for Health, Environment & Justice (CHEJ) was founded by Lois Gibbs, who in connection with the Love Canal toxic waste site awakened the nation to the link between toxic wastes and public health. CHEJ was instrumental in establishing some of the first national policies critical to protecting community health like the Superfund Program.

http://chej.org/

e.hormone

e.hormone is hosted by the Center for Bioenvironmental Research at Tulane and Xavier Universities in New Orleans. It provides accurate, timely information about environmental hormones and their impacts.

http://e.hormone.tulane.edu/

The J. Craig Venter Institute

The J. Craig Venter Institute, formed in October 2006, is a world leader in genomics research, including the effort to create synthetic cells.

www.jcvi.org/cms/

The La Hague Nuclear Reprocessing Plant

The AREVA NC La Hague site, located on the western tip of the Cotentin Peninsula in Normandy, reprocesses spent power reactor fuel to recycle reusable energy materials—uranium and plutonium—and to condition the waste into suitable final form.

www.areva.com/EN/operations-1118/areva-la-hague-recycling-spent-fuel.html

The Silicon Valley Toxics Coalition

The Silicon Valley Toxics Coalition (SVTC) was formed to engage in research, advocacy, and organizing associated with environmental and human problems caused by the rapid growth of the high-tech electronics industry.

www.etoxics.org/

Superfund

The U.S. Environmental Protection Agency provides a great deal of information on the Superfund program, including material on environmental justice.

www.epa.gov/superfund/

Toxic Chemicals

A great many of today's environmental issues have to do with industrial development, which expanded greatly during the twentieth century.
Just since World War II, many thousands of synthetic chemicals—
pesticides, plastics, and antibiotics—have flooded the environment. We
have become dependent on the production and use of energy, particularly
in the form of fossil fuels. We have discovered that industrial processes
generate huge amounts of waste, much of it toxic. Air and water pollution
have become global problems. And we have discovered that our actions
may change the world for generations to come. The following issues by no
means exhaust the possible topics for study and debate.

- Should Society Impose a Moratorium on the Use and Release of "Synthetic Biology" Organisms?

- Do Environmental Hormone Mimics Pose a Potentially Serious Health Threat?

- Should the Superfund Tax Be Reinstated?

- Should the United States Reprocess Spent Nuclear Fuel?

ISSUE 16

Should Society Impose a Moratorium on the Use and Release of "Synthetic Biology" Organisms?

YES: Jim Thomas, Eric Hoffman, and Jaydee Hanson, from "Offering Testimony from Civil Society on the Environmental and Societal Implications of Synthetic Biology" (May 27, 2010)

NO: Gregory E. Kaebnick, from "Testimony to the U.S. House of Representatives Committee on Energy and Commerce Hearing on Developments in Synthetic Genomics and Implications for Health and Energy" (May 27, 2010)

Learning Outcomes

After reading this issue, you should be able to:

- Explain what "synthetic biology" and its potential benefits are.
- Explain why people are concerned about the impact of "synthetic biology" on the environment.
- Describe measures for controlling the potential risks of "synthetic biology."
- Discuss the difficulty of preventing all the potential risks of a new technology.

ISSUE SUMMARY

YES: Jim Thomas, Eric Hoffman, and Jaydee Hanson, representing the Civil Society on the Environmental and Societal Implications of Synthetic Biology, argue that the risks posed by synthetic biology to human health, the environment, and natural ecosystems are so great that Congress should declare an immediate moratorium on releases to the environment and commercial uses of synthetic organisms and require comprehensive environmental and social impact reviews of all federally funded synthetic biology research.

NO: Gregory E. Kaebnick of the Hastings Center argues that although synthetic biology is surrounded by genuine ethical and moral concerns—including risks to health and environment—which warrant discussion, the potential benefits are too great to call for a general moratorium.

In the past century, biologists have learned an enormous amount about how the cell—the basic functional unit of all living things—works. By the early 1970s, they were beginning to move genes from one organism to another and dream of designing plants and animals (including human beings) with novel combinations of features. By 2002, with Defense Department funding, Jeronimo Cello, Aniko Paul, and Eckard Wimmer were able to construct a live poliovirus from raw laboratory chemicals. This feat was a long way from constructing a bacterium or animal from raw chemicals, but it was enough to set alarm bells of many kinds ringing. Some people thought this work challenged the divine monopoly on creation. Others feared that if one could construct one virus from scratch, one could construct others, such as the smallpox virus, or even tailor entirely new viruses with which natural immune systems and medical facilities could not cope. Some even thought that the paper was irresponsible and should not have been published because it pointed the way toward new kinds of terrorism. See Michael J. Selgelid and Lorna Weir, "Reflections on the Synthetic Production of Poliovirus," *Bulletin of the Atomic Scientists* (May/June 2010).

In 2010, the next step was taken. Craig Venter's research group announced that they had successfully synthesized a bacterial chromosome (the set of genes that specifies the function and form of the bacterium) and implanted it in a bacterium of a different species whose chromosome had been removed. The result was the conversion of the recipient bacterium into the synthesized chromosome's species. See Daniel G. Gibson et al., "Creation of a Bacterial Cell Controlled by a Chemically Synthesized Genome," *Science* (July 2, 2010). The report received a great deal of media attention, much of it saying that Venter's group had created a living cell, even though only the chromosome had been synthesized. The chromosome's biochemically complex container—a cell minus its chromosome—had not been synthesized.

The goal of this work is not the creation of life, but rather the ability to exert unprecedented control over what cells do. In testimony before the House Committee on Energy and Commerce Hearing on Developments in Synthetic Genomics and Implications for Health and Energy (May 27, 2010), Venter said, "The ability to routinely write the 'software of life' will usher in a new era in science, and with it, new products and applications such as advanced biofuels, clean water technology, food products, and new vaccines and medicines. The field is already having an impact in some of these areas and will continue to do so as long as this powerful new area of science is used wisely." See also Pamela Weintraub, "J. Craig Venter on Biology's Next Leap: Digitally Designed Life-Forms that Could Produce Novel Drugs, Renewable Fuels, and Plentiful

Food for Tomorrow's World," *Discover* (January/February 2010); and Michael A. Peters and Priya Venkatesan, "Bioeconomy and Third Industrial Revolution in the Age of Synthetic Life," *Contemporary Readings in Law and Social Justice* (vol. 2, no. 2, 2010). However, the ETC Group, which anticipated a synthetic organism in 2007, condemns the lack of rules governing synthetic biology, calls it "a quintessential Pandora's box moment," and calls for a global moratorium on further work; see "Synthia Is Alive . . . and Breeding: Panacea or Pandora's Box?" ETC Group News Release (May 20, 2010) (www.etcgroup.org/en/node/5142).

Researchers had been working on synthetic biology for a number of years, and well before Craig Venter's group announced their accomplishment, prospects and consequences were already being discussed. Michael Specter, "A Life of Its Own," *New Yorker* (September 28, 2009), describes progress to date and notes "the ultimate goal is to create a synthetic organism made solely from chemical parts and blueprints of DNA." If this sounds rather like manipulating living things the way children manipulate Legos, Drew Endy of MIT and colleagues created in 2005 the BioBricks Foundation to make that metaphor explicit. See also Rob Carlson, *Biology Is Technology: The Promise, Peril, and Business of Engineering Life* (Harvard University Press, 2010). David Deamer, "First Life and Next Life," *Technology Review* (May/June 2009), notes that the next step is to create entire cells, not just a single bacterial chromosome. Charles Petit, "Life from Scratch," *Science News* (July 3, 2010), describes the even more ambitious work of Harvard's Jack Szostak, who is trying to understand how life began by constructing a pre-cell just sophisticated enough to take in components, grow, divide, and start evolving. Szostak expects to succeed within a few years. Such efforts, say Steven A. Benner, Zunyi Yang, and Fei Chen, "Synthetic Biology, Tinkering Biology, and Artificial Biology: What Are We Learning?" *Comptes Rendus Chimie* (April 2011), will drive a better understanding of biology in ways that mere analysis cannot.

Immediately after the Venter group's announcement of their accomplishment, Vatican representatives declared that synthetic biology was "a potential time bomb, a dangerous double-edged sword for which it is impossible to imagine the consequences" and "pretending to be God and parroting his power of creation is an enormous risk that can plunge men into barbarity"; see "Vatican Greets First Synthetic Cell with Caution," *America* (June 7–14, 2010). Chuck Colson, "Synthetic Life: The Danger of God-Like Pretensions," *Christian Post* (June 16, 2010), says, "God-like control [of risks] isn't only hubris, it's pure fantasy. The only real way to avoid the unthinkable is not to try and play God in the first place. But that would require the kind of humility that Venter and company reject out-of-hand." Nancy Gibbs, "Creation Myths," *Time* (June 28, 2010), says, "The path of progress cuts through the four-way intersection of the moral, medical, religious and political—and whichever way you turn, you are likely to run over someone's deeply held beliefs. Venter's bombshell revived the oldest of ethical debates, over whether scientists were playing God or proving he does not exist because someone re-enacted Genesis in suburban Maryland." The "playing God" objection seems likely to grow louder as synthetic biology matures, but it is also likely to fade just as it has done after previous advances such as in vitro fertilization and surrogate mothering. Henk van den Belt, "Playing God in Frankenstein's Footsteps: Synthetic Biology and the Meaning of Life," *NanoEthics* (December 2009), notes, "While syntheses of artificial life forms cause some vague uneasiness that life may lose its special meaning, most concerns turn out to be narrowly anthropocentric. As long as synthetic biology creates only new microbial life and does not directly affect human life, it will in all likelihood be considered acceptable."

What will be more significant will be discussions such as Gautam Mukunda, Kenneth A. Oye, and Scott C. Mohr, "What Rough Beast? Synthetic Biology, Uncertainty, and the Future of Biosecurity," *Politics and the Life Sciences* (September 2009). Mukunda et al., see synthetic biology as seeking "to create modular biological parts that can be assembled into useful devices, allowing the modification of biological systems with greater reliability, at lower cost, with greater speed, and by a larger pool of people than has been the case with traditional genetic engineering." It is thus a "dual-use" technology, meaning that it has both benign and malign applications. This has clear implications for national security, both offensive and defensive, but they find those implications least alarming in the short term. In the long term, the defensive implications are most important. Because the offensive implications are there, regulation and surveillance of research and development will be necessary in order to forestall terrorists and criminals. Mildred K. Cho and David A. Relman, "Synthetic 'Life,' Ethics, National Security, and Public Discourse," *Science* (July 2, 2010), caution that some concerns about biosecurity and ethics are real but some are imagined; being realistic and avoiding exaggeration are essential if the science is not to become a victim of public mistrust. Meera Lee Sethi and Adam Briggle, "Making Stories Visible: The Task for Bioethics Commissions," *Issues in Science and Technology* (Winter 2011), caution that the stories we tell ourselves about technology (such as "synthetic biology is like computers") may hide issues that warrant deep and careful thought.

In the following selections, Jim Thomas, Eric Hoffman, and Jaydee Hanson, representing the Civil Society on the Environmental and Societal Implications of Synthetic Biology, argue that the risks posed by synthetic biology to human health, the environment, and natural ecosystems are so great that Congress should declare an immediate moratorium on releases to the environment and commercial uses of synthetic organisms and require comprehensive environmental and social impact reviews of all federally funded synthetic biology research. Gregory E. Kaebnick of the Hastings Center argues that although synthetic biology is surrounded by genuine ethical and moral concerns—including risks to health and environment—which warrant discussion, the potential benefits are too great to call for a general moratorium.

YES

Jim Thomas, Eric Hoffman, and Jaydee Hanson

Offering Testimony from Civil Society on the Environmental and Societal Implications of Synthetic Biology

Last week, the J. Craig Venter Institute announced the creation of the first living organism with a synthetic genome claiming that this technology would be used in applications as diverse as next generation biofuels, vaccine production and the clean up of oil spills. We agree that this is a significant technical feat however; we believe it should be received as a wake-up call to governments around the world that this technology must now be accountably regulated. While attention this week has been on the activities of a team from Synthetic Genomics Inc, the broader field of synthetic biology has in fact quickly and quietly grown into a multi-billion dollar industry with over seventy DNA foundries and dozens of 'pure play' synthetic biology companies entering the marketplace supported by large investments from Fortune 500 energy, forestry, chemical and agribusiness companies. That industry already has at least one product in the marketplace (Du Pont's 'Sorona' bioplastic), and another recently cleared for market entry in 2011 (Amyris Biotechnology's 'No Compromise' biofuel) as well as several dozen near to market applications. We believe the committee should consider the implications of this new industry as a whole in its deliberations not just the technical breakthrough reported last week. Without proper safeguards in place, we risk introducing synthetically constructed living organisms into the environment, intentionally or inadvertently through accident and worker error, that have the potential to destroy ecosystems and threaten human health. We will see the widespread commercial application of techniques with grave dual-use implications. We further risk licensing their use in industrial applications that will unsustainably increase the pressure of human activities on both land and marine ecologies through the increased take of biomass, food resources, water and fertilizer or displacement of wild lands to grow feedstocks for biobased fuel and chemical production.

We call on Congress to:

1. Implement a moratorium on the release of synthetic organisms into the environment and also their use in commercial settings. This moratorium should remain in place until there is an adequate

U.S. House of Representatives Committee on Energy and Commerce, May 27, 2010.

scientific basis on which to justify such activities, and until due consideration of the associated risks for the environment, biodiversity,
and human health, and all associated socio-economic repercussions,
are fully and transparently considered.

2. As an immediate step, all federally funded synthetic biology research
should be subject to a comprehensive environmental and societal impact review carried out with input from civil society, also
considering indirect impacts on biodiversity of moving synthetic
organisms into commercial use for fuel, chemicals and medicines.
This should include the projects that received $305 million from the
Department of Energy in 2009 alone.

3. All synthetic biology projects should also be reviewed by the
Recombinant DNA Advisory Committee.

On Synthetic Biology for Biofuels—Time for a Reality Check

Much of the purported promise of the emerging Synthetic Biology industry
resides in the notion of transforming biomass into next generation biofuels or bio-based chemicals where synthetic organisms work as bio-factories
transforming sugars to high value products. On examination much of this
promise is unrealistic and unsustainable and if allowed to proceed could
hamper ongoing efforts to conserve biological diversity, ensure food security
and prevent dangerous climate change. The sobering reality is that a switch
to a bio-based industrial economy could exert much more pressure on land,
water, soil, fertilizer, forest resources and conservation areas. It may also do
little to address greenhouse gas emissions, potentially worsening climate
change.

By way of an example, the team associated with Synthetic Genomics Inc
who have recently announced the creation of a synthetic cell have specifically
claimed that they would use the same technology to develop an algal species
that efficiently converts atmospheric carbon dioxide into hydrocarbon fuel,
supposedly addressing both the climate crisis and peak oil concerns in one fell
swoop. Yet, contrary to the impression put forth by these researchers in the
press, algae, synthetic or otherwise, requires much more than just carbon dioxide to grow—It also requires water, nutrients for fertilizer and also sunlight
(which therefore means one needs land or open ocean—this can't be done in
a vat without also consuming vast quantities of sugar).

In order for Synthetic Genomics or their partners to scale up algal biofuel production to make a dent in the fuel supply, the process would likely
exert a massive drain on both water and on fertilizers. Both fresh water and
fertilizer (especially phosphate-based fertilizers) are in short supply, both are
already prioritized for agricultural food production and both require a large
amount of energy either to produce (in the case of fertilizers) or to pump to
arid sunlight-rich regions (in the case of water). In a recent life-cycle assessment of algal biofuels published in the journal *Environmental Science and
Technology* researchers concluded that algae production consumes more water

and energy than other biofuel sources like corn, canola, and switch grass, and also has higher greenhouse gas emissions. "Given what we know about algae production pilot projects over the past 10 to 15 years, we've found that algae's environmental footprint is larger than other terrestrial crops," said Andres Clarens, an assistant professor in U.Virginia's Civil and Environmental Department and lead author on the paper. Moreover scaling-up this technology in the least energy-intensive manner will likely need large open ponds sited in deserts, displacing desert ecosystems. Indeed the federally appointed Invasive Species Advisory Committee has recently warned that non-native algal species employed for such biofuel production could prove ecologically harmful and is currently preparing a fuller report on the matter.

Meanwhile it is not clear that the yield from algal biofuels would go far to meeting our energy needs. MIT inventor Saul Griffiths has recently calculated that even if an algae strain can be made 4 times as efficient as an energy source than it is today it would still be necessary to fill one Olympic-size swimming pool of algae every second for the next twenty five years to offset only half a terawatt of our current energy consumption (which is expected to rise to 16 TW in that time period). That amounts to massive land use change. Emissions from land use change are recognized as one of the biggest contributors to anthropogenic climate change.

Moving Forward—Time for New Regulation

The rapid adoption of synthetic biology is moving the biotechnology industry into the driving seat of industrial production across many previously disparate sectors with downstream consequences for monopoly policy. Meanwhile its application in commercial settings uses a set of new and extreme techniques whose proper oversight and limits has not yet been debated. It also enables many more diverse living organisms to be produced using genetic science at a speed and volume that will challenge and ultimately overwhelm the capacity of existing biosafety regulations. For example, Craig Venter has claimed in press and in his patent applications that when combined with robotic techniques the technology for producing a synthetic cell can be perfected to make millions of new species per day. Neither the US government nor any other country has the capacity to assess such an outpouring of new synthetic species in a timely or detailed manner. The Energy and Commerce Committee urgently needs to suggest provisions for regulating these new organisms and chemicals derived from them under the Toxic Substances Control Act, Climate Change legislation and other legislation under its purview before allowing their release into the environment. It also needs to identify how it intends to ensure that the use of such organisms whether in biorefineries, open ponds or marine settings does not impinge on agriculture, forestry, desert and marine protection, the preservation of conservation lands, rural jobs or livelihoods.

To conclude, Congress must receive this announcement of a significant new lifeform as a warning bell, signifying that the time has come for governments to fully regulate all synthetic biology experiments and products.

It is imperative that in the pursuit of scientific experimentation and wealth creation, we do not sacrifice human health, the environment, and natural ecosystems. These technologies could have powerful and unpredictable consequences. These are life forms never seen on the planet before now. Before they are unleashed into the environment and commercial use, we need to understand the consequences, evaluate alternatives properly, and be able to prevent the problems that may arise from them.

Gregory E. Kaebnick  **NO**

Testimony to the House Committee on Energy and Commerce

. . . The ethical issues raised by synthetic biology are familiar themes in an ongoing conversation this nation has been having about biotechnologies for several decades. . . .

The concerns fall into two general categories. One has to do with whether the creation of synthetic organisms is a good or a bad thing in and of itself, aside from the consequences. These are thought of as intrinsic concerns. Many people had similar intrinsic concerns about reproductive cloning, for example; they just felt it was wrong to do, regardless of benefits. Another has to do with potential consequences—that is, with risks and benefits. The distinction between these categories can be difficult to maintain in practice, but it provides a useful organizational structure.

Intrinsic Concerns

I will start with the more philosophical, maybe more baffling, kind of concern—the intrinsic concerns. They are an appropriate place to start because the work just published by researchers at Synthetic Genomics, Inc., has been billed as advancing our understanding of these issues in addition to making a scientific advance.

This announcement is not the first time we have had a debate about whether biotechnology challenges deeply held views about the status of life and the power that biotechnology and medicine give us over it. There was a similar debate about gene transfer research in the 1970s and 1980s, about cloning and stem cell research in the 1990s, and—particularly in the last decade but also earlier—about various tools for enhancing human beings. They have been addressed by the President's Commission for the Study of Ethical Problems in Medicine and Biomedical and Behavioral Research in 1983, by President Clinton's National Bioethics Advisory Council, and by President Bush's President's Council on Bioethics. These concerns are related to even older concerns in medicine about decisions to withhold or withdraw medical treatment at the end of life.

The fact that we have had this debate before speaks to its importance. I believe the intrinsic concerns deserve respect, and with some kinds of biotechnology I think they are very important, but for synthetic biology, I do not think they provide a basis for decisions about governance.

U.S. House of Representatives Committee on Energy and Commerce, May 27, 2010.

Religious or Metaphysical Concerns

The classic concern about synthetic biology is that it puts human beings in a role properly held by God—that scientists who do it are "playing God," as people say. Some may also believe that life is sacred, and that scientists are violating its sacredness. Prince Charles had this in mind in a famous polemic some years ago when he lamented that biotechnology was leading to "the industrialisation of Life."

To object to synthetic biology along these lines is to see a serious moral mistake in it. This kind of objection may be grounded in deeply held beliefs about God's goals in creating the world and the proper role of human beings within God's plan. But these views would belong to particular faiths—not everybody would share them. Moreover, there is a range of opinions even within religious traditions about what human beings may and may not do. Some people celebrate human creativity and science. They may see science as a gift from God that God intends human beings to develop and use.

The announcement that Synthetic Genomics, Inc., has created a synthetic cell appears to some to disprove the view that life is sacred, but I do not agree. Arguably, what has been created is a synthetic genome, not a completely synthetic cell. Even if scientists manage to create a fully synthetic cell, however, people who believe that life is sacred, that it is something more than interacting chemicals, could continue to defend that belief. A similar question arises about the existence of souls in cloned people: If people have souls, then surely they would have souls even if they were created in the laboratory by means of cloning techniques. By the same reasoning, if microbial life is more than a combination of chemicals, then even microbial life created in the laboratory would be more than just chemicals. In general, beliefs about the sacredness of life are not undermined by science. Moreover, even the creation of a truly synthetic cell would still start with existing materials. It would not be the kind of creating with which God is credited, which is creating something from nothing—creation ex nihilo.

Concerns that Synthetic Biology Will Undermine Morally Significant Concepts

A related but different kind of concern is that synthetic biology will simply undermine our shared understanding of important moral concepts. For example, perhaps it will lead us to think that life does not have the specialness we have often found in it, or that we humans are more powerful than we have thought in the past. This kind of concern can be expressed without talking about God's plan.

Synthetic biology need not change our understanding of the value of life, however. The fact that living things are created naturally, rather than by people, would be only one reason for seeing them as valuable, and we could continue to see them as valuable when they are created by people. Further, in its current form, synthetic biology is almost exclusively about engineering single-celled organisms, which may be less troubling to people than engineering more complex organisms. If the work is contained within the laboratory and the factory, then it might not end up broadly changing humans' views of the value of life.

Also, of course, the fact that the work challenges our ideas may not really be a moral problem. It would not be the first time that science has challenged our views of life or our place in the cosmos, and we have weathered these challenges in the past.

Concerns about the Human Relationship to Nature

Another way of saying that there's something intrinsically troubling about synthetic biology, again without necessarily talking about the possibility that people are treading on God's turf, is to see it as a kind of environmentalist concern. Many environmentalists want to do more than make the environment good for humans; they also want to save nature from humans—they want to save endangered species, wildernesses, "wild rivers," old-growth forests, and mountains, canyons, and caves, for example. We should approach the natural world, many feel, with a kind of reverence or gratitude, and some worry that synthetic biology—perhaps along with many other kinds of biotechnology—does not square with this value.

Of course, human beings have been altering nature throughout human history. They have been altering ecosystems, affecting the survival of species, affecting the evolution of species, and even creating new species. Most agricultural crop species, for example, are dramatically different from their ancestral forebears. The issue, then, is where to draw the line. Even people who want to preserve nature accept that there is a balance to be struck between saving trees and harvesting them for wood. There might also be a balance when it comes to biotechnology. The misgiving is that synthetic biology goes too far—it takes human control over nature to the ultimate level, where we are not merely altering existing life forms but creating new forms.

Another environmentalist perspective, however, is that synthetic biology could be developed so that it is beneficial to the environment. Synthetic Genomics, Inc. recently contracted with Exxon Mobil to engineer algae that produce gasoline in ways that not only eliminate some of the usual environmental costs of producing and transporting fuel but simultaneously absorb large amounts of carbon dioxide, thereby offsetting some of the environmental costs of burning fuel (no matter how it is produced). If that could be achieved, many who feel deeply that we should tread more lightly on the natural world might well find synthetic biology attractive. In order to achieve this benefit, however, we must be confident that synthetic organisms will not escape into the environment and cause harms there.

Concerns involving Consequences

The second category of moral concerns is about consequences—that is, risks and benefits. The promise of synthetic biology includes, for example, better ways of producing medicine, environmentally friendlier ways of producing fuel and other substances, and remediation of past environmental damage. These are not morally trivial considerations. There are also, however, morally serious risks. These, too, fall into three categories.

Concerns About Social Justice

Synthetic biology is sometimes heralded as the start of a new industrial age. Not only will it lead to new products, but it will lead to new modes of production and distribution; instead of pumping oil out of the ground and shipping it around the world, we might be able to produce it from algae in places closer to where it will be used. Inevitably, then, it would have all sorts of large-scale economic and social consequences, some of which could be harmful and unjust. Some commentators hold, for example, that if synthetic biology generates effective ways of producing biofuels from feedstocks such as sugar cane, then farmland in poor countries would be converted from food production to sugar cane production. Another set of concerns arises over the intellectual property rights in synthetic biology. If synthetic biology is the beginning of a new industrial age, and a handful of companies received patents giving them broad control over it, the results could be unjust.

Surely we ought to avoid these consequences. It is my belief that we can do so without avoiding the technology. Also, traditional industrial methods themselves seem to be leading to disastrous long-term social consequences; if so, synthetic biology might provide a way toward better social outcomes.

Concerns About Biosafety

Another concern is about biosafety—about mechanisms for containing and controlling synthetic organisms, both during research and development and in industrial applications. The concern is that organisms will escape, turn out to have properties, at least in their new environment, different from what was intended and predicted, or maybe mutate to acquire them, and then pose a threat to public health, agriculture, or the environment. Alternatively, some of their genes might be transferred to other, wild microbes, producing wild microbes with new properties.

Controlling this risk means controlling the organisms—trying to prevent industrial or laboratory accidents, and then trying to make sure that, when organisms do escape, they are not dangerous. Many synthetic biologists argue that an organism that devotes most of its energy to producing jet fuel or medicine, that is greatly simplified (so that it lacks the genetic complexity and therefore the adaptability of a wild form), and that is designed to work in a controlled, contained environment, will simply be too weak to survive in the wild. For added assurance, perhaps engineering them with failsafe mechanisms will *ensure* that they are incapable of surviving in the wild.

Concerns About Deliberate Misuse

I once heard a well-respected microbiologist say that he was very enthusiastic about synthetic biology, and that the only thing that worries him is the possibility of catastrophe. The kind of thing that worries him is certainly possible. The 1918 flu virus has been recreated in the laboratory. In 2002, a scientist in New York stitched together stretches of nucleotides to produce a string of DNA that was equivalent to RNA polio virus and eventually produced the RNA virus using the DNA string. More recently, the SARS virus was also created in the laboratory. Eventually, it will almost certainly be possible to recreate bacterial pathogens like smallpox. We might also be able to enhance these pathogens. Some work in Australia on mousepox suggests ways of making smallpox more potent, for example. In theory, entirely new pathogens could be created. Pathogens that target crops or livestock are also possible.

Controlling this risk means controlling the people and companies who have access to DNA synthesis or the tools they could use to synthesize DNA themselves. There are some reasons to think that the worst will never actually happen. To be wielded effectively, destructive synthetic organisms would also have to be weaponized; for example, methods must be found to disperse pathogens in forms that will lead to epidemic infection in the target population while sparing one's own population. Arguably, terrorists have better forms of attacking their enemies than with bioweapons, which are still comparatively hard to make and are very hard to control. However, our policy should amount to more than hoping for the best.

Governance

In assessing these risks and establishing oversight over synthetic biology, we do not start from square one. There is an existing framework of laws and regulations, put into action by various agencies and oversight bodies, that will apply to R&D and to different applications. The NIH is extending its guidelines for research on genetic engineering to ensure that they are applicable to research on synthetic biology. These Guidelines are enforced by the NIH's Recombinant DNA Advisory Committee and a network of Institutional Biosafety Committees at research institutions receiving federal funding. Many applications would fall under the purview of various federal laws and the agencies that enforce them. For example, a plan to release synthetic organisms into the sea to produce nutrients that would help rebuild ocean food chains would have to pass muster with the EPA. The USDA and FDA also have regulatory authority over applications. The FBI and the NIH's National Science Advisory Board for Biosecurity are formulating policy to regulate the sale of synthetic DNA sequences that might pose a threat to biosecurity.

At the same time, the current regulatory framework may need to be augmented. First, there are questions about whether the existing laws leave gaps. Research conducted by an entirely privately funded laboratory might not be covered by the NIH's Guidelines, for example. Field testing of a synthetic organism—that is, release into the environment as part of basic research—might not be covered by the existing regulations of the EPA or the USDA. Questions about the adequacy of existing regulations are even more pointed when it comes to concerns about biosecurity, particularly if or when powerful benchtop synthesizers are available in every lab.

The other big question is whether the regulatory bodies' ability to do risk assessment of synthetic biology is adequate. Synthetic biology differs from older forms of genetic engineering in that a synthetic organism could combine DNA sequences found originally in many different organisms, or might even contain entirely novel genetic code. The eventual behavior of these organisms in new environments, should they accidentally end up in one, may therefore be hard to predict.

The synthetic biologists' goal of simplicity is crucial. One of the themes of traditional biology is that living things are usually more complex than they first appear. We should not assume at the outset that synthetic organisms will shed the unpredictability inherent to life. Life tends to find a way. As a starting assumption, we should expect that artificial life will try to find a way as well.

Another difficulty in assessing concerns about both biosafety and deliberate misuse is that, if the field evolves so that important and even innovative work could be done in small, private labs, even in homes, then it could be very difficult to monitor and regulate. The threats of biosafety and deliberate misuse would have to be taken yet more seriously.

Concluding Comments

I take seriously concerns that synthetic biology is bad in and of itself, and I believe that they warrant a thorough public airing, but I do not believe that they provide a good basis for restraining the technology, at least if we can be confident that the organisms will not lead to environmental damage. Better yet would be to get out in front of the technology and ensure that it benefits the environment. Possibly, some potential applications of synthetic biology are more troubling than others and should be treated differently.

Ultimately, I think the field should be assessed on its possible outcomes. At the moment, we do not understand the possible outcomes well enough. We need, I believe:

- more study of the emergence, plausibility, and impact of potential risks;
- a strategy for studying the risks that is multidisciplinary, rather than one conducted entirely within the field;
- a strategy that is grounded in good science rather than sheer speculation, yet flexible enough to look for the unexpected; and
- an analysis of whether our current regulatory framework is adequate to deal with these risks and how the framework should be augmented.

Different kinds of applications pose different risks and may call for different responses. Microbes intended for release into the environment, for example, would pose a different set of concerns than microbes designed to be kept in specialized, contained settings. Overall, however, while the risks of synthetic biology are too significant to leave the field alone, its potential benefits are too great to call for a general moratorium.

EXPLORING THE ISSUE

Should Society Impose a Moratorium on the Use and Release of "Synthetic Biology" Organisms?

Critical Thinking and Reflection

1. Do new technologies inevitably pose environmental risks?
2. Does the ability to engineer new forms of life offer at least potential solutions to environmental problems?
3. Should worrisome research be put on hold until all potential risks are analyzed and solutions devised?
4. Should the precautionary principle be applied to "synthetic biology"? In what ways?

Is There Common Ground?

As with many technologies, some people see mostly risks and would, if they could, stop the development of the technology. Others see mostly benefits and think that those benefits are worth putting up with the risks. A more nuanced approach is to determine which risks and benefits seem most likely and then to carefully weigh them against each other. This approach is known as risk–benefit or cost–benefit analysis, and it is used in medicine, engineering, business, and other areas.

1. What seem to be the most likely or worrisome environmental risks associated with "synthetic biology" technology?
2. What seem to be the most likely benefits associated with synthetic biology technology?
3. Do you think the benefits are worth the risks?

ISSUE 17

Do Environmental Hormone Mimics Pose a Potentially Serious Health Threat?

YES: **Michele L. Trankina**, from "The Hazards of Environmental Estrogens," *The World & I* (October 2001)

NO: **Michael Gough**, from "Endocrine Disrupters, Politics, Pesticides, the Cost of Food and Health," *Daily Commentary* (December 15, 1997)

Learning Outcomes

After reading this issue, you should be able to:

- Explain what endocrine disruptors are.
- Explain how endocrine disruptors may affect normal bodily function.
- Describe how people are exposed to endocrine disruptors.
- Explain what is being done to evaluate and prevent the possible problems posed by endocrine disruptors.

ISSUE SUMMARY

YES: Professor of biological sciences Michele L. Trankina argues that a great many synthetic chemicals behave like estrogen, alter the reproductive functioning of wildlife, and may have serious health effects—including cancer—on humans.

NO: Michael Gough, a biologist and expert on risk assessment and environmental policy, argues that only "junk science" supports the hazards of environmental estrogens.

F ollowing World War II, there was an exponential growth in the industrial use and marketing of synthetic chemicals. These chemicals, now known as "xenobiotics" because they are foreign to living things, were used in numerous products, including solvents, pesticides, refrigerants, coolants, and raw materials for plastics. This resulted in increasing environmental contamination. Many of these chemicals, such as DDT, PCBs, and dioxins, proved to be highly resistant to degradation in the environment; they accumulated in wildlife and were serious contaminants of lakes and estuaries. Carried by winds and ocean currents, these chemicals were soon detected in samples taken from the most remote regions of the planet, far from their points of introduction into the ecosphere.

Until very recently, most efforts to assess the potential toxicity of synthetic chemicals to living things, including human beings, focused almost exclusively on their possible role as carcinogens. This was because of legitimate public concern about rising cancer rates and the belief that cancer causation was the most likely outcome of exposure to low levels of synthetic chemicals.

Some environmental scientists urged public health officials to give serious consideration to other possible health effects of xenobiotics. They were generally ignored because of limited funding and the common belief that toxic effects other than cancer required larger exposures than usually resulted from environmental contamination.

In the late 1980s, Theo Colborn, a research scientist for the World Wildlife Fund who was then working on a study of pollution in the Great Lakes, began linking together the results of a growing series of isolated studies. Researchers in the Great Lakes region, as well as in Florida, the West Coast, and Northern Europe, had observed widespread evidence of serious and frequently lethal physiological problems involving abnormal reproductive development, unusual sexual behavior, and neurological problems exhibited by a diverse group of animal species, including fish, reptiles, amphibians, birds, and marine mammals. Through Colborn's insights, communications among these researchers, and further studies, a hypothesis was developed that all of these wildlife problems were manifestations of abnormal estrogenic activity. The causative agents were identified as more than 50 synthetic chemical compounds that have been shown in laboratory studies either to mimic the action or to disrupt the normal function of the powerful estrogenic hormones responsible for female sexual development and many other biological functions. Theo Colborn, Dianne Dumanoski, and John Peterson Myers, *Our Stolen Future: Are We Threatening Our Fertility, Intelligence, and Survival?—A Scientific Detective Story* (Dutton, 1996), find the evidence that extensive damage is being done to wildlife by synthetic estrogenic chemicals convincing and say it is likely that humans are experiencing similar health problems.

Concern that human exposure to these ubiquitous environmental contaminants may have serious health repercussions was heightened by a widely publicized European research study which concluded that male sperm counts had decreased by 50 percent over the past several decades (a result that is disputed by other researchers) and that testicular cancer rates have tripled. Some scientists have also proposed a link between breast cancer and estrogen disruptors.

Stephen H. Safe's "Environmental and Dietary Estrogens and Human Health: Is There a Problem?" *Environmental Health Perspectives* (April 1995), is often cited to support the contention that there is no causative link between environmental estrogens and human health problems. He draws a cautious conclusion, calling the link "implausible" and "unproven." Some people actually claim that the battle against environmental estrogens is motivated by environmentalist ideology rather than facts; see Angela Logomasini, "Chemical Warfare: Ideological Environmentalism's Quixotic Campaign Against Synthetic Chemicals," in Ronald Bailey, ed., *Global Warming and Other Eco-Myths: How the Environmental Movement Uses False Science to Scare Us to Death* (Prima, 2002). Some caution is certainly warranted, for the complex and variable manner by which different compounds with estrogenic properties may affect organisms makes projections from animal effects to human effects risky.

Sheldon Krimsky, in "Hormone Disruptors: A Clue to Understanding the Environmental Cause of Disease," *Environment* (June 2001), summarizes the evidence that many chemicals released to the environment affect—both singly and in combination or synergistically—the endocrine systems of animals and humans and may threaten human health with cancers, reproductive anomalies, and neurological effects. He cautions that the regulatory machinery is likely to move very slowly, adding that we cannot wait for scientific certainty about the hazards before we act. See also M. Gochfeld, "Why Epidemiology of Endocrine Disruptors Warrants the Precautionary Principle, *Pure & Applied Chemistry* (December 1, 2003); Rebecca Renner, "Human Estrogens Linked to Endocrine Disruption," *Environmental Science and Technology* (January 1, 1998); and Ted Schettler et al., *Generations at Risk: Reproductive Health and the Environment* (MIT Press, 1999). Julia R. Barrett, "Phthalates and Baby Boys,"*Environmental Health Perspectives* (August 2005), reports effects of phthalate plastic softeners on the development of the human male reproductive system. Laboratory work has found that at least one phthalate reduces the number of sperm-generating cells in fetal testis tissue; see Romain Lambrot et al., "Phthalates Impair Germ Cell Development in the Human Fetal Testis in Vitro Without Change in Testosterone Production," *Environmental Health Perspectives* (January 2009). When researchers added synthetic estrogen to a Canadian lake, they found that the reproduction of fathead minnows was so severely affected that the population nearly died out; see Karen A. Kidd et al., "Collapse of a Fish Population After Exposure to a Synthetic Estrogen," *Proceedings of the National Academy of Sciences of the United States of America* (May 22, 2007). In 1999, the National Research Council published *Hormonally Active Agents in the Environment* (National Academy Press), in which the Council's Committee on Hormonally Active Agents in the Environment reports on its evaluation of the scientific evidence pertaining to endocrine disruptors. The National Environmental Health Association has called for more research and product testing; see Ginger L. Gist, "National Environmental Health Association Position on Endocrine Disruptors," *Journal of Environmental Health* (January–February 1998).

Elisabete Silva, Nissanka Rajapakse, and Andreas Kortenkamp, in "Something from 'Nothing'—Eight Weak Estrogenic Chemicals Combined at Concentrations Below NOECS Produce Significant Mixture Effects," *Environmental Science and Technology* (April 2002), find synergistic effects of exactly the kind dismissed by Gough. Also, after reviewing the evidence, the U.S. National Toxicity Program found that low-dose effects had been demonstrated in animals; see Ronald Melnick et al., "Summary of the National Toxicology Program's Report of the Endocrine Disruptors Low-Dose Peer-Review," Environmental *Heal Perspectives* (April 2002). Some researchers say there is reason to think endocrine disruptors are linked to the epidemic of obesity in the United States and to a decline in the proportion of male births in the United States and Japan; Julia R. Barrett, "Shift in the Sexes," *Environmental Health Perspectives* (June 2007). Retha R. Newbold et al., "Effects of Endocrine Disruptors on Obesity," *International Journal of Andrology* (April 2008), review the literature on the obesity link and conclude that the evidence is strong enough to warrant expanding "the focus on obesity [and consequent diabetes and heart disease] from intervention and treatment to include prevention and avoidance of these chemical modifiers."

The view that environmental hormone mimics or disruptors have potentially serious effects continues to gain support. Evidence continues to accumulate, and changes are beginning to happen. The U.S. Congress passed legislation requiring that all pesticides be screened for estrogenic activity and the Environmental Protection Agency (EPA) develop procedures for detecting environmental estrogenic contaminants in drinking water supplies; see the EPA's Endocrine Disruptor Screening Program Web site at http://www.epa.gov/scipoly/oscpendo/index.htm. In April 2009, the EPA released its first list of chemicals for initial screening; in November 2010, a second list was released. Medical plastics, softened by phthalates, are being reformulated; see "Medical Devices to Get Phthalate Substitution Rule," *European Environment & Packaging Law Weekly* (June 1, 2007). Arnold Schechter et al., "Bisphenol A (BPA) in U.S. Food," *Environmental Science and Technology* (December 15, 2010), found BPA (an endocrine disruptor) in numerous commercially available foods. The authors note that BPA "is a chemical used for lining metal cans and in polycarbonate plastics, such as baby bottles. In rodents, BPA is associated with early sexual maturation, altered behavior, and effects on prostate and mammary glands. In humans, BPA is associated with cardiovascular disease, diabetes, and male sexual dysfunction in exposed workers. Food is a major exposure source." Plastic bottles, which contain BPA, are already being replaced by metal bottles; see Nancy Macdonald, "Plastic Bottles Get the Eco-Boot," *Maclean's* (October 15, 2007). Janet Raloff, "Concerns over Plastics Chemical Continue to Grow," *Science News* (July 18, 2009), reports that the House Committee on Energy and Commerce has asked the FDA to "reconsider the Bush Administration's position that BPA is safe at current estimated exposure levels." Europe and many U.S. states have already banned many uses of BPA.

In the following selections, professor of biological sciences Michele L. Trankina argues that a great many synthetic chemicals behave like estrogen, alter the reproductive functioning of wildlife, and may have serious health effects, including cancer, on humans. Regulatory agencies must minimize public exposure. Michael Gough, a biologist and expert on risk assessment and environmental policy, argues that only "junk science" supports the hazards of environmental estrogens. Expensive testing and regulatory programs can only drive up the cost of food, which will make it harder for the poor to afford fresh fruits and vegetables. Health protection will not be increased.

YES

Michele L. Trankina

The Hazards of Environmental Estrogens

What do Barbie dolls, food wrap, and spermicides have in common? And what do they have to do with low sperm counts, precocious puberty, and breast cancer? "Everything," say those who support the notion that hormones are disrupting everything from fish gender to human fertility. "Nothing," counter others who regard the connection as trumped up, alarmist chemophobia. The controversy swirls around the significance of a number of substances that behave like estrogens and appear to be practically everywhere—from plastic toys to topical sunscreens.

Estrogens are a group of hormones produced in both the female ovaries and male testes, with larger amounts made in females than in males. They are particularly influential during puberty, menstruation, and pregnancy, but they also help regulate the growth of bones, skin, and other organs and tissues.

Over the past 10 years, many synthetic compounds and plant products present in the environment have been found to affect hormonal functions in various ways. Those that have estrogenic activity have been labeled as environmental estrogens, ecoestrogens, estrogen mimics, or xenoestrogens (*xenos* means foreign). Some arise as artifacts during the manufacture of plastics and other synthetic materials. Others are metabolites (breakdown products) generated from pesticides or steroid hormones used to stimulate growth in livestock. Ecoestrogens that are produced naturally by plants are called phytoestrogens (*phyton* means plant).

Many of these estrogen mimics bind to estrogen receptors (within specialized cells) with roughly the same affinity as estrogen itself, setting up the potential to wreak havoc on reproductive anatomy and physiology. They have therefore been labeled as disruptors of endocrine function.

Bizarre Changes in Reproductive Systems

Heightened concern about estrogen-mimicking substances arose when several nonmammalian vertebrates began to exhibit bizarre changes in reproductive anatomy and fertility. Evidence that something was amiss came serendipitously in 1994, from observations by reproductive physiologist Louis Guillette of the University of Florida. In the process of studying the decline of alligator populations at Lake Apopka, Florida, Guillette and coworkers noticed that many male alligators had smaller penises than normal. In addition, females superovulated, with multiple nuclei in some of the surplus ova. Closer scrutiny linked these findings to a massive spill of DDT (dichloro-diphenyl-trichloroethane) into Lake Apopka in 1980. Guillette concluded that the declining alligator population was related to the effects of DDT exposure on the animals' reproductive systems.

Although DDT was banned for use in the United States in the early 1970s, it continues to be manufactured in this country and marketed abroad, where it is sprayed on produce that is then sold in U.S. stores. The principal metabolite derived from DDT is called DDE, a xenoestrogen that lingers in fat deposits in the human body for decades. Historically, there have been reports of estrogen-mimicking effects in various fish species, especially in the Great Lakes, where residual concentrations of DDT and PCBs (polychlorinated biphenyls) are high. These effects include feminization and hermaphroditism in males. In fact, fish serve as barometers of the effects of xenoestrogen contamination in bodies of water. An index of exposure is the presence of vitellogenin, a protein specific to egg yolk, in the blood of male fish. Normally, only females produce vitellogenin in their livers, upon stimulation by estrogen from the ovaries.

Ecoestrogens are further concentrated in animals higher up in the food chain. In the Great Lakes region, birds including male herring gulls, terns, and bald eagles exhibit hermaphroditic changes after feeding on contaminated fish. Increased embryo mortality among these avians has also been observed. In addition, evidence from Florida links sterility in male and female panthers to their predation on animals exposed to pesticides with estrogenic activity.

The harmful effects of DDT have also been observed in rodents. Female rodents treated with high concentrations of DDT become predisposed to mammary tumors, while males tend to develop testicular cancer. These observations raise the question of whether pharmacological (that is, low) doses of substances with estrogenic activity translate into physiological effects. Usually they do not, but chronic exposure may be enough to trigger such effects.

Dangers to Humans

If xenoestrogens cause such dramatic reproductive effects in vertebrates, including mammals, what might be the consequences for humans? Nearly a decade ago, Frederick vom Saal, a developmental biologist at the University of Missouri Columbia, cautioned that mammalian reproductive mechanisms are similar enough to warrant concern about the effects of hormone disruptors on humans.

A 1993 article in *Lancet* looked at decreasing sperm counts in men in the United States and 20 other countries and correlated these decreases with the growing concentration of environmental estrogens. The authors—Niels Skakkenbaek, a Danish reproductive endocrinologist, and Richard Sharpe, of the British Medical Research Council Reproductive Biology Unit in Scotland—performed a meta-analysis of 61 sperm-count studies published between 1938 and 1990 to make their connection.

"Nonbelievers" state that this interpretation is contrived. Others suggest alternative explanations. For instance, the negative effects on sperm counts could have resulted from simultaneous increases in the incidence of venereal diseases. Besides, there are known differences in steroid hormone metabolism between lower vertebrates (including nonprimate mammals) and primates, so one cannot always extrapolate from the former group to the latter.

Even so, the incidence of testicular cancer, which typically affects young men in their 20s and 30s, has increased worldwide. Between 1979 and 1991, over 1,100 new cases were reported in England and Wales—a 55 percent increase over previous rates. In Denmark, the rate of testicular cancer increased by 300 percent from 1945 to 1990. Intrauterine exposure to xenoestrogens during testicular development is thought to be the cause.

Supporting evidence comes from Michigan, where accidental contamination of cattle feed with PCBs in 1973 resulted in high concentrations in the breast milk of women who consumed tainted beef. Their sons exhibited defective genitalia. Furthermore, observers in England have noted increased incidences of cryptorchidism (undescended testes), which results in permanent sterility if left untreated, and hypospadias (urethral orifice on the underside of the penis instead of at the tip). The one area of agreement between those who attribute these effects to ecoestrogens and those who deny such a connection is that more research is needed.

Perhaps one of the most disturbing current trends is the alarming increase in breast cancer incidence. Fifty years ago, the risk rate was 1 woman in 20; today it is 1 in 8. Numerous studies have implicated xenoestrogens as the responsible agents. For instance, high concentrations of pesticides, especially DDT, have been found in the breast tissue of breast cancer patients on Long Island. In addition, it has long been known that under certain conditions, estrogen from any source can be tumor promoting, and that most breast cancer cell types have estrogen receptors.

Precocious Puberty

If that were not enough, the growing number of estrogen mimics in the environment has been linked to early puberty in girls. The normal, average age of onset is between 12 and 13. A recent study of 17,000 girls in the United States indicated that 7 percent of white and 27 percent of black girls exhibited physical signs of puberty by age seven. For 10-year-old girls, the percentages increased to 68 and 95, respectively. Studies from the United Kingdom, Canada, and New Zealand have shown similar changes in the age of puberty onset.

It is difficult, however, to elucidate the exact mechanisms that underlie these trends toward precocious puberty. One explanation, which applies especially to cases in the United States, points to the increasing number of children who are overweight or obese as a result of high-calorie diets and lack of regular exercise. Physiologically, an enhanced amount of body fat implies reproductive readiness and signals the onset of puberty in both boys and girls.

For girls, more body fat ensures that there is enough stored energy to support pregnancy and lactation. Young women with low percentages of body fat caused by heavy exercise, sports, or eating disorders usually do not experience the onset of menses until their body composition reflects adequate fat mass.

Many who study the phenomenon of premature puberty attribute it to environmental estrogens in plastics and secondhand exposure through the meat and milk of animals treated with steroid hormones. An alarming increase in the numbers of girls experiencing precocious puberty occurred in the 1970s and '80s in Puerto Rico. Among other effects, breast development occurred in girls as young as one year. Premature puberty was traced to consumption of beef, pork, and dairy products containing high concentrations of estrogen.

Another study from Puerto Rico revealed higher concentrations of phthalate —a xenoestrogen present in certain plastics—in girls who showed signs of early puberty, compared with controls.

It may be that excess body fat and exposure to estrogenic substances operate in concert to hasten puberty. Body fat is one site of endogenous estrogen synthesis. Exposure to environmental estrogens may add just enough exogenous hormone to exert the synergistic effect necessary to bring on puberty, much like the last drop of water that causes the bucket to overflow.

Although most xenoestrogens produce detrimental effects, at least one subgroup—the phytoestrogens—includes substances that can have beneficial effects. Phytoestrogens are generally weaker than natural estrogens. They are found in various foods, such as flax seeds, soybeans and other legumes, some herbs, and many fruits and vegetables. Some studies suggest that soy products may offer protection against certain cancers, including breast, prostate, uterus, and colon cancers. On the other hand, high doses of certain phytoestrogens— such as coumestrol (in sunflower seeds and alfalfa sprouts)—have been found to adversely affect the fertility and reproductive cycles of animals.

Unlike artificially produced xenoestrogens, phytoestrogens are generally not stored in the body but are readily metabolized and excreted. Their health effects should be evaluated on a case-by-case basis, considering such factors as the individual's age, medical and family history, and potential interactions with medications or supplements.

Plastics, Plastics Everywhere

It has been estimated that perhaps 100,000 synthetic chemicals are registered for commercial use in the world today and 1,000 new ones are formulated every year. While many are toxic and carcinogenic, little is known about the chronic effects of the majority of them. And there is growing concern about their potential hormone-disrupting effects. The problem of exposure is complicated by numerous carrier routes, including air, food, water, and consumer products.

Consider certain synthetics that turn up in familiar places, including food and consumer goods. Such seemingly inert products as plastic soda and water bottles, baby bottles, food wrap, Styrofoam, many toys, cosmetics, sunscreens, and even spermicides either contain or break down to yield xenoestrogens. In addition, environmental estrogens are among the byproducts created by such processes as the incineration of biological materials or industrial waste and chlorine bleaching of paper products.

In April 1999, Consumers Union confirmed information previously reported by the Food and Drug Administration regarding 95 percent of baby bottles sold in the United States. The bottles, made of a hard plastic known as polycarbonate, leach out the synthetic estrogen named bisphenol-A, especially when heated or scratched. Studies verifying the estrogenic activity of bisphenol-A were published in *Nature* in the 1930s, but it did not arouse much concern then. A 1993 report published in *Endocrinology* showed that bisphenol-A produced estrogenic effects in a culture of human breast cancer cells.

Additional studies published by vom Saal in 1997 and '98 have shown that bisphenol-A stimulates precocious puberty and obesity in mice. Others have detected leaching of bisphenol-A from polycarbonate products such as plastic tableware, water cooler jugs, and the inside coatings of certain cans (used for some canned foods) and bottle tops. Autoclaving in the canning process causes bisphenol-A to migrate into the liquid in cans.

Spokespersons from polycarbonate manufacturers have stated that they cannot replicate vom Saal's results, but he counters that industry researchers have not done the experiments correctly.

DEHA (di-[2-ethylhexyl] adipate) is a liquid plasticizer added to some plastic food wraps made of polyvinyl chloride (PVC). Scientific studies have shown that the fat-soluble DEHA can migrate into foods, especially luncheon meats, cheese, and other products with a high fat content. For a 45-pound child eating cheese wrapped in such plastic, the limit of safe intake is 1.5 ounces by European standards or 2.5 ounces by Environmental Protection Agency criteria. Studies conducted by Consumers Union indicate that DEHA leaching from commercial plastic wraps is eight times higher than European directives allow. Fortunately, alternative wraps—such as Handi-Wrap™ and Glad Cling Wrap—are made of polyethylene; chemicals in them do not appear to leach into foods.

Barbie dolls manufactured in the 1950s and '60s are made of PVC containing a stabilizer that degrades to a sticky, estrogen-mimicking residue and accumulates on the dolls' bodies. This phenomenon was noted by Danish museum officials in August 2000. Yvonne Shashoua, an expert in materials preservation at the National Museum of Denmark, warns that young children who play with older Barbie and Ken dolls expose themselves to this estrogenic chemical, and she suggests enclosing the dolls with xenoestrogen-free plastic wrap. Storing the dolls in a cool, dark place also helps prevent the harmful stabilizer from oozing out. Who would have thought that these models of glamour would become health hazards?

Another consumer nightmare is a group of chemical plasticizers known as phthalates. They have been associated with various problems, including testicular depletion of zinc, a necessary nutrient for spermatogenesis. Zinc deficiency results in sperm death and consequent infertility. Many products that previously contained phthalates have been reformulated to eliminate them, but phthalates continue to be present in vinyl flooring, medical tubing and bags, adhesives, infants' toys, and ink used to print on food wrap made of plastic and cardboard. They have been detected in fat-soluble foods such as infant formula, cheese, margarine, and chips.

Some environmental estrogens can be found in unusual places, such as contraceptive products containing the spermicide nonoxynol-9. This chemical degrades to nonylphenol, a xenoestrogen shown to stimulate breast cancer cells. Nonylphenol and other alkylphenols have been detected in human umbilical cords, plastic test tubes, and industrial detergents.

Various substances added to lotions (including sunscreens) and cosmetics serve as preservatives. Some of them, members of a chemical family called parabens, were shown to be estrogen mimics by a study at Brunel University in the United Kingdom. The researchers warned that "the safety of these chemicals should be reassessed with particular attention being paid to . . . levels of systemic exposure to humans." Officials of the European Cosmetic Toiletry and Perfumery Association dismissed the Brunel study as "irrelevant," on the grounds that parabens do not enter the systemic circulation. But they ignored the possibility of transdermal introduction.

Additional questions have been raised about the safety—in terms of xenoestrogen content—of recycled materials, especially plastics and paper. Because it is unlikely that a moratorium on chemical synthesis will occur anytime soon, such questions will continue to surface until the public is satisfied that regulatory agencies are doing all they can to minimize exposure. Fortunately, organizations such as the National Institutes of Health, the National Academy of Sciences, the Environmental Protection Agency, the Centers for Disease Control, many universities, and other institutions are involved in efforts to monitor and minimize the effects of environmental estrogens on wildlife and humans.

Michael Gough

 **NO**

Endocrine Disrupters, Politics, Pesticides, the Cost of Food and Health

Environmentalists and politicians and federal regulators have added environmental estrogens or endocrine disrupters to the "concerns" or scares that dictate "environmental health policy." That policy, from its beginning, has been based on ideology, not on science. To provide some veneer to the ideology, its proponents have spawned bad science and junk science that claims chemicals in the environment are a major cause of human illness. There is no substance to the claims, but the current policies threaten to cost billions of dollars in wasted estrogen testing programs and to drive some substantial proportion of pesticides from the market.

Rachel Carson, conjuring up a cancer-free, pre-industrial Garden of Eden launched the biggest environmental scare of all in the 1960s. She charged that modern industrial chemicals in the environment caused human cancers. It mattered not at all to her or to her readers that cancers are found in every society, pre-industrial and modern. What mattered were opinions of people such as Umberto Safiotti of the National Cancer Institute, who wrote:

> I consider cancer as a social disease, largely caused by external agents which are derived from our technology, conditioned by our societal lifestyle and whose control is dependent on societal actions and policies.

When Saffioti said "societal actions and policies," he meant government regulations.

By 1968, environmental groups and individuals—including some scientists—appeared on TV and on the floors of the House and Senate to say, over and over again, "The environment causes 90 percent of cancers." They didn't have to say "environment" meant pollution from modern industry and chemicals—especially pesticides—everyone already knew that. Saffioti and others had told them.

In the 1970s, the National Cancer Institute [NCI] released reports that blamed elevated rates for all kinds of cancers on chemicals in the workplace or in the environment. The institute did not have evidence to link those exposures to cancer. It didn't exist then, and it doesn't, except for a limited number of high exposures in the workplace, exist now. So what? The reports were gobbled up by the press, politicians, and the public.

In our ignorance of what causes most cancers, the "90 percent" misstatement provided great hope. If the carcinogenic agents in the environment could be identified and eliminated, cancer rates should drop. NCI scientists said so, and they said success was just around the corner if animal tests were used to identify carcinogens. Congress responded. It created the Environmental Protection Agency [EPA] and the Occupational Safety and Health Administration [OSHA]. Both agencies have lots of tasks, but both place an emphasis on controlling exposures to carcinogens. Congress passed and amended law after law. The Clean Air Act, the Clean Water Act, the Safe Drinking Water Act, amendments to the Fungicide, Insecticide, and Rodenticide Act—the euphonious FIFRA—the Toxic Substances Control Act, and the Resource Recovery and Control Act poured forth from Capitol Hill.

And in return, EPA and OSHA, to justify their existence, generated scare after scare. They are aided by all kinds of people eager for explanations about their health problems or for government grants and contracts for research or other work or for money to compensate for health effects or other problems that could be blamed on chemicals.

In 1978, we had the occupational exposure scare. Astoundingly, according to a government report, six workplace substances caused 38 percent of all the cancers in the United States. It was nonsense, of course, and many scientists ridiculed the report, but the government never retracted it. The government scientists who contributed to it never repudiated it.

At about the same time, wastes disposed in Love Canal near Niagara Fall, NY, spewed liquids and gases into a residential community. The chemicals were blamed as the cause of cancers, birth defects, miscarriages, skin diseases, you name it. None of it was true, but waste sites around the country were routinely identified as "another Love Canal" or a "Love Canal in the making," and Congress gave the nation the Superfund Law. Since its passage, Superfund has enriched lawyers and provided secure employment to thousands who wear moon suits and dig up, burn, and rebury wastes, and done nothing for the nation's health. For those who doubt the importance of politics in the environmental health saga, it's worth recalling that every state had two waste sites on the first list of sites slated for priority cleanup under Superfund.

By the 1980s, EPA was chucking out the scares. We had the 2,4,5-T scare, the dioxin scare, the 2,4-D scare, the asbestos in schools scare, the radon in homes scare, the Alar scare, the EMF scare. I've left some out, but the common thread that linked the scares together was cancer. Each scare prompted investigations by affected industries and non-government scientists. Each scare fell apart, revealed as a house of cards jerry-rigged from bad science, worse interpretations of the science, and terrible policy.

In fact, by the late 1970s, there was ample evidence that the much-talked about "cancer epidemic" and the 90 percent statement were simply wrong. Cancer rates were not increasing and rates for some cancers were higher in industrial countries and rates for others were higher in non-industrial countries. The U.S. fell in the middle of countries when ranked by cancer rates.

Sure, there are some carcinogenic substances in the nation's workplaces, but the best estimates are that they cause four percent or less of all cancers, and the percentage is decreasing because the biggest occupational threat, asbestos, is gone. Environmental exposures might cause two or three percent of cancer—on the outside—and they might cause much less.

The research into causes of cancer—not stories designed to bolster the chemicals cause cancer myth—did reveal that there are preventable causes of cancer. Not smoking is a good idea, as is eating lots of fruits and vegetables, not gaining too much weight, restricting the number of sexual partners, and, for those who are fair-skinned, being careful about sun exposures. It's not a lot different from what your mother or grandmother told you.

The government can take a nanny role in urging us to behave, but that's not where the big bucks are. The big bucks are in regulation, and regulation doesn't seem to have much to do with cancer.

In any case, cancer death rates began to fall in 1990, they've fallen since, and the fall is growing steeper. Maybe that information is blunting the cancer scare. I somehow doubt it. I think that the public has become numb to the cancer scare or that it fatalistically accepts the notion that "everything causes cancer." In any case, the environmentalists and the regulators needed a new scare.

The collapse of the cancer scare wasn't good news to everyone. Government bureaucrats and scientists in the anti-carcinogen offices and programs at EPA and elsewhere have secure jobs. Congress easily finds the will to write laws establishing environmental protection activities, but it lacks the will or patience to examine those activities to see if they've accomplished anything. And, let's face it, Congress doesn't eliminate established programs. But the growth of programs slows, and money can become scarce, and that can squeeze researchers who depend on EPA grants and contracts to fund their often senseless surveys and testing programs. Moreover, the fading of scares doesn't benefit environmental organizations that utter shrill cries about scares and coming calamities in their campaigns for contributions.

Here's an example of just how disappointed some people can be that cancer isn't on the rise. Dr. Theo Colborn, a wildlife biologist working for the Conservation Foundation in the late 1980s, was convinced that the chemicals in the Great Lakes were causing human cancer. She set out to prove it by reviewing the available literature about cancer rates in that region. She couldn't. In fact, she found that the rates for some cancers in the Great Lakes region were lower than the rates for the same cancers in other parts of the United States and Canada. . . .

Failing to find cancer slowed her down but didn't stop her. She knew that those chemicals were causing something. All she had to do was find it.

And find it, she did. She collected every paper that described any abnormality in wildlife that live on or around the Great Lakes, and concluded that synthetic chemicals were mimicking the effects of hormones. They were causing every problem in the literature, whether it was homosexual behavior among gulls, crossed bills in other birds, cancer in fish, or increases or decreases in any wildlife population.

The chemicals that have those activities were called "environmental estrogens" or "endocrine disrupters." There was no more evidence to link them to every abnormality in wildlife than there had been in the 1960s to link every human cancer to chemicals. The absence of evidence wasn't much of a problem. Colborn and her colleagues believed that chemicals were the culprit, and the press and much of the public, nutured on the idea that chemicals were bad, didn't require evidence.

Even so, Colborn had a problem that EPA faced in its early days. Soon after EPA was established, the agency leaders realized that protecting wildlife and the environment might be a good thing, but that Congress might not decide to lavish funds on such activities. They were sure, however, that Congress would throw money at programs that were going to protect human health from environmental risks. Whether Colborn knew that history or not, she apparently realized that any real splash for endocrine disrupters depended on tying them to human health effects.

Using the same techniques she'd used to catalogue the adverse effects of endocrine disrupters on wildlife, she reviewed the literature about human health effects that some way or another might be related to disruption of hormone activity. The list was long, including cancers, birth defects, and learning disabilities, but the big hitter on the list was decreased sperm counts. According to Colborn and others' analyses of sperm counts made in different parts of the world under different conditions of nutrition and stress and at different time periods, sperm counts had decreased by 50 percent in the post–World War II period.

If there's anything that catches the attention of Congress, it's risks to males. Congress banned leaded gasoline after EPA released a report that said atmospheric lead was a cause of heart attacks in middle-aged men. The reported decrease in sperm counts leaped up for attention, and attention it got. Congressional hearings were held, magazine articles were written, experts opined about endocrine disrupters and sexual dysfunctions.

And then it fell apart. Scientists found large geographical variations in sperm counts that have not changed over time. Those geographical variations and poor study designs accounted for the reported decrease. That scare went away, but endocrine disrupters were here to stay.

Well-organized and affluent women's groups are convinced that breast cancer is unusually common on Long Island. . . . We know that obesity, estrogen replacement therapy, and late child-bearing or no child-bearing, all of which are more common in affluent women, are associated with breast cancer. Nevertheless, from the very beginning, environmental chemicals have been singled out as the cause of the breast cancer excess. The insecticide aldicarb, which is very resistant to degradation was blamed, but subsequent studies failed to confirm the link. A well-publicized study found a link between DDT and breast cancer, but larger, follow-up studies failed to confirm it. But there're lots of environmental chemicals, and no evidence is required to justify a suggestion of a link between the chemicals and cancer.

Senator Al D'Amato is from Long Island, and he shares his constituents' concerns. During a hearing about the Clean Water Act, Senator D'Amato heard testimony by Dr. Anna Soto from Tufts University about her "E-Screen." According to Dr. Soto, her quick laboratory test could identify chemicals that behave as environmental estrogens or endocrine disrupters for $500 a chemical. Since environmental estrogens seem to some people to be a likely cause of breast cancer, Soto's test appeared to be a real bargain.

Senator D'Amato pushed for an amendment to require E-Screen testing of chemicals that are regulated under the Clean Water Act, but he was unsuccessful. Later in 1995, a senior Senate staffer, Jimmy Powell, took the E-Screen amendment to a very junior Senate staffer and told her to incorporate it into the Safe Drinking Water Act as an "administrative amendment." She did, it passed the Senate, and, for the first time, there was a legislative requirement for endocrine testing.

In the spring of 1996, House Committees were considering legislation to amend the Safe Drinking Water Act and new legislation related to pesticides in food. Aware of the Senate's action, some members of the House committees were eager to include endocrine testing in their legislation, but there was resistance as well. Chemical companies viewed the imposition of yet another test as certain to be an expense, unlikely to cost as little as $500 a chemical, and bound to raise new concerns about chemicals that would require far more extensive tests and research to understand or discount.

Furthermore, so far as food was concerned, there was a general conviction that all the safety factors built into the testing of pesticides and other chemicals that might end up in food provided adequate margins of protection. That conviction was shattered by rumors that reached the House in May 1996. According to the rumors, Dr. John McLachlan and his colleagues at Tulane University had shown that mixtures of pesticides and other environmental chemicals such as PCBs were far more potent in activating estrogen receptors, the first step in estrogen modulation of biochemical pathways than were single chemicals. In the most extreme case, two chemicals at concentrations considered safe by all conventional toxicity tests were 1600 times as potent as estrogen receptor activators as either chemical by itself. The powerful synergy raised new alarms.

In May, everyone concerned about pesticides knew that EPA had a draft of the Tulane paper, and EPA staff were drifting around House offices, but they refused to answer questions about the Tulane results. The silence signaled the expected significance of the paper. A month later, in June, the paper appeared in *Science*. It was a big deal. *Science* ran a news article about the research with a picture of the Tulane researchers. It also ran an editorial by a scientist from the National Institutes of Health who offered some theoretical explanations for how combinations of pesticides at very low levels could affect cells and activate the estrogen receptors. *The New York Times, The Washington Post,* other major newspapers, and news magazines and TV reported the news. If there was ever any doubt that FQPA [Food Quality Protection Act] would require tests for endocrine activity, the flurry of news about the Tulane results erased them.

While the House was drafting the FQPA, Dr. Lynn Goldman, an assistant administrator at EPA, established a committee called the "Endocrine Disrupter

Screening and Testing Advisory Committee" (EDSTAC) [which is now] considering tests for all of the 70,000 chemicals that it estimates are present in commerce, and it's not limiting its recommendations to tests for estrogenic activity. It's adding tests for testosterone and thyroid hormone activity as well as for anti-estrogenic, anti-testosterone, and anti-thyroid activity. The relatively simple E-Screen, which is run on cultured cells, is to be supplemented by some whole animal tests. Tests on single compounds will have to be complemented by tests of mixtures of compounds. The FQPA requires that "valid" tests be used. None of the tests being considered by EDSTAC has been validated; many of them have never been done.

EDSTAC's estimate of 70,000 chemicals in commerce is on the high side—some of those chemicals are used in such small amounts and under such controlled conditions that there's no exposure to them. Dr. Dan Byrd has estimated that 50,000 is a more realistic number. He's also looked at the price lists from commercial testing laboratories to see how much they would charge for a battery of tests something like EDSTAC is considering. Some of the tests haven't been developed, but assuming they can be, Dr. Byrd estimates testing each chemical will cost between $100,000 and $200,000. The total cost would be between $5 and $10 billion. . . .

The Tulane results played some major role in the passage of FQPA, the focus on endocrine disrupters, and Dr. Goldman's establishment of EDSTAC. The Tulane results are wrong. Several groups of scientists tried to replicate the Tulane results. None could. At first, Dr. McLachlan insisted his results were correct. He said that the experiments he reported required expertise and finesse and suggested that the scientists who couldn't repeat his findings were at fault, essentially incompetent. That changed. In July 1997, just 13 months after he published his report, he threw in the towel, acknowledging that neither his laboratory nor anyone else had been able to produce the results that had created such a stir.

Whether the initial results were caused by a series of mistakes or a willful desire to show, once and for all, that environmental chemicals, especially pesticides are bad, bad, bad, we don't know. We do know that the results were wrong.

No matter, EPA now assumes as a matter of policy that synergy occurs. Good science, repeatable science that showed the reported synergy didn't occur has been brushed aside. In its place, we have bad science or junk science. If the Tulane results were the products of honest mistakes, they're bad science; if they flowed from ideology, they're junk science. The effect is the same, but the reasons are different.

The estrogenic disrupter testing under FQPA is going to cost a lot of money and cause a lot of mischief. But the effects of that testing are off somewhere in the future. More immediately, a combination of ideology-driven science and congressional misreading of that science threatens to drive between 50 and 80 percent of all pesticides from the market.

In 1993, a committee of the National Research Council spun together the facts that childhood development takes place at specific times as an infant matures into a toddler and then a child, that infants, toddlers, and children eat, proportionally, far larger amounts of foods such as apple juice and apple sauce and orange juice than do adults, and that pesticides can be present in those processed foods. From those three observations, the committee concluded that an additional safety factor should be included in setting acceptable levels for pesticides in those foods. Left out from the analysis was any evidence that current exposures cause any harm to any infants, toddlers, or children. No matter.

Most people who worry about pesticides expected EPA and the Food and Drug Administration to react to the NRC recommendation by reducing the allowable levels of pesticides on foods that are destined for consumption by children. Maybe they would have. We'll never know. In the FQPA, Congress directed that a new ten-fold safety factor be incorporated into the evaluation of the risks from pesticides.

Safety factors are a fundamental part in the evaluation of pesticide risks. Pesticides are tested in laboratory animals to determine what concentrations cause effects on the nervous, digestive, endocrine, and other systems. At some, sufficiently low dose that varies from pesticide to pesticide, the chemical does not cause those adverse effects. That dose, called the "No Observed Adverse Effect Level" (or NOAEL), is then divided by 100 to set the acceptable daily limit for human ingestion of the chemical. The FQPA requires division by another factor of 10, so the acceptable daily limit will be the NOAEL divided by 1000 instead of 100. Acceptable limits will be ten-fold less.

Dr. Byrd has estimated that up to 80 percent of all currently permitted uses of pesticides would be eliminated by an across the board application of the 1000-fold safety factor. He cites another toxicologist who estimates that 50 percent of all pesticides would be eliminated from the market. The extent to which these draconian reductions will be forced remains to be seen, but pesticide manufacturers and users can look forward to a period of even-greater limbo as EPA sorts through it new responsibilities and decides how to implement FQPA.

There's no convincing evidence that pesticides in food contribute to cancer causation and none that they cause other adverse health effects. Restrictions on pesticides in food will not have a demonstrable effect on human health. On the other hand, the estrogen testing program and the new safety factor will drive pesticide costs up and pesticide availability down.

Some manufacturers may lose profitable product lines; some may even lose their businesses. Farmers will pay more. They will pass those costs onto middlemen and processors, who, in turn, will pass them onto consumers. Increases in the costs of fruits and vegetables won't change the food purchasing habits of the middle class, but they may and probably will affect the purchases of the poor. The poor are already at greater risks because of poor diets, and the increased costs can be expected to further decrease their consumption of fresh fruits and vegetables.

EXPLORING THE ISSUE

Do Environmental Hormone Mimics Pose a Potentially Serious Health Threat?

Critical Thinking and Reflection

1. The possible health effects of exposure to synthetic chemicals include cancer, reproductive malfunctions, and neurological problems. Which effects are most worthy of public concern?
2. Should we refrain from evaluating and regulating possibly toxic chemicals because the result may be more expensive food?
3. Most of the thousands of chemicals in use by industry have never been thoroughly tested for toxicity. Should they be?
4. Should the precautionary principle be applied to endocrine disruptors? In what ways?

Is There Common Ground?

Few people disagree that human health is worth protecting. They do, however, disagree on what constitutes acceptable evidence that human health is threatened. The endocrine disruptors debate began when signs were detected that synthetic chemicals were affecting wildlife. Effects on humans remain debatable.

1. Research the history of concerns over DDT (and other pesticides). Did wildlife effects play a role here?
2. What is the justification for animal testing of drugs (and other consumer products)? (Start here: http://www.fda.gov/Cosmetics/ProductandIngredientSafety/ProductTesting/ucm072268.htm)
3. In what ways is looking at effects of chemicals on wildlife similar to testing those effects on lab animals? In what ways are they different?

ISSUE 18

Should the Superfund Tax Be Reinstated?

YES: Stephen Lester and Anne Rabe, from *Superfund: In the Eye of the Storm* (Center for Health, Environment & Justice, June 2010) (included in testimony by Lois Gibb before Senate Committee on Environment & Public Works, Subcommittee on Superfund, Toxics and Environmental Health, hearing on "Oversight of the Environmental Protection Agency's Superfund Program," June 22, 2010)

NO: J. Winston Porter, from Testimony before the Senate Committee on Environment & Public Works, Subcommittee on Superfund, Toxics and Environmental Health, Hearing on "Oversight of the Environmental Protection Agency's Superfund Program," June 22, 2010

Learning Outcomes

After reading this issue, you should be able to:

- Explain the need for the Comprehensive Environmental Response, Compensation, and Liability Act of 1980 (CERCLA), commonly called "Superfund."
- Explain how Superfund cleanups have been funded in the past.
- Explain the purpose of "polluter pays" taxes.
- Discuss the relative merits of paying for cleanups using "polluter pays" taxes and general-fund revenues.

ISSUE SUMMARY

YES: Stephen Lester and Anne Rabe argue that because toxic waste cleanup is complicated by extreme weather events, corporations dodge their cleanup and payment obligations, and the taxpayer is left with the bill, Congress must reinstate the "polluter pays" fees.

NO: J. Winston Porter argues that Superfund cleanup efforts can be made much more efficient and that "polluter pays" taxes are unfair. The primary funder of cleanup work should be the people responsible for the problems. Taxpayers should foot the bill as a matter of last resort.

The potentially disastrous consequences of ignoring hazardous wastes and disposing of them improperly burst upon the consciousness of the American public in the late 1970s. The problem was dramatized by the evacuation of residents of Niagara Falls, New York, whose health was being threatened by chemicals leaking from the abandoned Love Canal, which was used for many years as an industrial waste dump. Awakened to the dangers posed by chemical dumping, numerous communities bordering on industrial manufacturing areas across the country began to discover and report local sites where chemicals had been disposed of in open lagoons or were leaking from disintegrating steel drums. Such esoteric chemical names as dioxins and PCBs have become part of the common lexicon, and numerous local citizens' groups have been mobilized to prevent human exposure to these and other toxins.

The expansion of the industrial use of synthetic chemicals following World War II resulted in the need to dispose of vast quantities of wastes laden with organic and inorganic chemical toxins. For the most part, industry adopted a casual attitude toward this problem and, in the absence of regulatory restraint, chose the least expensive means of disposal available. Little attention was paid to the ultimate fate of chemicals that could seep into surface water or groundwater. Scientists have estimated that less then 10 percent of the waste was disposed of in an environmentally sound manner.

The magnitude of the problem is mind-boggling: Over 275 million tons of hazardous waste is produced in the United States each year; as many as 10,000 dump sites may pose a serious threat to public health, according to the federal Office of Technology Assessment; other government estimates indicate that more than 350,000 waste sites may ultimately require corrective action at a cost that could easily exceed $500 billion.

Congressional response to the hazardous waste threat is embodied in two complex legislative initiatives. The Resource Conservation and Recovery Act (RCRA) of 1976 mandated action by the Environmental Protection Agency (EPA) to create "cradle to grave" oversight of newly generated waste. The Comprehensive Environmental Response, Compensation, and Liability Act of 1980 (CERCLA), commonly called "Superfund," gave the EPA broad authority to clean up existing hazardous waste sites. The implementation of this legislation has been severely criticized by environmental organizations, citizens' groups, and members of Congress who have accused the EPA of foot-dragging and a variety of politically motivated improprieties. Less than 20 percent of the original $1.6 billion Superfund allocation was actually spent on waste cleanup. In 2009, the U.S. Inspector General issued a report accusing the EPA of mismanagement; see "Inspector General's Report Faults EPA for Management of 819 Superfund Accounts," *BNA's Environmental Compliance Bulletin* (April 6, 2009) (for EPA commentary, visit http://www.epa.gov/oig/reports/2009/20090318-09-P-0119_glance.pdf).

Amendments designed to close RCRA loopholes were enacted in 1984, and the Superfund Amendments and Reauthorization Act (SARA) added $8.6 billion to a strengthened cleanup effort in 1986 and an additional $5.1 billion in 1990. Superfund cleanups, when those responsible for contaminated sites could not pay or could not be found (there are many such "orphan" sites), were to have been funded by taxes on industry (e.g., the Crude Oil Tax, the Chemical Feedstock Tax, the Toxic Chemicals Importation Tax, and the Corporate Environmental Income Tax, also known as "polluter pays" fees or taxes).

The Competitive Enterprise Institute objects that taxes such as the Superfund taxes are an assault on consumer pocketbooks, as is CERCLA's "joint and several liability" clause, which can make minor contributors to toxics sites liable for large cleanup costs even when they acted according to all laws and regulations in force at the time. However, in May 2009, the U.S. Supreme Court ruled that some contributors may not be liable for cleanup costs; see "US Supreme Court Ruling Limits Superfund Liability," *Oil Spill Intelligence Report* (May 7, 2009). While acknowledging some improvement, both environmentalists and industrial policy analysts remain very critical about the way that both RCRA and Superfund/SARA are being implemented. Efforts to reauthorize and modify both of these hazardous waste laws have been stalled in Congress since the early 1990s. But the work went on. The Superfund program continued to identify hazardous waste sites that warranted cleanup and to clean up sites; see http://www.epa.gov/superfund/ for the latest news. In 2004, EPA even declared that the infamous Love Canal site was finally safe.

The "polluter pays" taxes expired in 1995, and Congress has so far refused to reauthorize them, but the EPA has said they are necessary (Jay Landers, "EPA Calls on Congress to Reinstate Superfund Taxes," *Civil Engineering*, August 2010), and Representative Earl Blumenauer (D-OR) and Senator Frank R. Lautenberg (D-NJ) have proposed bills that would reinstate them. Support has come from the Obama administration (see "Obama Budget Would Reinstate Superfund Tax," *Chemical Week*, March 2, 2009), the American Society of Civil Engineers, and others. Opposition has come from the chemical industry (see David J. Hanson, "Waste Site Cleanup Tax Opposed," *Chemical & Engineering News*, April 19, 2010), which has been hiring lobbyists to fight the bills (see Kevin Bogardus, "Superfund Tax Push Spurs Rush for New Lobbyists," *The Hill*, June 23, 2010). At the Senate Committee on Environment & Public Works, Subcommittee on Superfund, Toxics and Environmental Health, hearing on "Oversight of the Environmental Protection Agency's Superfund Program," held June 22, 2010, Senator James M. Inhofe (R-OK) testified that cleanups are necessary but the EPA needs to improve efficiency (in part by not wasting money on public-relations campaigns). Since responsible parties already pay for most cleanups, "polluter pays" taxes are an unnecessary burden on small businesses.

Meanwhile, the hazardous waste problem takes new forms. Even in the 1990s, an increasingly popular method of disposing of hazardous wastes was to ship them from the United States to "dumping grounds" in developing countries. Iwonna Rummel-Bulska's "The Basel Convention: A Global Approach for the Management of Hazardous Wastes," *Environmental Policy and*

Law (vol. 24, no. 1, 1994), describes an international treaty designed to prevent or at least limit such waste dumping. But 8 years later, in February 2002, the Basel Action Network and the Silicon Valley Toxics Coalition (http://svtc.org) published *Exporting Harm: The High-Tech Trashing of Asia*. This lengthy report documents the fate of electronics wastes, including defunct personal computers, monitors, and keyboards, as well as circuit boards and other products rich in lead, beryllium, cadmium, mercury, and other toxic materials. Some 50–80 percent of the "e-waste" collected for recycling in the western United States is shipped to destinations such as China, India, and Pakistan, where crude recycling and disposal methods lead to widespread human and environmental contamination. February 2004 saw an updated version of this report, *Poison PCs and Toxic TVs: E-Waste Tsunami to Roll Across the US: Are We Prepared?* In August 2005, Greenpeace released K. Brigden et al., *Recycling of Electronic Wastes in China and India: Workplace and Environmental Contamination* (www.greenpeace.org/eastasia/publications/reports/toxics/2005/recycling-of-electronic-wastes/). Currently the problem of electronic waste continues to grow; see Elisabeth Jeffries, "E-Wasted," *World Watch* (July/August 2006).

Among the solutions that have been urged to address the hazardous waste problem are "take-back" and "remanufacturing" practices. Gary A. Davis and Catherine A. Wilt, in "Extended Product Responsibility," *Environment* (September 1997), urge such solutions as crucial to the minimization of waste and describe how they are becoming more common in Europe. Brad Stone, "Tech Trash, E-Waste By Any Name, It's an Issue," *Newsweek* (December 12, 2005), describes their appearance in the United States, where some states are making take-back programs mandatory for computers, televisions, and other electronic devices. After *Exporting Harm* was published and drew considerable attention from the press, some industry representatives hastened to emphasize such practices as Hewlett Packard's recycling of printer ink cartridges. See Doug Bartholomew's "Beyond the Grave," *Industry Week* (March 1, 2002), which also stressed the need to minimize waste by intelligent design. The Institute of Industrial Design published in its journal *IIE Solutions*, Brian K. Thorn's and Philip Rogerson's "Take It Back" (April 2002), which stressed the importance of designing for reuse or remanufacturing. Anthony Brabazon and Samuel Idowu, in "Costing the Earth," *Financial Management (CIMA)* (May 2001), note that "take-back schemes may [both] provide opportunities to build goodwill and [help] companies to use resources more efficiently."

In the following selections, Stephen Lester and Anne Rabe argue that because toxic waste cleanup is complicated by extreme weather events, corporations dodge their cleanup and payment obligations, and the taxpayer is left with the bill, Congress must reinstate the "polluter pays" fees. J. Winston Porter argues that Superfund cleanup efforts can be made much more efficient and that "polluter pays" taxes are unfair. The primary funder of cleanup work should be the people responsible for the problems. Taxpayers should foot the bill as a matter of last resort.

YES

Stephen Lester and Anne Rabe

Superfund: In the Eye of the Storm

The federal Superfund program was created in 1980 in response to serious threats across the country from toxic waste sites such as the infamous Love Canal landfill in Niagara Falls, NY. Since then, the U.S. Environmental Protection Agency (EPA) has completed the cleanup of more than 1000 of the nation's worst toxic waste sites, protecting hundreds of communities and drinking water supplies.

Today our nation faces a new threat to the health and safety of the American people—disruption and damage at Superfund sites caused by extreme weather conditions brought on by climate change. Hazardous waste sites can discharge and release large quantities of toxic substances when subject to flooding, tornadoes, and hurricanes. The increased costs from cleanup and disruption caused by extreme weather events place a tremendous financial burden on the already financially ailing Superfund program.

Furthermore, some large Fortune 500 corporations are declaring bankruptcy to avoid the cost of cleaning up their site and walking away. American taxpayers are then left holding an enormous cleanup bill which is paid for by Superfund, a program funded entirely with taxpayer dollars.

There is only one solution—Congress must reinstate the polluter pays fees. Without corporate fees to replenish Superfund, there is simply not enough money to do the critical job of cleaning up hundreds of toxic waste sites. Given the poor economic climate, it is unfair to expect the American taxpayers to pay for 100% of the annual costs of this program. Corporate polluters must once again contribute to the costs of cleaning up these contaminated sites.

Hurricanes and Tornadoes Impact Superfund Sites

As the climate-change related extreme weather events are becoming more frequent and more intense, they are posing a significant threat to the future integrity of many Superfund sites. The strong winds of hurricanes and tornadoes can cause significant damage such as disrupting contaminated soils and moving waste barrels long distances, or damaging protective liners covering dangerous toxic waste dumps. Flooding can dislodge buried waste, displace chemicals stored above ground, and spread contamination in soil.

From *Superfund: In the Eye of the Storm,* June 2010, pp. 4–8. Copyright © 2010 by Center for Health, Environment & Justice (CHEJ). Reprinted by permission.

Extreme weather conditions that have impacted Superfund sites include Hurricanes Ike in 2008, Katrina and Rita in 2005, and Ivan in 2004; tornadoes in Oklahoma and Iowa in 2008 and related flooding in Iowa, Kansas, and Missouri in 2008.

In the Gulf Coast region alone, 56 Superfund sites were impacted by hurricanes from 2004 to 2008. This region is one of the most heavily industrialized and polluted areas in the nation. Hurricane force winds and floodwaters stirred up toxic chemicals, oil, and pesticides and dispersed them across the region.

EPA staff have tested soil, sediment, air, and water for chemicals, removed barrels of hazardous substances, and investigated Superfund sites following extreme weather events such as hurricanes, tornadoes, and flooding. In the case of Hurricane Ike, for example, EPA used GPS tracking devices to identify over 18,000 containers suspected of containing chemicals that had been dispersed by the high winds and flood waters. The cost of these unplanned emergency response actions was a sizeable burden to an already financially ailing Superfund program.

Increased toxic contamination was found at several Superfund sites after hurricanes, tornadoes, and floods, such as the following sites in Louisiana and Oklahoma.

Agriculture Street Landfill in New Orleans, LA

The flooding of this Superfund site, containing large amounts of toxic industrial wastes, raised many questions about the release of hazardous contaminants into the neighborhood after Hurricane Katrina. EPA tests found sediment on the landfill that was deposited by the receding flood waters that contained toxic chemicals, one of which was almost three times higher than the state cleanup standard.

Oklahoma Tar Creek Superfund Site

A tornado slammed into the mining town of Picher and one of the country's largest Superfund sites leveling over 200 homes and creating a major public health hazard as lead-contaminated mining waste piles were dispersed throughout the community. Within a few weeks, EPA spent $8 million to buy-out and relocate 800 residents away from the toxic mess.

Corporate Bankruptcies Further Threaten Superfund

Another threat to the Superfund program is the eye of the economic storm, corporate bankruptcies, where polluting companies are allowed to avoid the costs of cleaning up their site by declaring bankruptcy. A potentially large number of sites could end up in Superfund if corporate polluters succeed in this exit strategy—declare bankruptcy and leave the American taxpayers to pay for the cleanup costs.

One 2007 study identified six companies connected to roughly 120 Superfund sites in 28 states that filed for bankruptcy in the last decade. Four of the companies, Bethlehem Steel Corp., Eagle-Picher Industries, Inc, Kaiser Aluminum Corp., and Polaroid Corp., avoided over half a billion dollars in cleanup costs by declaring bankruptcy.

Today, the American Smelting and Refining Company (Asarco) appears to be using the same exit strategy. The company filed for Chapter 11 reorganization starting a process that could result in the largest, most environmentally significant bankruptcy in America's history. The Asarco bankruptcy will impact an estimated 90 communities—many of them living with federal Superfund sites—and claims total over $25.2 billion. As more and more businesses struggle in today's weak economy, there is the potential for more sites to end up in Superfund as polluters declare bankruptcy and leave the taxpayers to pay for the cleanup costs.

As Superfund is saddled with major costs from corporate bankruptcies, this in combination with the increased financial burden of cleaning up sites damaged by hurricanes, tornadoes, and floods, poses significant new burdens for an already financially ailing program.

Funding Shortfall

Where will the money come from?

When Superfund was created in 1980 through the Comprehensive Environmental Response, Compensation and Liability Act, a Trust Fund was set up with approximately $1.6 billion to pay for the cleanup of any site where a polluter could not be located, was bankrupt, or refused to take action. Superfund was financed by fees from companies responsible for hazardous chemical releases, called polluter pays fees.

In 1995, Superfund had accumulated nearly $4 billion. However, the authorization to collect these fees sunset that year and was not reauthorized by Congress. Consequently, by 2003 the program ran out of money and the entire financial burden of paying for the cleanup of the worst toxic sites in America fell to the taxpayers.

For the past five years, Congress has annually allocated approximately $1.2 billion of general revenues—taxpayer's money—to Superfund since both Congress and President Bush failed to reinstate the fees.

The program's funding has been greatly reduced ever since the industry fees lapsed over a decade ago. The lack of polluter pays fees and the dependency on taxpayer revenues has led to a funding shortfall. In 2004, the House Energy and Commerce Committee found an estimated shortfall of $263 million dollars, and 19 sites were not cleaned up due to the lack of funds. In 2008, ten unfunded sites were not cleaned up.

Every year, the EPA makes roughly the same budget request and yet, from 2004 to 2008, Congress provided less than EPA requested. During this time, EPA requested an average of $1.69 billion and Congress appropriated an average of $1.26 billion, roughly 25% less than requested. A recent report by the Congressional Research Service suggests Congress could increase Superfund appropriations to address funding shortfalls, but notes "this could prove difficult in light of current interest in deficit reduction."

Superfund Slowdown

The decreased funding has led to a dramatic reduction in the number of sites cleaned up.

- From 1997 to 2000, EPA averaged 87 completed cleanups a year.
- In 2001, the number of cleanups dropped drastically to 47 sites.
- In 2002, the number dropped to 42 sites.
- From 2003 to 2006, 40 sites were cleaned up each year.
- In 2007, only 24 sites were cleaned up.
- In 2008, 30 sites were cleaned up and 20 sites in 2009.

From 2001 to 2009, there has been more than a 50% decrease in the pace of site cleanups from the late 1990's.

The agency has also started fewer cleanups since the Trust Fund ran out of polluter pays fee money. There were three times fewer cleanups started in the six year period from 2001 to 2007, compared to the previous six years. At sites targeted for cleanup, there were delays and decisions made to use cheaper, less effective remedies at some sites, according to EPA officials.

Compounding the Superfund slowdown problem is the addition of new sites every year. In past surveys EPA identified over 47,000 potentially hazardous waste sites and continues to discover new sites. Today, approximately 1600 known Superfund toxic waste sites are poisoning drinking water, land and air with chemicals that cause cancer, birth defects and other health problems. . . .

It's Time to Refinance Superfund

Superfund faces new threats as more money is needed to clean up sites impacted by hurricanes, tornadoes, and flooding, while bankrupt polluters continue to try to unload their cleanup costs on the program. At the same time, decreased funding and the Superfund slowdown have resulted in increased toxic exposures and health threats to communities across America. Stable and equitable funding is long overdue for this critically important pollution prevention program. It is time for Congress to reinstate the polluter pays fees. Without industry fees to replenish Superfund, there is simply not enough money to do the critical job of cleaning up hundreds of toxic waste sites and the American taxpayers are unfairly burdened by paying 100% of the annual costs.

Superfund was founded on the principle that those companies most closely associated with creating toxic waste sites and generating hazardous waste should bear the financial burden of cleaning them up. American taxpayers are unfairly bearing the full burden of paying for abandoned site cleanups. It is essential that industry fees are reinstated to replenish the ailing Superfund and get it back on the cleanup track. We can solve the problem by restoring the stable funding source of polluter pays fees which were the financial backbone of Superfund for more than 20 years. They included assessments on crude oil, chemical feedstock, imported chemical derivatives, and a corporate environmental income tax.

Except for President George W. Bush, the Superfund polluter pays fees have benefited from broad bipartisan presidential support. President Jimmy Carter signed the 1980 original law, President Ronald Reagan signed the 1986 law to expand the fees, and President George H.W. Bush signed a 1990 law renewing the fees. In 1995, President Clinton proposed Superfund fee reauthorization, but Congress did not approve it. President George W. Bush consistently opposed reinstatement of the polluter pays fees thus forcing American taxpayers to pay the bill when the program went bankrupt in 2003. Numerous bills have been introduced to reinstate Superfund's polluter pays fees and shift the cleanup expense burden back on polluting industries, but none have yet passed.

The Center for Health, Environment & Justice (CHEJ), Environment America, Sierra Club, and hundreds of state and local environmental, health, and community groups have waged a campaign to refinance Superfund over the years. CHEJ Executive Director Lois Gibbs was a leader of the successful community fight to relocate over 800 families away from the Love Canal toxic waste dump in Niagara Falls, NY which led to the creation of the Federal Superfund in 1980. After years of delay, Ms. Gibbs urges policymakers to take action on this critical environmental health problem.

"Congress should restore the hazardous waste fees on polluting industries and enable Superfund to move forward and respond to new toxic threats. The core principle of the Superfund program is that polluters, not taxpayers, should pay to clean up these deadly toxic waste sites. In addition to providing funding, the polluter pays principle creates a powerful disincentive against the reckless dumping of toxic waste."

Testimony before the Senate Committee on Environment & Public Works, Subcommittee on Superfund, Toxics and Environmental Health, Hearing on "Oversight of the Environmental Protection Agency's Superfund Program

. . . .

Background and Recommendations

Briefly, the current status of EPA's Superfund program is that about two-thirds of the 1500 national priority list sites have reached the construction completion (remedy installed) phase, about 370 sites are in the remedy design or construction phases, and approximately 120 sites are in the study phase.

In addition, many thousands of "emergency removals" have been conducted at Superfund sites in order to directly and cost effectively deal with obvious problem areas. This program has been perhaps Superfund's biggest success story.

In addition to the EPA, both the Departments of Energy and Defense have major Superfund-related programs underway. The DOE work primarily involves a few dozen very large facilities, most of which have been components of the nuclear weapons program. The DOD sites are much more numerous, although usually less complex, and include both Superfund and base closure activities.

So, a large amount of work is underway or has been completed by dedicated federal and state personnel as well as potentially responsible parties (PRPs) and various private contractors. For the remaining work it is important to improve program efficiency, as site study and remedial activities often take too long and cost too much.

In order to complete the remaining Superfund sites, the following **recommendations** are made to improve program efficiencies:

1. The focus of the Superfund program should increasingly be on the completion of existing sites. The various administrative and support

U.S. Senate, June 22, 2010.

services should be reduced as sites are completed in order to provide more funding for site completion work.

2. Consideration should be given to designation of a senior member of the assistant administrator's office to oversee and promote site completions.

3. Completion dates should be set for all current study and cleanup work. A "culture of completion" should replace the current "culture of deliverables." Program reports and other paperwork should be streamlined.

4. Some Superfund sites have been completed in a timely and cost effective manner. It is suggested that a sampling of such sites be identified and used to inform the timely completion of other sites.

Perhaps the most dramatic use of target setting has been the DOE Rocky Flats Closure Project, near Denver. For this site the "completion contractor," Kaiser-Hill, and the DOE agreed upon a 2005 target date for all study and remedy implementation work to be completed. If successful, the contractor was to receive a completion bonus. Not only was the project completed on time, but billions of dollars and many decades of time were saved. This work, of course, required good cooperation among the DOE, EPA, the State of Colorado, local stakeholders, and the contractor. The firm completion target date greatly focused this cooperation.

I will now provide more detailed comments or recommendations on the three major Superfund components, study, remedy selection, and construction phases.

The Study Phase

While the study projects related to Superfund sites are a decreasing part of the overall program, such activities are still very important to overall program success. Superfund projects usually begin with a "remedial investigation/feasibility study" (RI/FS). This complex study process is described in some detail in Superfund's primary regulation—the National Contingency Plan.

Very briefly, the RI portion calls for characterization of the site in terms of its natural features, as well as the amount and location of contamination and likely risks of such contamination to public health and the environment. The FS part involves identification of alternative remedial actions, and then comparison of such alternatives against a set of nine remedy selection criteria.

Based on the RI/FS process, as well as substantial stakeholder input, EPA then selects a remedy for the site through a "record of decision" (ROD) process.

In general, the RI/FS process has become steadily more complex and lengthy over the years, for almost all types of sites. My recommendations for conducting faster, less costly, and more technically sound RI/FSs are as follows:

1. Most importantly, timeframes for completing the study phase should be agreed to by the EPA and other key participants, such as States and PRPs.

Unfortunately, at many sites the study work simply meanders around for many years without much focus on alternative remedies, leading to wasted time and money, and, in some cases, an unimaginative or non-cost-effective remedy selection. Frankly, part of this lengthy process has to do with the fact that Superfund has become a lucrative source of work for various consultants, lawyers, and other participants. All of these specialists are needed, but their work needs to be more directed toward the timely selection of a sound site remedy rather than complex and lengthy work processes and paperwork.

Stated differently, there is often little sense of urgency in completing the study phase due, in part, to the lack of a senior "champion(s)" to complete the work. This is, or course, very frustrating to the communities involved. I would like to see such "completion champions" developed in both the governmental and private sectors at Superfund sites.

Some very complex federal and private sites will require longer study periods, but for most sites about 2–3 years should be adequate to produce a sound RI/FS.

To improve matters, early in the RI/FS process the EPA, PRPs, and other relevant organizations, should work together to set a clear goal to complete the study activities. This end date can be modified if necessary, but it is important for all to understand that, like almost every other type of engineering project, schedule and budget are key factors and should be adhered to.

2. When the RI/FS process begins one of the first orders of business should be to use experienced staff and key stakeholders to quickly identify about 4–7 major remedial action alternatives.

During this phase use should be made of EPA's list of "presumptive remedies" for many types of problems, as well as experience gained at similar Superfund sites.

The selected set of alternatives can always be modified during the study phase, but the current process which often involves "taking data" for many years before detailed focus on remedial options often leads to overly costly information, much of which may not be needed. Also, since the data collection is often not focused on comparing alternative remedies, the key information to compare such alternatives is sometimes missing.

An iterative approach should be used where information collection and analysis of remedial alternatives work cooperatively to achieve sound comparisons of options, leading to good remedy selections.

Even more importantly, the identification of key options early in the study process allows the decision-makers and stakeholders to begin their dialogue on non-technical factors which are contained in the remedy selection criteria. These include such items as cost-effectiveness, implementability, and state and community acceptance. Many times these types of factors are at least as important as the strictly technical matters, such as very precise measurement of numerous contaminants, many of which are present at near-zero levels.

3. Significantly streamline the process for developing the myriad of deliverables at Superfund sites.

While certain documents are clearly needed to guide the RI/FS activities, the long, tedious process of developing complex draft and final work plans, for example, should be expedited. This is also true of dozens of other "deliverables" which take so much time at Superfund sites, many of which should be quite standard by now. It might be helpful to revisit the need, or at least the complexity, of such deliverables.

There are several perverse effects which have led to such lengthy periods for document development and review. One has to do with the fact that Superfund is about the only federal environmental program where responsible parties have to pay for additional oversight beyond that which salaried regulators normally provide. Thus, if a group of PRPs are forced to give EPA, say, $ 3–5 million for oversight, then EPA can retain contractors to provide hundreds of pages of "comments" on such items as the aforementioned work plans. So, we now have dueling contractors battling over many pages of detailed text, before work can even begin.

One near term answer would be for review periods and oversight dollars to be reduced substantially, so participants can focus more on results than elaborate processes. PRPs should usually be encouraged to conduct the RI/FSs themselves with their own contractors and under EPA's overall supervision.

While this concept has been largely accepted and successfully promoted by the EPA, more could be done to encourage PRPs to do the study work, particularly where PRPs would commit to shorter timeframes than EPA often takes for its own studies.

A key aspect of PRP-conducted studies has to do with selection of appropriate consulting firms to conduct the necessary RI/FS activities. Such contractors have a difficult role in that they need to be responsive to their client, the PRPs, but must also provide the objective and professional work needed by EPA to allow selection of a sound and cost-effective remedy for the site in question.

The key is for the EPA, the relevant State, and the PRPs and their consultants to develop a cooperative and results-oriented relationship for the site work.

The Selection of Remedy Phase

The RI/FS process discussed above presents the decision-maker with detailed comparisons of alternative remedial actions, from which this person must select a remedy, present it to the public for comment, and make a final determination. The selection of protective, cost-effective remedies is, of course, a key to the overall success of the Superfund program. My suggestions in this area are as follows:

1. The decision-maker should be a very senior EPA official who can oversee all of the considerations which go into remedy selection.

As noted earlier, technical factors are very important in this process, but nontechnical factors are also key. For example, if there is very strong community opposition to a particular remedial action, or if a remedial option is not cost-effective, such factors must be considered by the decision-maker.

During my tenure as an EPA assistant administrator I made a number of ROD decisions, mainly at "nationally significant sites." Most decisions I delegated to the ten EPA regional administrators (RAs). However, over the years the

ROD decision responsibility has, in most cases, been delegated further down the line in the EPA regions.

My own view is that the RA should usually be the decision-maker in this important process since he or she is the one who can speak for the region and has the position and stature to consider all aspects of the problem, while "pushing" the staff to provide the necessary information to complete remedy selection expeditiously.

2. The role of expected land use should be an important factor in selecting a remedy.

While all remedies should be protective, it does not make much sense to demand that a cleanup be sufficient for, say, a children's daycare center, when the site is slated for use as a golf course, or a factory, or a wildlife preserve. All of these uses have their own requirements, so we do not need a one-size-fits-all approach to waste sites. The goal should be for a site to always be protective, so the remedial action may need to be modified at a later date if the site use changes significantly.

During Superfund's history one of the better examples of the role of land use in remedy selection had to do with the DOD's Rocky Mountain Arsenal in Colorado. For this site, the DOD decided ultimately that the land use would be for a wildlife refuge, not residential housing. Once this decision was made the DOD, Shell Oil, EPA, and the state and local stakeholders worked together to select the remedy and move quickly into the implementation phase, and an important wildlife refuge is the result.

Another DOD example may also be instructive with respect to the land use issue. This has to do with the DOD's Superfund-related remediation sites versus those conducted under the base closure program. Simply stated, the base closure cleanups, including the selection of remedy, seem to proceed much faster than those related to Superfund. One of the reasons, I believe, has to do with the fact that local communities and others are usually highly motivated to finish base closure cleanups in order to bring the affected land into productive use. The same time pressure often does not exist with Superfund remedial activities.

The Construction Phase

As noted earlier, the major activity these days has to do with the construction phase at Superfund sites. About 370 sites are in the phase where the selected remedy is being either designed or constructed. Currently, this is also the most controversial phase in that EPA may not have sufficient funds to expeditiously complete all of the construction work now planned.

This is particularly true for so-called fund-financed sites where EPA must install the remedy itself as there are insufficient willing and able PRPs to conduct this work at some sites.

The following are my recommendations on these construction-phase issues:

1. The roughly $1.2 billion dollars which is annually appropriated to EPA by Congress should be looked at very carefully by EPA senior

management to ensure that the highest priority is given to protecting human health and the environment by ensuring that Superfund sites are completed.

2. If Congress is satisfied that EPA has done all it can do to squeeze out funding for as many construction sites as possible, then it might consider a supplemental appropriation to EPA to focus on additional construction activities.

3. The EPA might selectively revisit the ROD decisions made at selected sites to see if some savings can be made based on new information or technology.

4. Although I suspect that this is already being done, that portion of the site which may provide actual, near term risk to the community should receive very high priority for funding.

5. While aiming at the highest risks is always the most important priority, I personally believe that where sites can be finished for modest sums of money, such funding should be considered, as there are usually site "carrying charges" which can then be reduced.

6. The EPA and others should be creative in finding non-federal funds for completing sites. In some cases, local developers or others may be so interested in having access to a completed site that they may be interested in helping financially. This type of financial driver has, of course, been instrumental in dealing with brownfields sites, which can often be very valuable when cleanup measures are completed.

7. Other creative measures should be pursued in the future to minimize costs and to develop more creative financing. A good example is the joint EPA and Army Corps of Engineers eight pilot programs referred to as the "urban rivers restoration initiative." In this program the EPA and the Corps, along with state and other agencies, work together to achieve a better and more cost-effective restoration program than by using Superfund alone.

8. Finally, it was mentioned earlier in this testimony that the emergency removal program has been one of Superfund's major successes. This program can deal with obvious contamination problems anytime during the Superfund process, with much less process costs than the remediation program. Given, this program's success, Congress might consider allowing EPA to spend more than the current limit on individual removal actions.

Implicit in all the above is the fact that I don't believe that the chemical and petroleum feedstock taxes should be renewed on Superfund. These taxes are unfair in that they target only two industries, which together account for much less than half of Superfund's contamination problems. Also, Superfund sites are a broad societal problem which has been created by many types of industries; local, state, and federal agencies; and even individuals.

Therefore, I believe the current process of requiring directly responsible parties at a site to fund the necessary work at that site is the best approach. For those sites, where responsible parties are not available to conduct the work general revenues are the most equitable approach, given the widely varied causes of contamination at such sites. EPA also has strong legal authorities to seek reimbursement from known responsible parties who are able, but not willing, to do the work in question.

EXPLORING THE ISSUE

Should the Superfund Tax Be Reinstated?

Critical Thinking and Reflection

1. In the past many industrial chemical wastes were disposed of according to the "best practice" of the time (as defined by laws and regulations). Later it was realized that "best practice" was not very good. Should companies be held liable for public health effects in such cases?
2. In the past many industrial chemical wastes were disposed of in the cheapest possible way, without regard to potential public health effects. Should companies be held liable for public health effects in such cases?
3. Orphan waste sites are those for which whatever company or person deposited waste at the sites can not be found. Who should pay for cleaning up such sites?
4. In what way is a tax imposed on all members of an industry (such as a "polluter pays" tax) like an insurance premium?

Is There Common Ground

Even those who object to "polluter pays" taxes agree that toxic waste sites need to be cleaned up. The debate is over exactly how to pay for the cleanup. There is also agreement that it would be good to generate less waste in the first place, and it has proven fruitful to consider the definition of a waste as "a resource out of place."

1. What does "resource out of place" mean?
2. What is a "waste broker"?
3. Can all wastes be useful to someone?

ISSUE 19

Should the United States Reprocess Spent Nuclear Fuel?

YES: Kate J. Dennis, Jason Rugolo, Lee T. Murray, and Justin Parrella, from "The Case for Reprocessing," *Bulletin of the Atomic Scientists* (November/December 2009)

NO: David M. Romps, Christopher D. Holmes, Kurt Z. House, Benjamin G. Lee, and Mark T. Winkler, from "The Case Against Nuclear Reprocessing," *Bulletin of the Atomic Scientists* (November/December 2009)

Learning Outcomes

After reading this issue, you should be able to:

- Explain why fear of nuclear proliferation has kept the United States from developing nuclear fuel processing facilities.
- Explain how nuclear fuel reprocessing can help deal with the nuclear waste problem.
- Explain how nuclear fuel reprocessing supports expended use of nuclear power.
- Explain why it will be difficult to find a publicly acceptable site for a nuclear fuel reprocessing plant.

ISSUE SUMMARY

YES: Kate J. Dennis, Jason Rugolo, Lee T. Murray, and Justin Parrella argue that nuclear fuel reprocessing extracts more energy from nuclear fuel and reduces the amount of nuclear waste to be disposed of. "If the United States truly wants to proceed with nuclear energy as a viable, low-carbon emitting source of energy, it should pursue reprocessing in combination with the development of fast reactors. Once such a decision is made, the debate should turn to how best to develop cheaper and safer reprocessing options, rather than denying its general benefit."

NO: David M. Romps, Christopher D. Holmes, Kurt Z. House, Benjamin G. Lee, and Mark T. Winkler argue that reprocessing is both dangerous and unnecessary. "It is in the best interests of the United States—from the perspective of waste management, national security, economics, and environmental protection—to maintain its de facto moratorium on reprocessing and encourage other countries to follow suit."

As nuclear reactors operate, the nuclei of uranium-235 atoms split, releasing neutrons and nuclei of smaller atoms called fission products, which are themselves radioactive. Some of the neutrons are absorbed by uranium-238, which then becomes plutonium. The fission product atoms eventually accumulate to the point where the reactor fuel no longer releases as much energy as it used to. It is said to be "spent." At this point, the spent fuel is removed from the reactor and replaced with fresh fuel.

Once removed from the reactor, the spent fuel poses a problem. Currently, it is regarded as high-level nuclear waste that must be stored on the site of the reactor, initially in a swimming pool-sized tank and later, once the radioactivity levels have subsided a bit, in "dry casks." Until 2009, there was a plan to dispose of the casks permanently in a subterranean repository being built at Yucca Mountain, Nevada, but the Obama administration has proposed to end funding for Yucca Mountain (see Dan Charles, "A Lifetime of Work Gone to Waste?" *Science*, March 20, 2009).

It is worth noting that spent fuel still contains useful components. Not all the uranium-235 has been burned up, and the plutonium created as fuel is burned can itself be used as fuel. When spent fuel is treated as waste, these components of the waste are discarded. Early in the Nuclear Age, it was seen that if these components could be recovered, the amount of waste to be disposed of could be reduced. The fuel supply could also be extended, and in fact, since plutonium is made from otherwise useless uranium-238, new fuel could be created. Reactors designed to maximize plutonium creation, known as "breeder" reactors because they "breed" fuel, were built and are still in use as power plants in Europe. In the United States, breeder reactors have been built and operated only by the Department of Defense, for plutonium extracted from spent fuel is required for making nuclear bombs. They have not seen civilian use in part because of fear that bomb-grade material could fall into the wrong hands.

The separation and recycling of unused fuel from spent fuel is known as reprocessing. In the United States, a reprocessing plant operated in West Valley, NY, from 1966 to 1972 (see "Plutonium Recovery from Spent Fuel Reprocessing by Nuclear Fuel Services at West Valley, New York, from 1966 to 1972," DOE, 1996; www.osti.gov/opennet/forms.jsp?formurl=document/purecov/nfsrepo.html). After the Nuclear Nonproliferation Treaty went into force in 1970, it became United States policy not to reprocess spent nuclear fuel and thereby to limit the availability of bomb-grade material. As a consequence, spent fuel was not recycled, it was regarded as high-level waste to be disposed of, and the waste continued to accumulate.

Despite the proposed termination of the Yucca Mountain nuclear waste disposal site, the nuclear waste disposal problem is real and it must be dealt with. If it is not, we may face the same kinds of problems created by the former Soviet Union, which disposed of some nuclear waste simply by dumping it at sea. For a summary of the nuclear waste problem and the disposal controversy, see Michael E. Long, "Half Life: The Lethal Legacy of America's Nuclear Waste," *National Geographic* (July 2002). The need for care in nuclear waste disposal is underlined by Tom Carpenter and Clare Gilbert, "Don't Breathe the Air," *Bulletin of the Atomic Scientists* (May/June 2004); they describe the Hanford Site in Hanford, Washington, where wastes from nuclear weapons production were stored in underground tanks. Leaks from the tanks have contaminated groundwater, and an extensive cleanup program is under way. But cleanup workers are being exposed to both radioactive materials and toxic chemicals, and they are falling ill. And in June 2004, the U.S. Senate voted to ease cleanup requirements. Per F. Peterson, William E. Kastenberg, and Michael Corradini, "Nuclear Waste and the Distant Future," *Issues in Science and Technology* (Summer 2006), argue that the risks of waste disposal have been sensibly addressed and we should be focusing more attention on other risks (such as those of global warming). Behnam Taebi and Jan Kloosterman, "To Recycle or Not to Recycle? An Intergenerational Approach to Nuclear Fuel Cycles," *Science and Engineering Ethics* (June 2008) argue that the question of whether to accept reprocessing comes down to choosing between risks for the present generation and risks for future generations.

In November 2005, President Bush directed the Department of Energy to start work toward a reprocessing plant; see Eli Kintisch, "Congress Tells DOE to Take Fresh Look at Recycling Spent Reactor Fuel," *Science* (December 2, 2005). By April 2008, Senator Pete Domenici of the U.S. Senate Energy and Natural Resources Committee was working on a bill that would set up the nation's first government-backed commercialized nuclear waste reprocessing facilities. Reprocessing spent nuclear fuel will be expensive, but the costs may not be great enough to make nuclear power unacceptable; see "The Economic Future of Nuclear Power," University of Chicago (August 2004) (http://www.ne.doe .gov/np2010/reports/NuclIndustryStudy-Summary.pdf). Sarah Widder, "Benefits and Concerns of a Closed Nuclear Fuel Cycle," *Journal of Renewable and Sustainable Energy* (November 2010), notes that the present once-through fuel cycle is not sustainable, but reprocessing remains controversial.

In February 2006, the United States Department of Energy announced the Global Nuclear Energy Partnership (GNEP), to be operated by the United States, Russia, Great Britain, and France. It would lease nuclear fuel to other nations, reprocess spent fuel without generating material that could be diverted to making nuclear bombs, reduce the amount of waste that must be disposed of, and help meet future energy needs. See Stephanie Cooke, "Just Within Reach?" *Bulletin of the Atomic Scientists* (July/August 2006), and Jeff Johnson, "Reprocessing Key to Nuclear Plan," *Chemical & Engineering News* (June 18, 2007). Critics such as Karen Charman ("Brave Nuclear World, Parts I and II," *World Watch* (May/June and July/August 2006), insist that nuclear power is far too expensive and carries too serious risks of breakdown and exposure to wastes to rely upon, especially when cleaner, cheaper, and less dangerous alternatives exist. Early in 2009, the Department of Energy announced it was closing down the GNEP.

It is an unfortunate truth that the reprocessing of nuclear spent fuel does indeed increase the risks of nuclear proliferation. Both nations and terrorists itch to possess nuclear weapons, whose destructive potential makes present members of the "nuclear club" tremble. Can the risks be controlled? John Deutch, Arnold Kanter, Ernest Moniz, and Daniel Poneman, in "Making the World Safe for Nuclear Energy," *Survival* (Winter 2004/2005), argue that present nuclear nations could supply fuel and reprocess spent fuel for other nations; nations that refuse to participate would be seen as suspect and subject to international action. Nuclear physicist and Princeton professor Frank N. von Hippel, "Rethinking Nuclear Fuel Recycling," *Scientific American* (May 2008) argues that reprocessing nuclear spent fuel is expensive and emits lethal radiation. There is also a worrisome risk that the increased availability of bomb-grade nuclear materials will increase the risk of nuclear war and terrorism. Prudence demands that spent fuel be stored until the benefits of reprocessing exceed the risks (if they ever do). See also Rodney C. Ewing and Frank N. von Hippel, "Nuclear Waste Management in the United States—Starting Over," *Science* (July 10, 2009).

In the summer of 2009, 11 Harvard graduate students and postdocs went to France to investigate nuclear fuel reprocessing. During the trip, they divided into two groups, one for reprocessing and one against (with two undecided). After the trip, they wrote "Should the United States Resume Reprocessing? A Pro and Con" for the *Bulletin of the Atomic Scientists*. In the following selections, from that article, Kate J. Dennis, Jason Rugolo, Lee T. Murray, and Justin Parrella argue that nuclear fuel reprocessing extracts more energy from nuclear fuel and reduces the amount of nuclear waste to be disposed of. "If the United States truly wants to proceed with nuclear energy as a viable, low-carbon emitting source of energy, it should pursue reprocessing in combination with the development of fast reactors. Once such a decision is made, the debate should turn to how best to develop cheaper and safer reprocessing options, rather than denying its general benefit." David M. Romps, Christopher D. Holmes, Kurt Z. House, Benjamin G. Lee, and Mark T. Winkler argue that reprocessing is both dangerous and unnecessary. "It is in the best interests of the United States—from the perspective of waste management, national security, economics, and environmental protection—to maintain its de facto moratorium on reprocessing and encourage other countries to follow suit."

YES

Kate J. Dennis et al.

The Case for Reprocessing

The United States should reconsider reprocessing its spent nuclear fuel to obtain the highest efficiency and lowest waste. With the correct and necessary guidelines, closing the nuclear fuel cycle by allowing waste to be turned back into fuel is a viable option and should be a goal for the country.

Critics argue that the high cost of reprocessing spent fuel and fabricating mixed-oxide (MOX) fuel rods—a mixture of uranium and plutonium oxides— far outweighs the benefits. According to MIT's 2003 "Future of Nuclear Power" report, the so-called once-through fuel cycle, where spent fuel is directly deposited in geologic repositories, is four to five times less expensive than the costs associated with reprocessing. We do not argue that reprocessing is cheaper in the short-term, but that it is extremely difficult, if not impossible, to compare the short-term costs associated with reprocessing with the benefits that would occur 50–100 years hence. (A 2006 study by the Boston Consulting Group does find the long-term cost of reprocessing to be almost equivalent to once-through fuel management.) These longer-term benefits include a twofold volumetric reduction in nuclear waste, conservation of uranium resources, and a reduction in the environmental impact of uranium mining. Additionally, if the United States considers building fast reactors in the future, reprocessing becomes a necessary step to remove the plutonium that these reactors generate. Fast reactors consist of a core of fissile plutonium or highly enriched uranium surrounded by a blanket of uranium 238, which captures neutrons escaping from the core and partially transmutes into plutonium 239. The result is a reactor that produces or "breeds" a surplus of fuel. This blanket material, however, must be reprocessed to recover the generated plutonium. The development of such breeder reactors would increase the energy output for a given amount of nuclear fuel 60–100 times, according to a range of estimates, and therefore reduce consumption of natural uranium ore.

Currently, it is unclear if the United States will continue to pursue fast reactors as outlined by former President George W. Bush in the 2006 Global Nuclear Energy Partnership (GNEP), which encouraged the use of nuclear energy abroad and the restart of U.S. reprocessing. But even without GNEP, the Obama administration has continued funding the Advanced Fuel Cycle Initiative, a research project within the Energy Department that is focused on "proliferation-resistant fuel cycles and waste reduction strategies." This seems to indicate research into reprocessing will continue.

From *Bulletin of the Atomic Scientists,* November/December 2009, pp. 30–36. Copyright © 2009 by Bulletin of the Atomic Scientists. Reprinted by permission of Sage Publications via Rightslink.

Until early 2009, the U.S. nuclear waste plan was to dispose of the country's spent fuel in a permanent geologic site at Yucca Mountain in Nevada. In the 2010 Energy budget, however, the Obama administration effectively shut down Yucca Mountain, stating its funding would be "scaled back to those costs necessary to answer inquiries from the Nuclear Regulatory Commission, while the administration devises a new strategy toward nuclear waste disposal." As a result, we are left with a dispersed, decentralized patchwork of highly radioactive waste sites (interim storage pools and dry-cask storage at nuclear power plants) with the hope that a long-term repository will be opened somewhere, sometime in the future. Otherwise, an alternative waste-disposal strategy must be established, which is no easy task.

To give an idea of what regulatory hurdles any alternate waste-disposal strategy faces, it is helpful to look back at the history of Yucca Mountain. Until recently, the repository was scheduled to open in 2017, but that date itself was delayed numerous times from 1998, the original year it was supposed to open and start accepting U.S. spent fuel. Even if the Obama administration had not ended the project, it was unclear whether Energy's Office of Civilian Radioactive Waste Management would have been able to meet the 2017 deadline, especially with political opposition from key congressional leaders. The Democratic Senate majority leader, Nevada Sen. Harry Reid, for example, is strongly opposed to Yucca Mountain; he has argued instead for on-site, dry-cask storage as a more viable solution for the country's spent nuclear fuel. Although we disagree with the long-term viability of dry-cask storage as a solution (not only does on-site, dry-cask storage create a patchwork of nuclear waste, but it also pushes long-term decision making on nuclear waste strategy to the next generation), the fact remains that even without President Barack Obama's recent cuts, congressional support for the repository site is lacking.

Cost estimates for Yucca Mountain also have risen significantly over time. Estimates to complete the project were $79.3 billion in 2008, much higher than the 1998 estimate of $11.6 billion (in 2000 dollars). Plus, it is increasingly obvious that Yucca Mountain would not have been large enough to accept all of the country's current and future nuclear waste. As of 2008, the United States had generated 58,000 tons of civilian spent fuel, and along with the 13,000 tons of government-sourced spent fuel and high-level waste, the repository's 70,000-ton capacity would have been exceeded on the day it opened. If waste continues to be generated at the rate of approximately 2,500 tons per year, another permanent geologic disposal site, or an increase in the capacity of Yucca Mountain, would have been required. If there are additional nuclear plants built in the future, this problem only will be exacerbated. Although it remains unclear if the Obama administration will support extensive nuclear power investment, Energy Secretary Steven Chu has repeatedly come out in favor of nuclear power. In addition, the 2005 Energy Policy Act provided incentives for the nuclear industry, including tax credits and up to $18.5 billion in loan guarantees for new U.S. reactors. Although we are not aware of any allocations of these funds, it again suggests that there is interest in expanding the civilian nuclear energy industry.

Since 1983, $10.3 billion has been spent to manage and dispose of the country's nuclear waste. That amount has been taken out of the Nuclear Waste Fund, which was established with a $0.001 per kilowatt-hour surcharge on nuclear power generating utilities. The total amount in the fund was $29.6 billion at the end of 2008. With the majority of the $10.3 billion spent developing Yucca Mountain as a repository, it is unclear how much more money will be needed before a nuclear waste storage site is decided upon, built, and opened.

Reprocessing saves valuable repository space. For long-term geologic storage, reductions in waste volume are important. But it is not just the space that the waste would physically take up that is vital, the heat output of the waste also must be taken into consideration, as does the space between waste packages necessary to prevent overheating in the repository. While it is true that high-level waste from reprocessing is hotter than non-reprocessed spent fuel, this does not completely nullify the decrease in waste volume achieved by reprocessing. The heat emitted from post-reprocessing waste decreases by approximately 70 percent during its first 30 years. In other words, such waste initially can be stored either aboveground in well-ventilated storage buildings (as Areva does), or it can be stored in geologic repositories with space between packages left empty and then filled over the years as heat output decreases.

In contrast, spent fuel rods that are directly disposed in repositories cool more slowly and require larger geologic repositories. One estimate, which appears in the book *Megawatts and Megatons* by Richard Garwin and Georges Charpak, suggests that even with the increased heat output of high-level wastes from reprocessing, the amount of space required for a geologic repository can be reduced by *one-half* if the waste is reprocessed. Overall, Garwin and Charpak argue against reprocessing but acknowledge several benefits that we believe outweigh the economic burdens, the most important being that reprocessing can effectively double the capacity of a Yucca Mountain-sized permanent repository.

Reprocessing reduces radioactivity of waste. Reprocessing also reduces the radiotoxicity of high-level waste by one-half to one-tenth when compared with direct burial, and the waste decays to the radioactivity of natural uranium in 10,000 years versus 100,000 years. With the advent of fast reactors, coupled with reprocessing, radiotoxicity of waste would be further reduced with radioactivity reaching the level of natural uranium in only 1,000 years. Given the necessity of any nuclear waste strategy's long-term viability, these reductions are significant advantages for reprocessing.

Our discussion of waste management would not be complete without acknowledging that after reprocessing spent fuel and fabricating MOX fuel rods, the spent MOX fuel rods present a unique problem when dealing with their final disposal. Spent MOX fuel has higher contents of plutonium (plutonium 238 and plutonium 241), americium, and curium than conventional low-enriched uranium (LEU) spent fuel rods, and as a result, the management of spent MOX fuel is more challenging due to cooling and criticality concerns. For interim storage, spent MOX fuel can be dispersed among LEU spent fuel resulting in no change in storage requirements. But in a geologic repository, according to a 2003 International Atomic Energy Agency report on MOX fuel technology, spent MOX fuel would need three times as much space as spent LEU fuel, or require interim storage aboveground for 150 years to reach the same thermal output and then be able to occupy the same amount of space. If we are to assume fast reactors are the long-term goal of the nuclear industry, the optimal and safest use of MOX fuel rods would be to continue recycling them in fast reactors. Yet without that option available, we must acknowledge that some of the gains made by reprocessing are lost in the storage of spent MOX fuel.

Reprocessing does not pose a proliferation threat. A frequent criticism of reprocessing is that separating pure plutonium from spent fuel creates a proliferation and theft risk. Specifically, critics say that spent nuclear fuel without reprocessing is too radioactive to be stolen easily and thus is self protecting. Therefore, some suggest, all spent fuel should remain unreprocessed. Currently, Areva, the only large-scale operator of reprocessing plants in the world, uses the PUREX technique, which separates spent fuel rods into individual streams of uranium, plutonium, and high-level fission products. Although it is true that PUREX results in a pure plutonium stream, the separated plutonium is not considered weapon-grade. Weapon-grade plutonium contains more than 93 percent plutonium 239, while reprocessed fuel contains closer to 50 percent with the remainder being oxygen and other plutonium isotopes, including approximately 15 percent plutonium 241. A nuclear explosive device that used 50 percent plutonium 239 would have an expected yield of about 1 kiloton (approximately 5 percent the power of a weapon-grade plutonium bomb). Even so, such an explosion could wipe out a handful of city blocks (in comparison, the nuclear weapon dropped on Hiroshima was 12–15 kilotons, destroying a radial area of approximately 1.6 kilometers) and is exactly why the implementation of effective security measures is paramount for safe reprocessing. According to an Areva representative, the company, along with French national security personnel, goes to great lengths to secure the transport of reactor-grade plutonium across France. In fact, during conversations with Areva we were told that if a truck were ever hijacked, "it would not get further than 100 meters." The representative was not able to further elaborate, stating that security protocol restricted his ability to clarify.

It is reasonable to worry about the risks presented by transporting nuclear materials and plutonium separated by reprocessing, and this is exactly why we stress that the United States should use and build reprocessing and fuel fabrication facilities in a single location if it chooses to restart its domestic reprocessing program. (Areva's facilities are on opposite ends of France.) Such a combined facility, with all elements in one location, would circumvent the security concerns associated with reprocessing. The United States is currently using this design for its Savannah River Site in Aiken, South Carolina, where it plans to decommission nuclear weapons. This facility will downblend surplus weapon-grade material and use that material in the fabrication of MOX fuel rods. Although the site does not currently reprocess civilian fuel, it seems logical that any civilian program would benefit from a similarly combined facility design.

To further prevent proliferation risks, the United States should develop advanced reprocessing technologies that have been researched under GNEP and will likely continue with the Advanced Fuel Cycle Initiative. These new reprocessing techniques (COEX, UREX+, and NUEX) avoid creating a pure plutonium stream. COEX extracts plutonium and uranium together, while UREX+ and NUEX extract plutonium with some combination of highly radioactive elements that are present in spent fuel. The inclusion of such transuranic elements increases the heat output and radioactive emission rate of the produced waste, necessitating robust radioactive shielding to safely manipulate or handle the material. Such advanced reprocessing techniques coupled with combined reprocessing and fabrication facilities would provide additional layers of security to address the major proliferation concerns associated with reprocessing.

The final word. Reprocessing allows the utilization of more available energy from nuclear fuel than is currently possible in the once-through fuel cycle. Reprocessing represents a path toward decreasing the current nuclear waste burden on geologic disposal sites and on future generations. If the United States truly wants to proceed with nuclear energy as a viable, low-carbon emitting source of energy, it should pursue reprocessing in combination with the development of fast reactors. Once such a decision is made, the debate should turn to how best to develop cheaper and safer reprocessing options, rather than denying its general benefit.

David M. Romps et al.

NO

The Case Against Nuclear Reprocessing

In France, as in the United States and other countries, reprocessing technology was originally pursued to produce plutonium for nuclear weapons. As reprocessing plants churned out their product, the growing stockpiles of plutonium cast a doomsday pall across the globe. Therefore, it is somewhat ironic that reprocessing became part of a utopian dream to provide the world with cheap and nearly limitless energy. The concept was dubbed the "plutonium economy." In this vision, a special kind of nuclear reactor—the fast "breeder" reactor—would burn plutonium to make electricity, but it also would convert non-fissile uranium 238 into plutonium 239, thereby creating or "breeding" more fuel. Before that new fuel could be used, however, it would have to be separated from less useful radioactive material in a nuclear reprocessing plant and then formed into fuel rods in a mixed-oxide (MOX) fuel fabrication plant.

In the 1970s, France embarked on an effort to make the plutonium economy a reality. To do so, it needed a breeder reactor, a reprocessing plant, and a MOX fuel fabrication plant. For the reprocessing plant, France pressed into service its military reprocessing facility at La Hague, blurring the line between its civilian and military nuclear programs. To turn the plutonium into ceramic MOX fuel pellets, it constructed the MELOX plant at Marcoule. All that was left to build was the commercial-scale fast reactor, dubbed Superphénix. Unfortunately the reactor's construction turned out to be much more complicated than anyone imagined, and Superphénix became a notorious flop.

In the aftermath of the failure of fast reactors (not just in France, but throughout the world), the plutonium economy died. This left the French with two very expensive nuclear facilities that had lost their *raison d'être*. Although the French nuclear industry still hopes that commercial fast reactors will someday become viable, they have been forced to argue the merits of reprocessing in the current world of light water reactors fueled by low-enriched uranium.

Today's reprocessing advocates have two main arguments: It reduces hazardous waste and conserves uranium resources. Both of these justifications are seriously flawed. It is also unlikely that fast reactors will be economically competitive in the next several decades, so reprocessing will remain a technology in search of a rationale for years to come. In the nearer term, the economic

From *Bulletin of the Atomic Scientists*, November/December 2009, pp. 36–41. Copyright © 2009 by Bulletin of the Atomic Scientists. Reprinted by permission of Sage Publications via Rightslink.

costs, environmental harm, and proliferation hazards overwhelm the supposed benefits of reprocessing.

Reprocessing does not save uranium or repository space. Reprocessing proponents claim that the process dramatically reduces the volume of nuclear waste. Indeed, spent nuclear fuel contains roughly 1 percent plutonium, 4 percent high-level radioactive waste, and 95 percent uranium. After reprocessing, the highly radioactive fuel waste and fuel-rod casing material occupy only 20 percent of the original spent-fuel volume. Aboveground, where air currents can cool the high-level waste, the space savings is significant. In fact, all of France's high-level waste sits in a modestly sized building at the La Hague complex. For final geological storage, however, the high temperature of the waste is a more restrictive design consideration than the waste's volume. In order to avoid damaging the geologic repository and risk releasing radioactivity, the high-level waste must be spaced at sufficient intervals to allow for cooling. Even if the reprocessed high-level waste is allowed to cool for 100 years before final disposal, it has been estimated that the repository volume only would be cut by one-half. Whether the United States chooses Yucca Mountain or another site for geologic storage, this modest space savings does not justify the additional costs and hazards of reprocessing because reducing the volume of the repository has an almost negligible effect on the storage costs. According to a recent report from Harvard University's Project on Managing the Atom, even a fourfold decrease in the repository volume would decrease storage costs by less than 15 percent.

The other major claim is that, without reprocessing, there is only enough low-cost uranium for 50 more years of nuclear power at current usage levels. Indeed, from 1965 to 2003, the Organisation for Economic Co-operation and Development's "Red Book" listed known conventional reserves of natural uranium recoverable at less than $130 per kilogram at 3–5 megatons. At the low end of this range, the current global consumption of 0.07 megatons per year would exhaust reserves in 50 years. Reserves, however, will continue to grow in the future, as they have in the past. As prices rise and extraction technology improves, extraction from difficult deposits will become more profitable, increasing the size of known reserves. For example, from 2005 to 2007, the known conventional reserves recoverable at $130 per kilogram increased by 20 percent to 5.5 megatons. Adding in "Red Book" estimates of so-far undiscovered resources brings the amount of uranium to roughly 17 megatons, which, if mined, would provide a 240-year supply. Reprocessing is a proliferation threat. Reprocessing supporters argue that plutonium from a reprocessing plant is not weapon-grade. The term "weapon-grade" refers to plutonium that is primarily plutonium 239, with less than 6 percent plutonium 240. Plutonium 240 has a high rate of spontaneous fission, which increases the odds of prematurely igniting a fission weapon during the detonation sequence, thereby decreasing its yield. In contrast to weapon-grade plutonium, the plutonium that comes out of France's La Hague facility has about 24 percent plutonium 240. The plutonium bombs detonated during the Trinity test and over Nagasaki used super-weapon-grade plutonium to produce an explosion equivalent to 20 kilotons TNT with a 1.6-kilometer destruction radius. If the same bomb were made with plutonium from the La Hague plant, its expected yield would be around 1 kiloton of TNT equivalent. Although a 1-kiloton explosion might sound less dangerous than a 20-kiloton explosion, its likely blast radius would still be one-half to four-fifths of a kilometer, enough to level most of down-town Boston. Such a device would be 90 times more powerful than the largest conventional weapon in the U.S. arsenal and 500 times more powerful than what destroyed the Alfred P. Murrah Federal Building in Oklahoma City. So while it is true that reactor-grade plutonium is not as powerful an explosive as weapon-grade plutonium, it certainly should be considered "terrorist-grade."

In France, reprocessed plutonium is shipped the entire length of the country from La Hague on the north coast to the MELOX fuel fabrication plant in Marcoule near the south coast. To avoid uncontrolled fission, the plutonium oxide—a fine, yellow powder—is divided into small, thermos-sized canisters. Each canister contains about 2.5 kilograms of material, and only three of these provide enough to make a bomb. Every week, Areva ships dozens of these canisters from La Hague to Marcoule in shielded trucks, present-ing an opportunity for theft. During our visit, we observed two such trucks on the highway guarded by four police cars. Given Areva's spin that this is non-weapon-grade plutonium, we wondered whether the police officers were aware that they were guarding enough material to destroy the downtown areas of 30 mid-sized cities.

Unfortunately, this is not the only chance someone has to steal the plu-tonium. Even with strict monitoring at reprocessing plants, it is not feasible to account for every single gram of plutonium produced. With the large volumes of plutonium being separated and the possibilities for measurement errors or poor bookkeeping, an insider could smuggle out a bomb's worth of plutonium in several months without anyone noticing the missing plutonium.

In addition to theft, there is the danger that reprocessing plants could be used by a host nation to initiate a nuclear weapons program. This has already happened in the case of India. The United States sold reprocessing technology to India with the understanding that it would only be used for civilian nuclear power. The plutonium product, however, was diverted for military purposes. After the detonation of an Indian bomb in 1974, the United States reversed its pro-reprocessing stance and put an end to domestic civilian reprocessing. Although reprocessing is no longer technically outlawed in the United States, a de facto ban has persisted.

In response to proliferation concerns, many reprocessing advocates rec-ommend a new method, called COEX, for future reprocessing facilities. In con-trast to the current PUREX process, which produces pure plutonium oxide, COEX would extract plutonium and uranium together to form a mixed pluto-nium-uranium oxide that can be directly fabricated into MOX fuel. Although this product cannot fuel bombs directly, a malicious organization could later extract the plutonium. Since none of the MOX components are highly radio-active, plutonium separation could be carried out safely in a standard chemi-cal laboratory with existing methods. Even more worrisome, the separation process need not be efficient to obtain a large quantity of material because the COEX product is composed of 50 percent plutonium. In addition, because

COEX has never been deployed on an industrial scale, the costs of developing it commercially could be massive. In combination with the unknown operating and environmental costs, COEX is a big gamble for little gain.

Reprocessing is not cost effective. In order for reprocessing to make sense economically, the price of a new MOX fuel rod must be competitive with the price of a new uranium fuel rod, which largely depends on the price of mined uranium. Several studies have concluded that the price of uranium would have to be in the range of $400–$700 per kilogram in order for reprocessed MOX to break even. But for the first half of 2009, the price of uranium oxide has hovered around $100 per kilogram. In fact, uranium prices have reached $300 per kilogram (in 2008 dollars) only twice in history—in the late 1970s during the energy crisis and briefly in the summer of 2007. In other words, uranium has never sustained a price that would make reprocessing profitable. And given the large estimated resources of uranium available at or below $130 per kilogram, it is unlikely that reprocessing will become cost competitive any time in the foreseeable future.

The high expense of reprocessing is rooted in the fact that chemically separating and processing spent nuclear fuel requires large, complex facilities that produce significant quantities of radioactive and chemical wastes. These facilities also must meet modern health and safety standards for dealing with highly toxic plutonium, which adds to the expense. In addition, these facilities produce weapons-usable plutonium and must be operated under military guard, which adds more costs to the process. At the same time, it is difficult to recoup all of these expenses when reprocessing yields so little usable product—only one MOX fuel rod is produced for every seven spent uranium fuel rods reprocessed. That means, for all of the investment and operating costs, reprocessing boosts the usable energy extracted from mined uranium only about 14 percent.

Reprocessing harms the environment. The reprocessing plants at Sellafield in Britain and La Hague in France are two of the largest anthropogenic emitters of radioactivity in the world. Both facilities intentionally discharge significant amounts of radioactive cesium, technetium, and iodine into the surrounding oceans, which show up in seafood harvested in the Irish and North seas. When it comes time to close a reprocessing plant, it requires decades to decontaminate facilities, soil, and groundwater. The United States currently spends billions of dollars each year to rehabilitate the reprocessing facilities at the Hanford site in Washington state and Britain will require similar resources to clean up Sellafield, making these among the most complicated and costly environmental cleanup projects in the world.

The final word. Contrary to the arguments given by its proponents, reprocessing is dangerous and unnecessary. Although nuclear power will remain a significant source of electricity for the coming decades, spent fuel reprocessing should have no part in this future. It is in the best interests of the United States—from the perspective of waste management, national security, economics, and environmental protection—to maintain its de facto moratorium on reprocessing and encourage other countries to follow suit.

EXPLORING THE ISSUE

Should the United States Reprocess Spent Nuclear Fuel?

Critical Thinking and Reflection

1. Why has fear of nuclear proliferation inhibited nuclear fuel reprocessing?
2. What are the advantages of putting off the decision to reprocess spent nuclear fuel until the necessary technology has been more fully developed?
3. What are the disadvantages of putting off the decision to reprocess spent nuclear fuel until the necessary technology has been more fully developed?
4. Economic incentives have been used to persuade people to accept nearby undesirable facilities such as dumps, power plants, and factories. Would it be ethical to do the same for a nuclear fuel reprocessing plant?

Is There Common Ground?

Both sides agree that nuclear fuel reprocessing can help support the use of nuclear power and reduce the nuclear waste problem. The major difference lies in whether the technology can be implemented immediately or must undergo further research and development. Present and under-development technologies are described on the website of the World Nuclear Association. Visit it at http://www.world-nuclear.org/info/inf69.html and answer the following questions:

1. What is the purpose of nuclear fuel reprocessing?
2. What is the dominant technology today?
3. What is transmutation and how does it reduce the nuclear waste problem?

ISSUE 20

Can "Green" Marketing Claims Be Believed?

YES: Jessica Tsai, from "Marketing the New Green," *Customer Relationship Management* (*CRM*) (April 2010)

NO: Richard Dahl, from "Green Washing: Do You Know What You're Buying?" *Environmental Health Perspectives* (June 2010)

Learning Outcomes

After reading this issue, you should be able to:

- Explain what greenwashing is and how it misleads consumers.
- Explain how to make "green" marketing claims without being guilty of greenwashing.
- Describe why corporations have in the past made excessive "green" advertising claims.
- Explain why better definitions of "green" advertising terms are needed.

ISSUE SUMMARY

YES: Jessica Tsai argues that even though marketing is all about gaining attention in a very noisy environment, it is possible to improve brand image by being environmentally responsible without being guilty of greenwashing.

NO: Richard Dahl argues that consumers are reluctant to believe corporate claims of environmental responsibility because in the past such claims have been so overblown as to amount to "greenwashing."

With the birth of the environmental movement, it quickly became apparent that people were interested in the environmentally related behavior of corporations. Initially, this interest took the form of criticism, some of which involved public protest, invasions of corporate offices, and even dumping barrels of toxic goop, collected from the pipes delivering a corporation's wastes to a local river or lake, on the carpet of the CEO's office. The point was to shame corporations into cleaning up their act. Sometimes it worked. More often, it took a while—even years—for corporations to realize that it was good public relations to look environmentally responsible. The result was more emphasis on landscaping around factories and office buildings, even to the extent of providing public walking trails, tree-planting projects, and more. As clean air and water laws were passed, corporations made much of the way they were cleaning up air and water pollution, usually without mentioning that they were being forced to do so by law. Product ingredients that had always been derived from tree barks and other natural sources were now touted as "natural." When sustainability became an environmental buzzword, it also became a PR buzzword whose definition was remarkably elastic. Is using coal more sustainable than using oil? Well, the supply of coal will certainly last longer, but the effect on the environment hardly fits the idea of meeting present needs without impairing the ability of future generations to meet theirs. Is "clean coal" green coal? Many think the term an oxymoron, coined mostly to clean up perceptions of an inescapably dirty energy technology.

The term that has come to be used for such behavior is "greenwashing," by analogy with "whitewashing," which means painting over dirt and blemishes in a fence or a reputation with whitewash (an early form of white paint). See Wendy Priesnitz, "Greenwash: When the Green Is Just Veneer," *Natural Life* (May/June 2008). Perhaps unfortunately, greenwashing works. Satyendra Singh, Demetris Vrontis, and Alkis Thrassou, "Green Marketing and Consumer Behavior: The Case of Gasoline Products," *Journal of Transnational Management* (2011), found that consumers preferentially seek out gasoline brands perceived as greener.

In 2010, the Business Reference and Services Section (BRASS) Program of the American Library Association's annual meeting discussed "Clean, Green, and Not So Mean: Can Business Save the World?" *Reference & User Services Quarterly* (Winter 2010). The theme was corporate social responsibility, of which environmental responsibility is one component, expressed as sustainable marketing, defined as "meeting the . . . definition of marketing in a way that meets both the organizational goals and customers' needs while preserving, benefiting, and replenishing both society and the environment."

Jason Daley, "Green Fallout," *Entrepreneur* (August 2010), notes the essential hypocrisy of many businesses when he describes the way the Deepwater Horizon oil spill in May 2010 tarnished the image of BP (or British Petroleum), the "company that had cultivated the greenest image in the oil industry."

> The BP blowout was the swan song of an old style of green marketing, one in which companies could make green claims and hope that no one would look over their shoulders. In the last five years, a new type of green marketing has taken hold, and it has high standards.
>
> It's no longer enough to say you're green in your advertising. It's not even enough to have one or two flagship green products in your line or to screw in a few compact fluorescents and send out a press release.

Not that BP was getting away with much before the oil spill, for *European Environment & Packaging Law Weekly* (May 15, 2009), could title an article "Shell and BP in Poll Position on 'Greenwash' Blacklist."

The basic problem has not yet vanished. According to Greenpeace:

> Corporations are falling all over themselves to demonstrate to current and potential customers that they are not only ecologically conscious, but also environmentally correct.
>
> Some businesses are genuinely committed to making the world a better, greener place. But for far too many others, environmentalism is little more than a convenient slogan. Buy our products, they say, and you will end global warming, improve air quality, and save the oceans. At best, such statements stretch the truth; at worst, they help conceal corporate behavior that is environmentally harmful by any standard.
>
> The average citizen is finding it more and more difficult to tell the difference between those companies genuinely dedicated to making a difference and those that are using a green curtain to conceal dark motives. Consumers are constantly bombarded by corporate campaigns touting green goals, programs, and accomplishments. Even when corporations voluntarily strengthen their record on the environment, they often use multi-million dollar advertising campaigns to exaggerate these minor improvements as major achievements.
>
> Sometimes, not even the intentions are genuine. Some companies, when forced by legislation or a court decision to improve their environmental track record, promote the resulting changes as if they had taken the step voluntarily. And at the same time that many corporations are touting their new green image (and their CEOs are giving lectures on corporate ecological ethics), their lobbyists are working night and day in Washington to gut environmental protections.

As Greenpeace notes, not every corporation that makes "green" claims is trying to fool the public. But some are, so that the question for the consumer becomes how to tell the difference. Awareness of the problem has grown, and the Federal Trade Commission (FTC) published the first of its "Guides for the Use of Environmental Marketing Claims" (or "Green Guides") in 1992. In October 2010, the FTC announced proposed revisions to the Green Guides; see Lynn L. Bergeson, "Selling Green: US FTC Releases Proposed Revisions to the 'Green Guides,'" *Environmental Quality Management* (Spring 2011). The proposed revisions address the use of product certifications and seals of approval and claims relating to "renewable energy," "renewable materials," and "carbon offsets." Before proposing the revisions, the FTC studied consumer perceptions of the terms often used in such claims (such as "biodegradable") to help draw the line between valid and misleading claims. Since perception plays an important role in this issue, some—including some in the public relations and advertising industry—have objected that the revisions do not go far enough. The Cone company filed comments saying that it "had hoped to see the Commission take a more definitive stance on general environmental benefit claims, perhaps even prohibiting the use of words such as 'sustainable' or 'earth friendly.'" It also wished for attention to the imagery used in ads, which can sometimes convey very strong messages that might not be defensible if put into words. See "Cone Files Comment to FTC in Response to Proposed Green Guides Revisions" at www.coneinc.com/ftcgreenguides. Jessica E. Fliegelman, "The Next Generation of Greenwash: Diminishing Consumer Confusion Through a National Eco-Labeling Program," *Fordham Urban Law Journal* (October 2010), calls for "a nationwide eco-labeling program [to] reduce consumer confusion." Such a program would be enforced by the FTC and the states. Robert B. White, "Preemption in Green Marketing: The Case for Uniform Federal Marketing Definitions," *Indiana Law Journal* (Winter 2010), calls for "federal definitions for green-marketing terms that have the force of law and expressly preempt all state definitions should be promulgated, thereby increasing green-marketing regulations' certainty and uniformity." The problem is similar in Britain; Josh Naish, "Lies . . . Damned Lies . . . and Green Lies," *Ecologist* (June 2008), notes that there the number of relevant regulations and definitions lead to consumer confusion. Simplification and even more government regulation are required.

In the following selections, Jessica Tsai argues that even though marketing is all about gaining attention in a very noisy environment, it is possible to improve brand image by being environmentally responsible without being guilty of greenwashing. Richard Dahl argues that consumers are reluctant to believe corporate claims of environmental responsibility because in the past such claims have been so overblown as to amount to "greenwashing."

YES

Jessica Tsai

Marketing the New Green

**Companies should be rightfully proud of their environmental improvements.
So why can't they market those achievements without seeming mercenary?**

Unless your company happens to be in an industry where the environment is literally part of the business model—solar-cell manufacturer, forestry-supply firm, waste-management consultancy—then it's unlikely to have had an enterprisewide sustainability process woven into its corporate DNA. In fact, most companies have never even used the word *sustainability,* and the others probably reverse-engineered some half-hearted "green" initiative as a mere afterthought. The reality is that, among existing companies, green processes are usually just a byproduct of cost-saving or efficiency-improving projects. According to Mark Smith, executive vice president at customer engagement specialist Portrait Software, the rare businesses that have truly tackled sustainability "got there as a side effect of their primary goal—making money or saving money. In terms of 'green marketing'? No one started there." Glitz and glamour, bigger and better: From a traditional marketer's perspective, Smith says, "it's all about gaining attention in a very noisy marketplace." Whether that involves toxic paint to ensure colors "pop" or nonbiodegradable packaging materials designed to endure a rough-and-tumble delivery, marketers have long embraced any and all environmental catastrophes that help get the message across. The recession, however, made marketing budgets themselves the scene of a catastophe—drastic cuts were the order of the day. In early 2009, 71 percent of respondents in Forrester Research's Global CMO Recession Online Survey reported reduced budgets compared to the year prior, with more than half of those bloodied reporting cuts of 20 percent or more. "The interesting side effect [of this] is that those big, grand [campaigns] have been cut back and reined in," Smith says. The added bonus? It's had a great effect on the environment as well. Still, sustainability has never been a top consideration for marketing, at least not on the departmental or corporate level. It's usually a single person or small group of people who have a personal stake in environmental activism.

That pretty much describes Seema Haji, the senior product marketing manager at Actuate, a provider of business intelligence solutions. Actuate's companywide green initiatives—and the eventual launch of its sustainability management product—can be traced back to the passion of Haji, who calls herself a "pseudo-environmentalist in a corporate world," having started her own blog, *World-Saving Tips for the Lazy (and Busy)*, a few years before joining Actuate. (You can find her efforts at bleedinggrass.blogspot.com.)

Founded in 1993, Actuate now has more than 4,400 customers worldwide, but it wasn't until 2008 that the company began thinking seriously about going green—and helping its customers do the same. "We were all brainstorming," Haji recalls. "My bosses turned to me and said, 'Oh, yeah—she's a hippie. Let her do this, she'll have fun with it.' And I really did."

Word had been trickling up from users that going green was not only emerging as a huge competitive advantage but as a way to cut energy costs and improve brand image. Despite the enthusiasm, though, users didn't know what steps to take.

SIX WAYS TO PROMOTE GREEN MARKETING

- **Integrating green initiatives into every aspect of the organization:** Companies are trying to link the corporate brand to efforts in social responsibility, Edwards says—and environmental stewardship can affect the bottom line as it improves customer relationships. At United Parcel Service (UPS), for example, new mapping systems enabled a "No Left Turn" rule to eliminate costly left turns from drivers' routes. According to *The New York Times*, UPS spokeswoman Heather Robinson reported that the company shortened delivery routes by 28.5 million miles, saving 3 million gallons of gas and reducing carbon emissions by 31,000 metric tons. "They're tying their brand image to efficiency and environmental savings," Edwards says.

- **Using ecolabels and ecologos on products or marketing materials:** Perhaps the most well-known ecolabel is the recycling symbol composed of chasing arrows, created in 1970 by Gary Anderson, who won a graphics and design competition hosted by the Container Corporation of America. Since then, a significant number of labels have popped up, some of which have contributed to an industry malfeasance known as "greenwashing.". . . Other widely recognized symbols include the USDA Organic, which signifies the use of organic ingredients in food; Forest Stewardship Council (FSC) indicates wood and paper products produced in methods that advocate responsible forestry; and Energy Star identifies home, building and construction, and electronics that are energy efficient.

  When adopting these labels, Edwards warns marketers to be careful—while it can help inform consumers, over-saturation of labels in the market has resulted in label blindness. For the most part, consumers today only recognize a handful of labels. Therefore, marketers must identify whether the logo: a) is credible; b) is meaningful and recognizable by the intended audience; and c) fits with the organization's message.

- **Engaging customers in green marketing:** Companies are looking to motivate consumers by encouraging them to participate and engage in the campaign or directly with the product. Marketers that send out direct mail pieces can put links directing marketers to participate in green programs online, or do something as simple as ask customers to recycle the mail after reading. Edwards has seen largely positive feedback from marketers who've attempted to bring customers into the mix; the number of those doing so is growing but still pretty small, Edwards admits. Only about 100 marketers have enlisted in the DMA's "Recycle Please" program—a nationwide public education campaign where DMA members are asked to display a "Recycle Please" logo in catalogues and direct mail pieces.

- **Asking and respecting customer choices and preferences:** Segmentation is a practice that goes back to Marketing 101. Companies that are leveraging customer data and respecting their preferences will inevitably have fewer unnecessary mailings. . . . In October 2007, the DMA launched its Commitment to Consumer Choice policy, which among other stipulations, requires all DMA members to provide existing and prospective customers and donors with notice of an opportunity to modify or opt out of commercial communications. By giving consumers this choice, companies are not only acting environmentally responsible, but also reinforcing their corporate responsibilities.
- **Adopting a lifecycle approach:** Companies are selecting green materials and products for their marketing materials and adopting a lifecycle approach that looks at the whole of the campaign, thereby foreseeing areas of potential waste. Edwards sees more marketers adopting recycled and FSC-approved papers and printing, vegetable and soy-based inks, smaller formats and trim sizes, and a reduction in paper use overall. Aromatherapy and skincare treatment provider Decleor now only uses Programme for the Endorsement of Forest Certification and FSC-certified paper, despite the fact that it's 3 percent to 7 percent more costly. This year, the company stopped printing its logo on gold foil and changed it to a deep, eggplant color, in order to ensure that its paper products are 100 percent recyclable. Moreover, the company only maintains relationships with FSC-certified printers and has actually stopped doing business with a printer that wasn't—until that printer came back six months later newly certified.
- **Shifting to the online space:** Digital marketing was projected to reach $25.6 billion in 2009, and reach $55 billion, 21 percent of all marketing spend, by 2014, according to Forrester Research's United States Interactive Marketing Spend report. Channels included in this report were mobile marketing, social media, email marketing, display advertising, and search marketing. More and more companies are requiring that employees remind email recipients to think about the environment before printing.

After discussions with customers and partners, Actuate developed a set of more than 100 metrics intended to help people understand what they should be tracking: carbon-credit usage, average water consumption by facility, average electrical consumption per employee, to name just a few. At the same time, Actuate wanted users to have the flexibility to define custom metrics according to the needs of their respective businesses and industries. The end result—Actuate for Sustainability Management—deploys interactive dashboards, sustainability scorecards, and strategy maps to help companies measure overall employee satisfaction, environmental impact, access to training and education, and community engagement.

At Portrait, going green wasn't a corporate mandate, but support spread like wildfire. "A year or so ago, we had a huge green initiative where we plastered stickers all over the place saying we were a green company," Smith says. The company installed recycling bins around the office; its United Kingdom branch composted food and waste; energy use for computers and lights was more closely monitored and regulated; new window shades were installed to help reduce the need for air conditioning.

"It all started as a grassroots thing," Smith says, "but then our CFO came around and said it was fantastic. Our electricity and power bills were lower— I don't think he was expecting it at all." Cost-cutting certainly wasn't the main objective, but the savings have earned the attention—and support—of senior-level executives. Even initiatives that have cost more—it's cheaper and easier to discard than to recycle—are still being encouraged as the company strives to be even more ecoconscious.

Still, other than looking at comparisons between one electric bill and the next, Portrait couldn't speak to a particular technology that measures a business's environmental impact—let alone any attendant cost savings. At press time, even Actuate was only just a few months into implementing its own sustainability application. "The stuff they ask us to measure is all about dollars, response rates, mailing volumes," Smith says. "Not many [companies] have particularly brought it to that next level."

For some, however, becoming a sustainable enterprise is simply motivated by the desire to do the right thing. Portrait hasn't yet utilized its green initiatives to enhance its brand image, although Smith admits it's something he'll definitely be considering in the future.

Meta Brophy, director of publishing operations for Consumers Union (CU), the parent of *Consumer Reports*, says that her company doesn't herald its green initiatives in marketing solicitations to its consumers, focusing instead on its industry peers. "CU has worked to lead by example . . . in hopes that other direct marketers will follow suit and help to build a more sustainable direct marketing community," Brophy says.

CU abides by the Direct Marketing Association's Green 15 Standards & Environmental Action Program, which has led to changes in the six principal areas of list hygiene and data management, design, paper procurement and usage, printing, recycling, and pollution reduction. For five years, CU has been a green advocate, sharing its progress and best practices at industry events such as National Postal Forums, New England Mail Expo, New York Nonprofit Conference, and the World Environment Center Roundtable. "We incorporate such initiatives because it's the right thing to do," she says, "and generally speaking we save money."

Gregory Unruh, author of *Earth, Inc.*, has been involved in bridging the gap between the business sector and the sustainability movement. As a professor of global business and director of the Lincoln Center for Ethics in Global Management at Thunderbird School of Global Management, Unruh has found that the only way to get companies to stick to sustainability is by proving that it's profitable. "The ultimate goal here is to embed sustainability and make it a standard business practice," Unruh says. "You start with one [practice], set up in a way that—after you complete that one—you create cost-reduction and profitability opportunities that then provide the momentum you need," he says. "If it makes a profit, companies will do it.". . .

Building-products provider BlueLinx has always had a focus on sustainability—by necessity, considering its business is heavily dependent on wood. BlueLinx long ago made the connection between practicing sustainable forestry and maintaining a sustainable—and profitable—business. Nevertheless, Shiloh Kelly, the firm's communications and national sustainability lead, contends that being green "wasn't something to try and sell an extra piece of wood over." With the launch of its virtual trade show, however, BlueLinx was able to promote a new sustainability program and brand of ecoproducts. . . .

"The whole aspect of sustainability is to get every last drop from everything you can," Kelly says. "Not only from a resource standpoint, but from a financial standpoint and a viral standpoint." To that end, Kelly admits she was especially blown away by the viral impact of the BlueLinx virtual event, and in particular its effectiveness around the company brand.

"Prior to . . . [our] rollout, our customers didn't know the depth of our resources or expertise," Kelly explains. "The calls concerning these matters were few and far between." After the virtual event, however, top vendors, customers, and media publications were pouring in, soliciting BlueLinx as a primary source of information around sustainable-building products and industry programs. . . .

Decleor, a French provider of aromatherapy and skincare treatments, has used only essential oils and plant-based ingredients since its founding 35 years ago—factors that Cindy Willette, the company's director of marketing, admits were taken for granted during most of that time. The company has had certain advantages simply by virtue of its products. For instance, essential oils corrode plastic, so glass bottles were the most natural packaging. However, it wasn't until the company saw its competitors taking away market share that Decleor realized the importance of strengthening its green marketing.

Similar to BlueLinx, Decleor realized that it, too, needed to be responsible for replenishing the resources its products consumed. In 2008, the company launched a three-year, responsible-development project in Madagascar, whereby Decleor would partner with Association Madagascar to contribute to the region's reforestation, install solar electrification in schools, and provide medical equipment to hospitals. (As of press time, 40,000 trees had been planted and a school with 600 students had been equipped with solar electrification.) As part of the promotion, the company sold 100-percent-natural shopping bags at spas and retail stores and promised to plant five trees for every bag purchased. These initiatives have captured the attention of both the consumer and trade audiences, Willette says, citing a 20 percent increase in inquiries coming through the company Web site last year.

And yet, despite the company's ecofriendly roots, some traditional practices remain difficult to weed out. "There are ways of doing things that get

THE 7 SINS OF GREENWASHING

In partnership with EcoLogo, a program that provides a certification mark approved by the Global Ecolabelling Network, environmental marketing firm TerraChoice released a report last April describing the seven sins of "greenwashing." TerraChoice defines this concept as "the act of misleading consumers regarding the environmental practices of a company or the environmental benefits of a product or service."

In order to avoid committing greenwashing—which can lead to consumer mistrust, delay true innovation, stir market skepticism, and, thus, damage credibility—the firm warns companies against the following "sins":

- **The Hidden Trade-Off:** Fully understand your product's environmental impact across its entire lifecycle and continue to improve it. Don't overemphasize one facet to hide the drawbacks of another.
- **Lack of Proof:** Back up your claims with scientific evidence and make that information readily accessible to the public.
- **Vagueness:** Clearly articulate how your product or service is environmentally beneficial in a language your customer can understand.
- **Worship of False Labels:** Select eco-labels that are from accredited programs, preferably ones that address a product's entire lifecycle.
- **Irrelevance:** Don't claim to be something you're not or to be something that all or most of your competition shares.
- **Preference for the Lesser of Two Evils:** Connect consumers with the right product rather than pitching a product that may still be harmful and unnecessary but may be more "green" than the next.
- **Dishonesty:** Don't lie—ever. Don't even exaggerate.

Sources: TerraChoice, EcoLogo

ingrained in corporate culture," Willette says. "We need to have this material, this piece, to support this launch in this way. As we started looking at the brand from a bigger picture . . . [we said], 'Let's take a harder look at what we're doing and decide if we need to do that anymore. Just because it's always been done, doesn't mean it needs to be done.'"

In scrutinizing its marketing and overall business processes, Decleor has not only changed its printing and paper policies . . ., but overall energy consumption fell by 10 percent in the last two years; industrial waste declined by 6.6 percent; paper usage dropped by 14 percent; gas usage was down 23 percent; water usage was down 21 percent; all while production increased 4.7 percent during the same time frame. Seeing the benefits of going green, Decleor is unlikely to revert back to the traditional ways. "Our mantra here is 'less is more," Willette says, "so rather than create more, we try to create better."

As a certain amphibian once said, it's not easy being green. A culture that's been trained to be high in both consumption and waste production took centuries to build—and will be difficult to tear down.

"Start small," Haji advises. "Pick the most important things that will make the most impact and start with that. Start in one department [and] spread this through the rest of the organization."

Reducing paper consumption may be the first and easiest step to take. Encourage your marketing and sales teams to send materials digitally through email or USB flash drives. (If you *really* want to go green, there are bamboo drives, wood drives, and biodegradable, leadfree drives available.)

Sure, you could say that even walking around isn't 100 percent green—the energy to fuel that movement came from food, which was prepared at a restaurant after being processed at (and shipped from) a factory, and so on. But Kelly says that perspective picks at lesser evils and distracts from the bigger issue, one that companies can finally tackle—and marketers can sincerely boast about.

"With the way technology and innovation has changed, there are huge differences between the evils now," Kelly says. "Anything you do out there has a sustainable effect. It's just whether you're going to ignore it, or do something about it."

Richard Dahl  **NO**

Green Washing: Do You Know What You're Buying?

In a United States where climate change legislation, concerns about foreign oil dependence, and mandatory curbside recycling are becoming the "new normal," companies across a variety of sectors are seeing the benefit of promoting their "greenness" in advertisements. Many lay vague and dubious claims to environmental stewardship. Others are more specific but still raise questions about what their claims really mean. The term for ads and labels that promise more environmental benefit than they deliver is "greenwashing." Today, some critics are asking whether the impact of greenwashing can go beyond a breach of marketing ethics—can greenwashing actually harm health?

Greenwash: Growing (Almost) Unchecked

Greenwashing is not a recent phenomenon; since the mid-1980s the term has gained broad recognition and acceptance to describe the practice of making unwarranted or overblown claims of sustainability or environmental friendliness in an attempt to gain market share.

Although greenwashing has been around for many years, its use has escalated sharply in recent years as companies have strived to meet escalating consumer demand for greener products and services, according to advertising consultancy TerraChoice Environmental Marketing. Last year TerraChoice issued its second report on the subject, identifying 2,219 products making green claims—an increase of 79% over the company's first report two years earlier. TerraChoice also concluded that 98% of those products were guilty of greenwashing. Furthermore, according to TerraChoice vice president Scot Case, the problem is escalating.

TerraChoice also measured green advertising in major magazines and found that between 2006 and 2009, the number mushroomed from about 3.5% of all ads to just over 10%; today, Case says, the number is probably higher still. Case says researchers are currently working on another update that will be released later this year, and he predicts the number of products making dubious green claims will double.

Compounding the problem is the fact that environmental advertising—in the United States, at least—is not tightly regulated. The Federal Trade Commission (FTC), the agency responsible for protecting the public from unsubstantiated or unscrupulous advertising, does have a set of environmental

Dahl, Richard. From *Environmental Health Perspectives,* June 2010, pp. A247–A252. Published in 2010 by National Institute of Environmental Health Sciences. http://ehp.niehs.nih.gov/

marketing guidelines known as the Green Guides. Published under Title 16 of the *Code of Federal Regulations*, the Green Guides were created in 1992 and most recently updated in 1998. According to Laura DeMartino, assistant director of the FTC Division of Enforcement, the proliferation of green claims in the marketplace includes claims that are not currently addressed in the Green Guides, and updated guidance currently is being developed.

The FTC originally planned to begin a review of the Green Guides in 2009, but the commission moved the schedule up, according to DeMartino, in response to a changing landscape in environmental marketing. "The reason, at least anecdotally, was an increase in environmental marketing claims in many different sectors of the economy and newer claims that were not common, and therefore not addressed, in the existing Guides," she says. "These are things like carbon offsets or carbon-neutrality claims, terms like 'sustainable' or 'made with renewable materials.'"

The FTC held a series of workshops in 2008, holding separate events for each of three areas: carbon-offset and renewable-energy claims, green packaging, and buildings and textiles. In association with each workshop, the FTC asked for comments to help shed light on consumer perception of green advertising, but DeMartino says the commission received very few. The FTC responded to this gap by commissioning a research firm, Harris Interactive, to provide that information. DeMartino says that research has been completed, and a report on it will accompany the revision announcement, which is expected soon.

How Updated Guidance Might Look

In aspiring to revise its environmental marketing guidelines, the FTC is following a trend that has been evident in other nations. In 2008, the Canadian Competition Bureau (a government agency similar in function to the FTC) updated its environmental marketing guidelines to reduce green misinformation, and the Australian Competition and Consumer Commission took a similar step. In March 2010, the U.K. Committee of Advertising Practice and Broadcast Committee of Advertising Practice announced an update to their codes of practice designed to curtail greenwashing.

All three updates are "remarkably similar," Case says, but he suggests the Canadian revisions might provide the best "sneak peak" at what the FTC might do because the two agencies have a long history of working together on cross-border consumer matters. Attorney Randi W. Singer, a litigation partner at New York's Weil Gotschal who has defended companies accused of false advertising, agrees the moves made in Canada but also in Australia and the United Kingdom may provide a good look at what is to come in the United States. Those changes, coupled with her own analysis of the FTC workshop discussions, provide the basis for her to make several predictions about what the new U.S. regulatory scheme might look like.

Singer predicts the revisions will probably contain new definitional language for terms such as "carbon neutral" and "sustainable." She also expects the FTC will address the issue of third-party certifications—that is, the plethora of green labels consumers see on their products. She says the workshop discussions included "a lot of talk about the need for standardization of certifications, a need to have a process for certifications so it's not just people registering themselves, a need to standardize the iconography and the testing."

According to Case, there are now more than 500 green labels in the United States, and some are "significantly more meaningful" than others. "I testified before Congress last summer and I pointed out a certain lawyer in Florida who set up a website and is 'certifying' products. He doesn't need to see the product, he doesn't need test results. He just needs to see your credit card number," Case says.

Meanwhile, the FTC has begun to step up enforcement regarding claims that it considers clear violations of the existing Green Guides, last year charging three companies with false and unsupportable claims that a variety of paper plates, wipes, and towels were biodegradable. "When consumers see a 'biodegradable' claim they think that product will degrade completely in a reasonably short period of time after it has been customarily disposed," DeMartino says. But for about 91% of the waste in the United States, the FTC wrote in its 2009 decisions, customary disposal means disposal in a landfill, where conditions prevent even a theoretically biodegradable item from degrading quickly.

In another instance, the FTC charged four sellers of clothing and other textiles with deceptively advertising and labeling various textile items as biodegradable bamboo that had been grown in a more sustainable fashion than conventional cotton, when, in fact, the items were rayon, a heavily processed fiber. In January 2010, the FTC sent letters to 78 additional sellers of clothing and textiles warning them they may be breaking the law by advertising and labeling textile products as bamboo.

The Health Impact of Greenwash

One major result of greenwashing, say Case and others, is public confusion. But can greenwashing also pose a threat to the environment and even to public health? Critics say greenwashing is indeed harmful, and they cite examples.

In 2008, the Malaysia Palm Oil Council produced a TV commercial touting itself in very general terms as eco-friendly; a voiceover stated "Malaysia Palm Oil. Its trees give life and help our planet breathe, and give home to hundreds of species of flora and fauna. Malaysia Palm Oil. A gift from narure, a gift for life." But according to Friends of the Earth and other critics of the ad, palm oil plantations are linked to rainforest species extinction, habitat loss, pollution from burning to clear the land, destruction of flood buffer zones along rivers, and other adverse effects. The U.K. Advertising Standards Authority agreed, declaring the ad in violation of its advertising standards; contrary to the message of the ad, the authority ruled, "there was not a consensus that there was a net benefit to the environment from Malaysia's palm oil plantations."

In 2008, the authority rebuked Dutch energy giant Shell for misleading the public about the environmental effects of its oil sands development project in Canada in the course of advertising its efforts to "secure a profitable and sustainable future." While acknowledging the term "sustainable" is "used

THE SEVEN SINS OF GREENWASHING

In the course of assessing thousands of products in the United States and Canada, TerraChoice Environmental Marketing categorized marketing claims into the following "seven sins of greenwashing":

1. Sin of the hidden trade-off: committed by suggesting a product is "green" based on an unreasonably narrow set of attributes without attention to other important environmental issues (e.g., paper produced from a sustainably harvested forest may still yield significant energy and pollution costs).
2. Sin of no proof: committed by an environmental claim that cannot be substantiated by easily accessible supporting information of by a reliable third-party certification (e.g., paper products that claim various percentages of postconsumer recycled content without providing any evidence).
3. Sin of vagueness: committed by every claim that is so poorly defined or broad that its real meaning is likely to be misunderstood by the consumer (e.g., "all-natural").
4. Sin of irrelevance: committed by making an environmental claim that may be truthful but is unimportant or unhelpful for consumers seeking environmentally preferable products (e.g., "CFC-free" is meaningless given that chlorofluorocarbons are already banned by law).
5. Sin of lesser of two evils: committed by claims that may be true within the product category, but that risk distracting the consumer from the greater health or environmental impacts of the category as a whole (e.g., organic cigarettes).
6. Sin of fibbing: committed by making environmental claims that are simply false (e.g., products falsely claiming to be Energy Star certified).
7. Sin of false labels: committed by exploiting consumers' demand for third-party certification with fake labels or claims of third-party endorsement (e.g., certification-like images with green jargon such as "eco-preferred").

Adapted from: The Seven Sins of Greenwashing: Environmental Claims in Consumer Markets.

and understood in a variety of ways by governmental and non-governmental organisations, researchers, public and corporate bodies and members of the public," the authority also noted that Shell provided no evidence backing up the "sustainability" of the oil sands project, which has been criticized widely for its environmental impact.

Case contends that makers of indoor cleaning products are among the worst greenwash offenders. "People are attempting to buy cleaning chemicals that have reduced environmental and health impacts, but [manufacturers] are using greenwashing to either confuse or mislead them," he says. "People aren't really well-equipped to navigate the eco-babble, and so they end up buying products that don't have the environmental or human-health performances that they expect."

TerraChoice's 2009 report concluded that of 397 cleaners and paper cleaning products assessed, only 3 made no unsubstantiated or unverifiable green claims. The report noted that cleaners, along with cosmetics and children's products, are particularly prone to greenwashing—a worrisome state, given that these items are "among the most common of products in most households."

While companies see consumers' growing demands for green products as an opportunity to increase sales by making perhaps dubious environmental claims, they may also be doing so in an attempt to avoid regulation, says Bruno. In addition to the FTC's promises to tightening up its rules on environmental advertising, broader governmental pressures increasingly place greater burdens on producers to ensure their products are environmentally sound.

"A single ad or ad campaign may be an attempt to sway a customer. But the preponderance of green image ads, many of which are not even attempting to sell a product, combined with lobbying efforts to avoid regulation, add up to a political project that I call 'deep greenwash,'" Bruno says. "Deep greenwash is the campaign to assuage the concerns of the public, deflect blame away from polluting corporations, and promote voluntary measures over bona fide regulation."

However, several corporate and marketing professionals warn that growing consumer cynicism about these kinds of general campaigns make them risky ventures for companies who engage in them. Keith Miller, manager of environmental initiatives and sustainability at 3M, last year addressed a seminar of The Conference Board, a business-management organization, about what his company does to avoid greenwashing allegations. Summarizing his presentation for the business blog CSR Perspective, Miller said that, based on 3M's experiences, he encouraged companies to avoid making "broad environmental claims" and that any claims made should be specific to products and backed up by "compelling" data.

Ogilvy & Mather advertising agency recently released a handbook designed to guide managers in how to avoid greenwashing charges and called upon them to adopt a policy of "radical transparency" in green advertising campaigns. Business for Social Responsibility, a consulting and research organization, has also published a handbook, *Understanding and Preventing Greenwash: A Business Guide*, which also emphasized the need for transparency as well as for bolstering any environmental claims with independent verification.

Reining in Greenwash

In the absence of a strong regulatory scheme, consumer and environmental groups have stepped into the vacuum to keep an eye on corporate use of greenwashing. Greenpeace was one of the first groups to do so, creating a separate anti-greenwash group, stopgreenwash.org, which monitors alleged greenwash ads and provides other information on identifying and combating greenwash. The University of Oregon School of Journalism and Communication and EnviroMedia Social Marketing operate greenwashingindex.com, where people may post suspected greenwash print or electronic ads and rank them on a scale of 1 to 5 (1 is "authentic," 5 is "bogus").

Claudette Juska, a research specialist at Greenpeace, also points to numerous antigreenwash blogs that have emerged. The result, she says, is that "there's been a lot of analysis of greenwashing, and the public has caught on to it. I think in general people have become skeptical of any environmental claims. They don't know what's valid and what isn't, so they disregard most of them."

Thomas P. Lyon, a business professor at the University of Michigan who has written and spoken extensively about greenwashing, agrees. He says companies are aware they may be criticized or mocked for making even valid claims, so they're starting to grow skittish about making green claims of any kind. "That's why companies, I think, want to see the FTC act—to give them some certainty," he says.

David Mallen, associate director of the National Advertising Division of the Council of Better Business Bureaus, the advertising industry's self-regulatory body, says companies are growing increasingly aware of the dangers of greenwashing. Although some of the matters his office handles are initiated by consumers, the large majority are prompted by companies disputing competitors' claims.

"We're definitely seeing a rise in challenges about the truth and accuracy of green marketing and environmental marketing," he says. "It's certainly taking up a greater percentage of the kinds of advertising cases that we look at. Because green advertising is so ubiquitous now, there's so much greater potential for confusion, misunderstanding, and uncertainty about what messages mean and how to substantiate them."

Typically, he says, a company will be attacked for making a broad or general claim about a product being environmentally friendly "based only on a single attribute, which might not even be a meaningful one." But he says many other cases focus on a competitor's use of a word such as "biodegradable" or "renewable." He adds, "We're also seeing these aggressive, competitive green advertisements where a company will say 'Not only are we green, not only are we making significant efforts toward sustainability, but our competitors aren't.'"

Lyon says he's found the companies that are most likely to engage in greenwashing are the dirtiest ones, because dirty companies know they have a bad reputation, so little is lost in making a green claim if the opportunity arises. At the same time, he and coauthor John W. Maxwell wrote in 2006, "[P]ublic outrage over corporate greenwash is more likely to induce a firm to become more open and transparent if the firm operates in an industry that is likely to have socially or environmentally damaging impacts, and if the firm is relatively well informed about its environmental social impacts."

"It's somewhat counterintuitive, but the clean guy is likely to shut up altogether," Lyon says. "The rationale is: if you're clean and people already think you're a green company, you don't need to bother touting it so much—and if touting it puts you at risk of being attacked, just shut up and let people think you're clean."

Making Green Claims Work

But when a clean company pulls in its horns over the risks of backlash from a cynical public, Lyon believes an opportunity has been lost. He suggests clean companies can be effective green marketers if they take certain steps. First, he says, they might incorporate a full-blown environmental management system (EMS), which would detail its full environmental program in a comprehensive manner. "When a company has an EMS in place, you have a greater expectation that they actually do know what their environmental results are," Lyon explains. EMSs themselves are supposed to meet an international standard called ISO 14001 developed by the nongovernmental International Organization for Standardization in Geneva, which sets out a variety of voluntary environmental standards.

Another step is to take part in the Global Reporting Initiative (GRI), an international organization that has pioneered the world's most widely used corporate sustainability reporting framework. The GRI was launched in 1997 by a nonprofit U.S. group called Ceres—a network of investors, environmental organizations and other public interest groups—in partnership with the United Nations Environment Programme. Lyon says the GRI can provide good green credibility at the company level.

The FTC's attention, however, is directed at products—not companies. And Lyon is one of several experts questioning how effective the looming changes to the Green Guides will be in modifying greenwashing. "Honestly, I don't think the FTC Green Guides are going to block much activity," he says. "All the FTC can do is force companies not to provide materially false information. They could potentially go into the domain of what's misleading as well, but that's very tricky. But they could . . . require companies to give you a more complete story."

To Lyon, the ideal system for regulating green marketing claims would entail comprehensive labeling and certification requirements. "You could picture a system that would be a little like the nutrition labeling that we get for food," he explains. "But whether or not that would be helpful is really unclear to me. From what I understand, there's not a lot of evidence that those nutrition labels have changed America's eating habits."

Among hundreds of green labels available today, a few are broadly recognized as highly reliable. One of them is Green Seal, which awards its seal to companies that meet standards that examine a product's environmental impact along every step of the production process, including its supply of raw materials. "It's a differentiator," says Linda Chipperfield, vice president of marketing and outreach at Green Seal. "If you're really walking the walk, you should be able to tell your customers about it."

Other labels are attained via self-certification—that is, if a company wants the label, they can buy it—and aren't so reliable. The Government Accountability Office (GAO) recently proved that in an investigation of Energy

Star, a joint program of the U.S. Environmental Protection Agency (EPA) and Department of Energy.

Energy Star provides labels to companies who submit data about products and seek the stamp of approval to place on their packages. "Currently, in a majority of categories [Energy Star] is a self-certification by the manufacturer, which leaves it vulnerable to fraud and abuse by unscrupulous companies," says Jonathan Meyer, an assistant director in the GAO's Dallas office. Indeed, over a nine-month period, GAO investigators gained Energy Star labels for 15 bogus products, including a gas-powered alarm clock the size of a portable generator. In addition, two of the bogus firms that GAO created as "manufacturers" of the products received phone calls from real companies that wanted to purchase products because the fake companies were listed as Energy Star partners.

The EPA and DOE subsequently issued a joint statement pledging to strengthen the program. The GAO report has also prompted responses from consumers and industry alike that a strong and reliable federal certification program is needed. In a story on the investigation *The New York Times* quoted the director of customer energy efficiency at Southern California Edison as saying industries affected by Energy Star hope the report will be "a wake-up call to whip [the program] into shape."

Case believes an improved regulatory scheme does require some kind of certification and labeling. "I think there is room for some kind of unifying green label," he says. "But I'm not sure if the government wants to get into the business of putting 'approved' stickers on good products." He proposes that the function of providing environmental labels be handled by a new office of the EPA. Under this plan, the EPA would combine several existing environmental labels (such as Energy Star and Green Seal) under a single brand to make it easier for consumers to identify more environmentally preferable goods and services. He points to the U.S. Department of Agriculture's affirming label on organic foods as a model.

Toward a Unified Approach

The growing demands of society for greener products and corporate America's desires to meet it and make a profit make for "a fascinating interaction with cultural change," says Lyon. "The norms have really started to shift. I think that's our hope for information and labeling—that it will create a new floor that keeps rising. I don't think we're anywhere close to that yet, but I think it's starting to happen."

Case says he is "somewhat hopeful" that all involved are moving toward a unified approach to solving the challenges posed by greenwashing. "The huge danger of greenwashing is if consumers get so skeptical that they don't believe any green claims," he says. "Then we've lost an incredibly powerful tool for generating environmental improvements. So we don't want consumers to get too skeptical."

EXPLORING THE ISSUE

Can "Green" Marketing Claims Be Believed?

Questions for Critical Thinking

1. Should corporations be allowed to claim they or their products are kind to the environment even when they are not?
2. Why are cleaning supplies particularly prone to greenwashing?
3. How can a consumer reliably choose environmentally benign products?

Is There Common Ground?

No one seems willing to defend the practice of greenwashing. The debate is between those who say green marketing cannot be trusted and those who say, "It can too! At least if it's done right." Yet it seems clear that greenwashing happens and one can easily suspect that there must be internal corporate documents that say something like "Go ahead and lie."

1. How might such internal documents be revealed?
2. Are there similar cases of corporate mendacity on record? See Rebecca Leung, "Battling Big Tobacco," *60 Minutes* (February 11, 2009) (www.cbsnews.com/stories/2005/01/13/60II/main666867.shtml). For a list of the top ten whistle-blowers, see www.toptenz.net/top-10-whistle-blowers.php.

Additional Resources

Greenpeace—Greenwashing

Greenpeace has long monitored corporate malfeasance, confronted polluters, and tried to stop environmental crimes whenever and wherever they occur. Its StopGreenwash.org website aims to confront deceptive greenwashing campaigns and give consumers, activists, and lawmakers the information and tools they need to confront corporate deception and hold corporations accountable for the impacts their core business decisions and investments are having on our planet.

http://stopgreenwash.org/

Federal Trade Commission

The Federal Trade Commission offers consumers help in interpreting what claims like "environmentally safe," "recyclable," "degradable" or "ozone friendly" really mean.

www.ftc.gov/bcp/edu/pubs/consumer/general/gen02.shtm

ISSUE 21

Does Designating "Wild Lands" Harm Rural Economies?

YES: Mike McKee, from testimony of Mike McKee before the House Natural Resources Committee hearing on "The Impact of the Administration's Wild Lands Order on Jobs and Economic Growth" (March 1, 2011)

NO: Robert Abbey, from testimony of Robert Abbey before the House Natural Resources Committee hearing on "The Impact of the Administration's Wild Lands Order on Jobs and Economic Growth" (March 1, 2011)

Learning Outcomes

After reading this issue, you should be able to:

- Explain how declaring public lands to be "wild lands" or wilderness can affect rural livelihoods.
- Explain why public involvement in land use planning is essential.
- Describe the various ways in which public lands are valuable to the nation.
- Describe how "wild lands" are designated.

ISSUE SUMMARY

YES: Mike McKee, Uintah County Commissioner, argues that the government's new "Wild Lands" policy is illegal, contradicts previously approved land use plans for public lands, and will have dire effects on rural economies based on extractive industries. It should be repealed.

NO: Robert Abbey, director of the Bureau of Land Management, argues that the government's new "Wild Lands" policy is legal, restores balance and clarity to multiple-use public land management, and will be implemented in collaboration with the public. Destruction of local extractive economies is by no means a foregone conclusion.

On March 30, 2009, President Obama signed the Omnibus Public Lands Management Act, protecting more than two million acres of wilderness in nine states. Stephen Trimble, "Wildeness: 'The System' Delivers: Hope Abounds for Wilderness Bills," *Wilderness* (2009/2010), notes that "As we search for a resilient ethic of sustainability that balances traditional American vitality with the stark demands of the 21st Century our solutions must balance wilderness and restoration, preservation and reconciliation, collaboration and core values, environmental and social justice." See also his book, *Bargaining for Eden: The Fight for the Last Open Spaces in America* (University of California Press, 2009). However, not everyone is interested in such balance. On December 23, 2010, Ken Salazar, Secretary of the United States Department of the Interior, issued an order proposing to designate public lands with wilderness characteristics as "Wild Lands" and then to manage them to protect their wilderness value. The order can be seen at www.blm.gov/pgdata/etc/medialib/blm/wo/Communications_Directorate/public_affairs/news_release_attachments.Par.26564.File.dat/sec_order_3310.pdf. Doc Hastings (R-WA), chairman of the House Natural Resources committee, called the order "a clear attempt to allow the Administration to create *de facto* Wilderness areas without Congressional approval." He added:

> "Designating land as Wilderness imposes the most restrictive land-use policies. Lands that are currently used for multiple-use—including recreation activities, agriculture, ranching, American energy production and other economic activities—are in danger of being placed off-limits.
> "This Secretarial Order will disproportionately impact rural communities, who depend on public lands for their livelihoods. These communities have already been hit hard by onerous existing federal restrictions and by the current economic crisis. They suffer from some of the highest unemployment rates in the country. The 'wild lands' order threatens to inflict further economic pain. This is just one more example of the onslaught of harmful actions that the Obama Administration is imposing on rural America."

This debate is not new. A great deal of the American West is public land, and a great many farmers, ranchers, and others have become accustomed to using (and misusing) it as if it were their own. When the federal government has attempted to rein in overgrazing and other misuses, protests have been loud and Western politicians have fought to protect their constituents' prior uses of public lands. The conflict is also seen in connection with water rights, as in California where the debate is over whether the waters of the San Joaquin River should be routed through the river to protect endangered species of fish or through pipes and canals to supply farmers with irrigation water and protect agricultural production and jobs. In that context, Representative Hastings has accused the Environmental Protection Agency and "the radical policies of the environmental left" of waging "an assault on rural America." For the history of the wilderness idea and the controversies surrounding it, see Michael P. Nelson and J. Baird Caldicott, eds., *The Wilderness*

Debate Rages On: Continuing the Great New Wilderness Debate (University of Georgia Press, 2008).

As is so often the case in environmental issues, the basic question is one of priorities. Do human needs—jobs, incomes, etc.—come before the needs of wildlife? Do vested interests—farmers, ranchers, miners, the oil industry, etc.—have a higher claim to environmental resources than do forests, rivers, and wildlife? Are short-term benefits (jobs, incomes) more important than long-term benefits (sustainability, and hence survival)? To people like Representative Hastings, the choice is obvious. So too is it for environmentalists. It is the job of institutions such as the Environmental Protection Agency and the Department of the Interior to balance the claims to the environment, and it is a safe bet that no such balancing will satisfy everyone. David N. Laband, "Regulating Biodiversity: Tragedy in the Political Commons," *Ideas on Liberty* (September 2001), argues that one basic problem is that those who demand protection of endangered species do not bear the costs (in jobs and economic activity) of that protection. On the other hand, Howard Youth, "Silenced Springs: Disappearing Birds," *Futurist* (July/August 2003), argues that the actions needed to protect biodiversity not only have economic benefits, but also are the same actions needed to ensure a sustainable future for humanity. According to Martin Jenkins, "Prospects for Biodiversity," *Science* (November 14, 2003), the consequences for human life of failing to protect the natural world are "unforeseeable but probably catastrophic." In the San Francisco Bay area, local land managers recognize this and are working to conserve wild lands; see Glen Martin, "Big Plans for Wild Lands," *Bay Nature* (April-June 2011). New York State recently bought a "conservation easement" on a tract of paper company land, committing it to sustainable use by timber interests, recreational users, and others; see Connie Prickett, "More than a Working Forest," *New York State Conservationist* (June 2011). However, Nick Salafsky, "Integrating Development with Conservation: A Means to a Conservation End, or a Mean End to Conservation?" *Biological Conservation* (March 2011), warns that trying to combine conservation and development often fails. If one cannot pursue conservation by itself, one must be aware of the inevitable trade-offs. See also Johan A. Oldekop, Anthony J. Bebbington, Dan Brockington, and Richard F. Preziosi, "Understanding the Lessons and Limitations of Conservation and Development," *Conservation Biology* (April 2010); and Thomas O. McShane, et al., "Hard Choices: Making Trade-Offs Between Biodiversity Conservation and Human Well-Being," *Biological Conservation* (March 2011).

In the following selections, Mike McKee, Uintah County Commissioner, argues that the government's new "Wild Lands" policy is illegal, contradicts previously approved land use plans for public lands, and will have dire effects on rural economies based on extractive industries. It should be repealed. Robert Abbey, director of the Bureau of Land Management, argues that the government's new "Wild Lands" policy is legal, restores balance and clarity to multiple-use public land management, and will be implemented in collaboration with the public. Destruction of local extractive economies is by no means a foregone conclusion.

GUIDE TO ACRONYMS

APD   Application for permit to drill
BLM   Bureau of Land Management
CEO   Chief Executive Officer
DOI   Department of the Interior
FLPMA   Federal Land Policy and Management Act
LWC   Lands with Wilderness Characteristics
NEPA   National Environmental Policy Act
RMP   Resource Management Plan
WSA   Wilderness Study Area

YES

Mike McKee

Testimony of Mike McKee before the House Natural Resources Committee hearing on "The Impact of the Administration's Wild Lands Order on Jobs and Economic Growth" (March 1, 2011)

T hank you for holding this hearing on the Wild Lands Policy and its negative impacts on my constituents. In Uintah County we are proud of our history, our heritage, and the multiple uses on our public lands from recreation to development of our natural resources.

Uintah County is the largest producer of natural gas in the state of Utah, with 63% of the State's natural gas coming from our County. Oil and gas have been produced in Uintah County since the early 1900's. We remain committed to responsible development of our public lands in an environmentally safe manner.

In Uintah County, only 15% of our land is privately owned. Policy changes during the past two years have had a chilling and detrimental effect on the economy of our County. In 2009, Uintah County lost 3,200 jobs in the mining and extraction industry. Many of our citizens are relocating to other states in order to retain employment and family members are left behind with the hope that the jobs will return. Jobs and the economy are not the only consequences of this administration's policy actions. Uintah County is concerned about homelessness, drug abuse, domestic violence, crime, and other social impacts. Jobs and economy are important to the citizens of Utah and Uintah County. In Uintah County, 50% of our jobs and 60% of our economy are tied to the extractive industry. This fact underscores the importance of sound policy and procedure on our public lands. The Wild Lands Policy issued by the Secretary will make all of these lands off limits in the predictable future for natural gas production, oil production, and shale oil, which are in such rich abundance.

Our community is suffering, and this suffering can be directly tied to policies of the Department of Interior.

Wild Lands Policy which the Interior Secretary signed on December 23, 2010 directly repudiates a Settlement Agreement signed by the State of Utah,

U.S. House of Representatives, March 1, 2011.

the Utah School and the Institutional Trust Lands Administration (SITLA), the Utah Association of Counties and Department of the Interior. The Interior Department incorrectly describes Wild Lands Policy as a revocation of the Norton no-more wilderness policy. The fact is that BLM adopted an instruction memorandum to implement an out-of-court settlement that resolved litigation between the state of Utah and the Department of the Interior.

Interior officials continue to say that there is no violation of this Settlement Agreement, presumably based on the incorrect premise that "Wild Lands" are different from "Wilderness Study Areas" or WSAs. But aside from the name, they are identical and are treated the same.

In the Settlement Agreement, the Department of the Interior committed to not manage public lands outside of WSAs as if they were WSAs. The Wild Lands Policy in fact manages non-WSA public lands under the same protective framework that DOI has applied to WSAs for more than 30 years. The Wild Lands Policy clearly violates the Utah Wilderness Settlement Agreement.

In the Settlement Agreement, the Department of the Interior also pledged not to create new WSAs. The Wild Lands Policy does just exactly that and changing the name does not make it any less of a violation.

No federal law gives the Interior Secretary the authority to implement Secretarial Order 3310, the Wild Lands Policy.

In addition to being poor policy, the Wild Lands Policy is illegal. Under the U.S. Constitution, Congress has the sole authority to regulate federal lands. For public lands, Congress delegates that authority to the Interior Secretary in a series of federal laws, including the Bureau of Land Management Organic Act or the Federal Land Policy and Management Act (FLPMA). For wilderness designation, Congress chose to retain the sole power to designate wilderness.

The Wild Lands Policy attempts to override the laws that apply to public lands in several key respects:

> The Wild Lands Policy declares protection of lands with wilderness character a management priority.
>
> FLPMA dedicates the public lands to multiple use, with principal emphasis on six multiple uses: including domestic livestock grazing, fish and wildlife development and utilization, mineral exploration and production, rights-of-way [including transmission lines and pipelines], outdoor recreation, and timber production.

FLPMA does not include the word 'wilderness' in its definition of multiple use. It defines 'wilderness' only with respect to the now-expired wilderness review program in Section 603.

The Wild Lands Policy attempts to revise federal law by changing land management priorities to promote wilderness protection over all of the other uses that, by federal law, apply to public lands. This contradicts FLPMA, which dedicates the public lands to other uses, several of which, like mineral exploration and development, conflict with wilderness management. It also contradicts the Wilderness Act, which reserves to the sole authority to designate wilderness only by Congress.

The Wild Lands Policy assumes that the Secretary can manage public lands to protect wilderness, although FLPMA provided for a single and limited wilderness review program. FLPMA defines wilderness solely in terms of Section 603, which prescribed a 15-year wilderness review period. It is widely accepted that the authority to study public lands for wilderness expired in 1991, 15 years after FLPMA was enacted. There is no new authority to manage public lands for wilderness protection without attempting to rewrite FLPMA, and only Congress can do so.

It is also worth pointing out that federal agencies must involve the public and local governments when making a significant public land management change. These procedures ensure that there is a robust discussion of the effects of a proposal, and in the case of federal lands, there is coordination with state and local governments. In his haste to issue this policy right before the Christmas holiday, the Interior Secretary ignored these procedural steps.

The Interior Department also ignored the significant adverse environmental impacts that will come from the Wild Lands Policy. Proponents of this policy forget that the Wild Lands Policy will also prohibit wind turbines and transmission lines that are necessary for the green energy promoted by the Interior Secretary. For two years we have heard how the Administration will fund and subsidize green energy for wind turbines, solar energy farms, and the transmission lines necessary to put these alternative energy projects into the electrical power grid. Many energy projects are proposed for public lands, without considering the fact that these structures will violate the Wild Lands Policy. The structures associated with wind and solar energy are prohibited as permanent development and cannot be said to conform to the visual standards applied to wild lands. These important impacts are entirely ignored in the discussion by the Interior Department. It also appears that the Energy Department, which is issuing millions of dollars in incentive grants and loans, is not coordinating with the Interior Department which has adopted a policy that will prohibit or certainly delay implementation of any project.

Since early 2009, DOI has imposed a *de facto* moratorium on drilling and leasing on these lands. Uintah County initiated litigation in October of 2010 because the management policies violated the Settlement Agreement, contradicted the approved land use plans for public lands, and also were harming the local economy.

The Wild Lands Policy could potentially close millions of acres to oil and gas leasing in the State of Utah. BLM previously studied the lands that were said to have wilderness character when it revised the land use plans between 2000 and 2008, so we know the scope of the lands which may be impacted in Utah. These lands do not meet the actual definition of wilderness but are being called wilderness even with dirt roads, livestock developments, oil and gas rigs, pipelines and transmission lines.

We are concerned that the Wild Lands Policy now creates defacto wilderness. In our County, this policy is already negatively affecting areas that were open for multiple use activity. Recently signed Resource Management Plans are being turned upside down by this policy. For example, current road improvement requests, oil and gas leases, and permits to drill are being affected based on Wild Lands Policy.

Historically, Uintah County, on behalf of its citizens, has fully participated in federal land management forums in numerous land management issues, including resource management plans, oil and gas leasing decisions, transportation corridors on Federal lands, and wilderness issues. The County has expended a tremendous amount of resources over the past 20 years to engage in these processes in a responsible manner and representing our constituents. When Secretary Salazar announced the Wild Lands Policy just two days before Christmas in 2010, it was not only a shock to our constituents but was clearly an effort to circumvent established public processes that have governed our federal lands. In an economy and energy situation that is already at rock bottom, this action is further proof that Secretary Salazar has little regard for jobs or energy security in the West.

Over the past decade, the BLM began a revision of the Resource Management Plan for Utah and the Uintah Basin. This process, governed by NEPA, was open to the public and Uintah County participated as a cooperating agency. Thousands of hours and well over a million dollars of tax payer funds were expended by Uintah County. Other entities participated to bring to fruition a management plan that takes a comprehensive look at all uses of public lands in Uintah County. Although long, sometimes painful, and certainly no one group liked everything in the plan; this is what NEPA contemplated. Concessions were made on all sides. Uintah County supports open, public processes where all views are heard and considered, and then the hard working professionals of the BLM make informed decisions. All of the issues the Secretary claims to address under the new Wild Lands Policy are addressed in the Resource Management Plan—the only difference is that the Secretary clearly disagrees with the outcome of this Plan. Instead of attempting to short circuit the NEPA process, we urge the Secretary to vigilantly defend the BLM's Resource Management Plans. We need to end the practice of settling claims with litigants for the sole purpose of setting new policy outside the bright light of public input. Simply, the Wild Lands Policy undermines the Resource Management Plans.

We also note that toward the conclusion of the Vernal Resource Management Plan process, alternative "E" was added. This alternative's sole purpose was to evaluate the full spectrum of potential wilderness and the management thereof. This process required an additional two years to complete. Director Bob Abbey, in a meeting recently held in Salt Lake City, Utah, stated that the reason for reanalyzing work that was already complete was because not enough wilderness was found. This continual upheaval, unrest, change of direction, and philosophy, is discouraging. Either the land has wilderness quality or it does not. Why, with the huge deficits of spending that the Government is going through, do we have the BLM redo that which they have already completed?

In real terms, this policy will make it economically less viable for natural resource developers to operate on federal lands in the West. The State of Utah processes applications for permit to drill (APD's) in 35 days, while BLM takes an average of one and a half years. The Wild Lands Policy will add years to the permitting process and effectively further reduce access to natural resource production. It will yet create another layer of unnecessary bureaucracy that will only result in the further loss of jobs in my County and in other public lands counties throughout the West. Moreover, Uintah County will be forced to spend precious tax payer dollars to fight our own government to try to force the Department of Interior to live by the law of the land.

The combination of regressive gas leasing policies and the new Wild Lands Policy will result in further job losses and economic impact in Uintah County and throughout the west. Recently, I visited with a local CEO whose business has a cutting edge technology in the natural gas industry, yet, he can see the writing on the wall with the current policies. He will likely move his headquarters. He just returned from Dhabi as an option. Why would a business owner even consider such an option with all the unrest in the Middle East? What is wrong with this picture? Is the business environment better in the Middle East than on our own public lands in Uintah County? Planned and balanced development of these resources takes years to move into production. Driving these companies overseas is detrimental to our economy and to our energy security.

Unfortunately, today's policies are stopping responsible development and endangering America's energy security. This is not a spigot you can simply turn on and off on a whim.

Many companies stand ready to invest large sums of money in our County over the next ten years. All told, these investments would exceed two billion dollars over a ten year period. However, the regulatory uncertainty and the adverse policies of the Department of Interior is keeping these companies from investing, and in many cases, driving them overseas where U.S. dollars are being invested in foreign economies.

Eastern Utah is a treasure chest of natural resources. Uintah County has a great opportunity to help America become energy independent. Utah has 6.7 trillion cubic feet of proven natural gas reserves, conventional oil reserves of 286 million barrels, much of these are found in Uintah County. According to a Rand Report, the Uintah Basin has a staggering amount of shale oil ranging from 56 billion barrels to 321 billion barrels.

Each morning our newspapers carry disturbing pictures of governmental unrest in the Middle East and news of more and larger oil supply disruptions. In less than a month, previously stable countries in northern Africa and the Middle East have erupted in violent demonstrations. The governmental overthrow of Tunisia and Egypt has gone viral in Yemen, Libya, Saudi Arabia, and Bahrain with new calls for changes in the governments of the region. These shifts in power will have profound changes for the future, especially for the United States that produces and transports oil from those regions to the United States.

The Wild Lands Policy threatens national security by sharply reducing the nation's energy independence. It applies equally to all sources of energy from public lands such that the country is made weaker at a time when it needs to be stronger and more self-sufficient.

In addition to this, the Wild Lands Policy will impact the education of our children. The State of Utah was granted upon statehood, school trust lands, which by State Constitution are mandated to generate income to fund schools in the State of Utah. These lands are interspersed with federal lands throughout the State of Utah and Uintah County. It is commercially unviable to develop these lands for natural resources without access to the surrounding lands. If the federal lands become off limits to development, state lands go undeveloped as well, and education suffers directly from the federal policies.

To sum it up, the Wild Lands Policy is a short-sighted initiative that undermines the interests of this country and its people. The Wild Lands Policy overreaches by revising federal law when only Congress can do so. We urge this Committee to take every action possible to repeal it.

Our natural resources should be responsibly developed pursuant to the laws of the land. We have a responsibility to carefully develop our resources for America, for energy security, for our economy, and jobs for our citizens. I commend the House for choosing to de-fund the Wild Lands Policy for this current fiscal year and I urge the Senate to follow your lead. The role of Congress is clear in terms of wilderness policy, and I urge this Congress to preserve its authority and reverse this policy to save my county and our country from further economic harm.

Robert Abbey  **NO**

Testimony of Robert Abbey before the House Natural Resources Committee hearing on "The Impact of the Administration's Wild Lands Order on Jobs and Economic Growth" (March 1, 2011)

The Wild Lands policy, established by Secretarial Order 3310, restores balance and clarity to the management of our public lands and follows clear legal direction. This order directs the BLM to work collaboratively with the public and local communities to determine how best to manage the public lands, taking into account all of their potential uses, including uses associated with the wilderness characteristics of certain public lands. It does not dictate the results of that planning process.

Section 102 of the Federal Land Policy and Management Act (FLPMA) declares that preservation and protection of public lands in their natural condition are part of the BLM's mission. Just as conventional and renewable energy production, grazing, mining, off-highway vehicle use, and hunting are considered in the development of the BLM's Resource Management Plans (RMPs), so too must the protection of wilderness characteristics be considered in the agency's land use plans.

Lands with wilderness characteristics are valued for their outstanding recreational opportunities (such as hunting, fishing, hiking, photography, or just getting outdoors) as well as for their important scientific, cultural, and historic contributions. Failing to consider protecting these wild places would undermine the careful balance in management mandated by law, a balance that we need on our public lands. Public lands provide billions of dollars in local economic benefits and they should be managed for multiple uses and many values, including energy production, recreation, and conservation.

U.S. House of Representatives, March 1, 2011.

The BLM's Multiple-Use Mission/Economic Contributions

I have worked for over 30 years in public service, 25 of those years as a career BLMer. I believe in, and am dedicated to, the BLM's multiple-use mission. This multiple-use mission is what makes the agency unique among Federal land management agencies, and it is what makes us welcome members of every community in which we work and live. However, multiple-use does not mean every use on every acre.

The BLM strives to be a good neighbor and a vital part of communities across America. Public lands managed by the BLM contribute significantly to the nation's economy and, in turn, often have a positive impact on nearby communities. The BLM's management of public lands contributes more than $100 billion annually to the national economy, and supports more than 500,000 American jobs.

A key component of these economic benefits is the BLM's contribution to America's energy portfolio. The BLM expects its onshore mineral leasing activities to contribute $4.3 billion to the Treasury in Fiscal Year 2012. The BLM currently manages more than 41 million acres of oil and gas leases, although less than 30 percent of that acreage is currently in production. More than 114 million barrels of oil were produced from BLM-managed mineral estate in Fiscal Year 2010 (the most since Fiscal Year 1997), and the almost 3 billion MCF (thousand cubic feet) of natural gas produced made 2010 the second-most productive year of natural gas production on record. The coal produced from nearly a half million acres of federal leases powers more than one-fifth of all electricity generated in the United States.

The BLM is also leading the nation toward the new energy frontier with active solar, wind, and geothermal energy programs. The BLM has proposed 24 Solar Energy Zones within 22 million acres of public lands identified for solar development, and in 2010 approved nine large-scale solar energy projects. These projects will generate more than 3,600 megawatts of electricity, enough to power close to 1 million homes, and could create thousands of construction and operations jobs. Development of wind power is also a key part of our nation's energy strategy for the future. The BLM manages 20 million acres of public lands with wind potential; currently, there is 437 MW of installed wind power capacity on the public lands. Geothermal energy development on the public lands, meanwhile, accounts for nearly half of U.S. geothermal energy capacity and supplies the electrical needs of about 1.2 million homes.

Energy production is not the only way in which the BLM contributes to local communities and the national economy. The combined economic impacts of timber-related activities on BLM-managed lands, grazing-related activities, and activity attributable to non-energy mineral production from BLM-managed mineral estate total more than $5 billion each year. Recreation on public lands also provides major economic benefits to local economies and communities. In 2010, more than 58 million recreational visits took place on BLM-managed lands and waters, contributing billions of dollars to the U.S. economy. The diverse recreational opportunities on BLM-managed lands draw crowds of backpackers, hunters, off-road vehicle enthusiasts, mountain bikers, anglers, and photographers.

In an increasingly urbanized West, these recreational opportunities are vital to the quality of life enjoyed by residents of western states, as well as national and international visitors. It should be noted that many of these recreationists are seeking the primitive experience available in BLM's wilder places.

The BLM's multiple-use mission is all about balancing public land management, and balancing all of the myriad resource values of this nation's great public lands. Wilderness character is one of these many resource values, and the BLM's new Wild Lands policy is a rational approach to ensuring that balance.

Secretarial Order 3310—Wild Lands Policy

The BLM's authority to designate new Wilderness Study Areas (WSAs) under section 603 of the FLPMA expired after President George H.W. Bush completed his recommendations for wilderness designation to Congress in January 1993. However, the BLM was still required to inventory and consider wilderness characteristics in the land use planning process.

Secretary of the Interior Gale Norton and the State of Utah entered into an out-of-court settlement agreement (the "Norton-Leavitt settlement") in 2003 that resulted in BLM rescinding the agency's then existing guidance on wilderness inventory. Since that time, the BLM has been without long-term national guidance on how to meet the FLPMA requirements to inventory and manage lands with wilderness characteristics. In 2008, the Ninth Circuit Court of Appeals in *Oregon Natural Desert Association v. BLM* stated that FLPMA's requirement that BLM maintain an inventory of public lands and their resources and other values includes inventory of wilderness values and that BLM must consider those values in its land use planning when they are present in the planning area. Secretarial Order 3310 and the related BLM manuals address that previous lack of direction on inventorying and managing lands with wilderness characteristics.

On December 23, 2010, I joined Secretary Salazar in announcing clear direction for implementing the BLM's mandate under FLPMA to conduct wilderness characteristics inventories and decide how best to manage those lands. The BLM also issued draft manuals that were recently finalized. This Wild Lands policy restores balance to the BLM's multiple-use management of the public lands in accordance with applicable law. It also provides the field with clear guidance on how to comply with FLPMA and more specifically how to take into account wilderness characteristics in the agency's planning process.

With this consistent guidance, we believe that the BLM will enhance its ability to sustain its land use plan and project level decisions. In the past, some of these decisions have been invalidated because the courts in the Ninth and Tenth Circuits have found the analysis of wilderness characteristics lacking.

Policy Implementation/BLM's Manuals

There has been a great deal of confusion about what this new policy does, and perhaps more importantly, what it does not do. Be assured that the new policy itself does not immediately change the management or status of the public lands. I would like to outline for you the facts about the new policy and its implementation. The BLM's new manuals set out a two-step process for inventorying and managing lands that may have wilderness characteristics. The first step is to maintain an inventory of Lands with Wilderness Characteristics (LWCs) as required by section 201 of FLPMA. The BLM's new manual on Wilderness Characteristics Inventory provides guidance on both updating existing inventory information and inventorying lands not previously assessed.

The manual carefully spells out the process for making these determinations, based on size, naturalness, and outstanding opportunities for solitude or a primitive and unconfined type of recreation—using the same Wilderness Act criteria the agency has always used. This process makes no determination about how the lands should be managed; it simply documents the current state of the lands.

Step two of the process, deciding how LWCs should be managed, is an open, public process undertaken through the BLM's land use planning process. Through this public process, a decision may be made to protect LWCs as "Wild Lands" or to manage them for other uses. For example, the BLM may determine that impairment of LWCs is appropriate for some areas due to other resource considerations, such as energy development. Other areas may be managed as Wild Lands with restrictions on surface disturbance and the construction of new structures. In addition, Wild Lands designations must be consistent with other applicable requirements of law. The BLM must consider these additional statutory requirements, where appropriate, in determining whether LWCs can be managed to protect their wilderness characteristics.

It is important to emphasize that if lands are designated as Wild Lands they are not wilderness and they are not WSAs. First, Wild Lands may only be designated administratively through an open, public planning process. The designation of Wild Lands may be revisited, as the need arises, through a subsequent public planning process. Second, allowed uses in Wild Lands may include some forms of motorized and mechanized travel. Allowed uses in each specific Wild Land will be determined by the land use plan governing those lands and will be accomplished through a process that allows the public and local communities full access to that decision-making. These decisions will be made locally, not in Washington, D.C. This policy doesn't change the delegation of authority for land use planning decisions. The BLM's state and field offices will continue to be responsible for those planning decisions.

The BLM regularly makes project-level decisions for activities on public lands. These decisions can involve a wide range of proposals such as locating roads and power lines, filming commercials and movies, and permitting mineral extraction activities. When considering these proposals, the BLM relies on existing land use plans, as well as any new information, to make a determination of how and if these projects can be accommodated within the BLM's multiple-use mission. This determination is necessarily a balancing act, taking into account all of the resources for which the BLM is responsible—including wilderness characteristics—as mandated by FLPMA.

A Wild Lands designation will be made and modified through an open public process, and therefore these designations differ from designated wilderness areas and WSAs. Wilderness areas can only be designated through an act of Congress and modified through subsequent legislation. The BLM manages WSAs to protect their wilderness characteristics until Congress designates them as wilderness or releases them from WSA status.

I have heard concerns that the new Wild Lands policy has put a halt to new projects and will prevent important economic activity in local communities. This claim is, simply put, false. A recent example involves a potash lease proposal in Utah that the BLM has approved through this new process. Through the NEPA process, the BLM has undertaken a review of a proposal to offer a competitive lease sale for potash on Sevier Lake, a dry lake bed in southwestern Utah. Following the issuance of the Secretarial Order roughly two months ago, the BLM completed an inventory of the lands involved and determined that the area does not meet the criteria for LWCs. The project is moving forward and it has been reported that it may result in as many as 300 permanent jobs in the local community.

Conclusion

The BLM is committed, and I am personally committed, to working with Congress and other key stakeholders to ensure that the Wild Lands policy works. My staff and I have spoken with many of you directly about the policy. In January, I traveled to Utah at the request of Governor Herbert, and participated in several meetings and forums on the policy. We have heard your concerns, and we are listening.

The BLM's Wild Lands policy affirms the agency's responsibility to take into account all of the public land resources for which the BLM is responsible. The policy provides local communities and the public with a strong voice in the decisions affecting the nation's public lands. Working cooperatively with our stakeholders, and being sensitive to local needs, we will ensure that all of the potential uses of the public lands and the BLM's multiple-use mission are taken into account when determining how best to manage the nation's public lands.

EXPLORING THE ISSUE

Does Designating "Wild Lands" Harm Rural Economies?

Questions for Critical Thinking

1. Do states with different amounts of publicly owned land warrant different land-management policies?
2. Why do many business interests oppose the concept of wilderness?
3. To what degree should public land management policy be left to the federal government?

Is There Common Ground?

The users of public lands fear the BLM's Wild Lands policy because of its potential effect on jobs and local economies. The BLM agrees that jobs and local economies are important but "preservation and protection of public lands in their natural condition are part of the BLM's mission." A crucial question is how to satisfy both the economic and protective demands. Public involvement in policy-making is one way to reduce conflicts; Stephen E. Decker and Alistair J. Bath, "Public versus Expert Opinions Regarding Public Involvement Processes Used in Resource and Wildlife Management," *Conservation Letters* (December 2010), note that "Successful public involvement efforts can reduce conflict and build trust between resource managers and the public."

1. Where else does public involvement in decision-making play a part?
2. Find a local example (perhaps on campus). Has it been successful at reducing conflict and building trust?
3. Sometimes "public involvement" is replaced by "stakeholder involvement" in decision-making. Who are the stakeholders in the Wild Lands debate?

Internet Resources

The Wilderness Society

The Wilderness Society is the leading American conservation organization working to protect the 635 million acres of government-managed public lands in the United States.

http://wilderness.org/

Contributors to This Volume

EDITOR

THOMAS A. EASTON is a professor of science at Thomas College in Waterville, Maine, where he has been teaching environmental science, science, technology, and society, emerging technologies, and computer science since 1983. He received a BA in biology from Colby College in 1966 and a PhD in theoretical biology from the University of Chicago in 1971. He writes and speaks frequently on scientific and futuristic issues. His books include *Focus on Human Biology*, 2nd ed., coauthored with Carl E. Rischer (HarperCollins, 1995), *Careers in Science*, 4th ed. (VGM Career Horizons, 2004), *Taking Sides: Clashing Views in Science, Technology and Society* (McGraw-Hill, 10th ed., 2012), *Taking Sides: Clashing Views in Energy and Society* (McGraw-Hill, 2nd ed., 2011), and *Classic Editions Sources: Environmental Studies* (McGraw-Hill, 4th ed., 2012). Dr. Easton is also a well-known writer and critic of science fiction.

AUTHORS

ROBERT ABBEY is Director of the Bureau of Land Management, United States Department of the Interior.

SENATOR LAMAR ALEXANDER (R-TN), Republican Senator from Tennessee, has been a Tennessee governor (1979–1987) and a United States Secretary of Education (1991–1993).

MARK ANSLOW is the *Ecologist*'s senior reporter.

DAVID ATTENBOROUGH is a British naturalist and broadcaster who has produced numerous popular wildlife documentaries.

RONALD BAILEY is a science correspondent for *Reason Magazine*. A member of the Society of Environmental Journalists, his articles have appeared in many popular publications, including the *Wall Street Journal*, *The Public Interest*, and *National Review*. He has produced several series and documentaries for PBS television and ABC News, and he was the Warren T. Brookes Fellow in Environmental Journalism at the Competitive Enterprise Institute in 1993. He is the editor of *Earth Report 2000: Revisiting the True State of the Planet* (McGraw-Hill, 1999) and the author of *Global Warming and Other Eco-Myths: How the Environmental Movement Uses False Science to Scare Us to Death* (Prima Publishing, 2002).

STEPHEN L. BAIRD is a technology education teacher for the Virginia Beach City Public School system and an adjunct assistant professor at Old Dominion University.

TOM BETHELL is a senior editor of *The American Spectator*.

PAUL CICIO is president of the Industrial Energy Consumers of America.

JAMIE CLARK is senior vice president for Conservation Programs, National Wildlife Federation.

BENEDICT S. COHEN is deputy general counsel for Environment and Installations, Department of Defense.

D. J. CONNOR is professor emeritus in the School of Agriculture and Food Systems, The University of Melbourne, Victoria, Australia.

KEN CUSSEN is with the Graduate School of the Environment, Macquarie University, Sydney, Australia.

RICHARD DAHL is a Boston freelance writer who writes periodically for MIT.

VIRGINIA H. DALE is a scientist in the Center for BioEnergy Sustainability at the Oak Ridge National Laboratory, Oak Ridge, Tennessee.

GIULIO A. DE LEO is a professor of ecology at the Università degli studi di Parma Italy.

KATE J. DENNIS is a PhD candidate at Harvard University's Department of Earth and Planetary Sciences.

C. JOSH DONLAN is a research biologist at Cornell University, founder and director of Advanced Conservation Strategies, an adviser to the Galapagos

National Park and to Island Conservation, and a senior fellow at the Robert and Patricia Switzer Foundation and Environmental Leadership Program.

AARON EZROJ is a Law Fellow at Adams Broadwell Joseph & Cardozzo. He studied market-based environmental mechanisms in Europe as a Fulbright Scholar.

TAKIS FOTOPOULOS is an economist and political philosopher. He founded the "inclusive democracy" movement and is the author of *Towards an Inclusive Democracy* (Cassell, 1997).

MARINO GATTO is a professor of applied ecology in the Dipartimento di Elettronica e Informazione at Politecnico di Milano in Milan, Italy. He is associate editor of *Theoretical Population Biology*.

THOMAS A. GAVIN is an associate professor in the Department of Natural Resources at Cornell University.

THE GLOBAL HUMANITARIAN FORUM was established to bring together a global community and create the conditions for dialogue and rapid action. Its mission was to inspire others and to connect them so that they can take action together. It closed in 2010 for lack of funds.

REBECCA L. GOLDMAN is a senior scientist for Board Relations in the central science division of The Nature Conservancy.

MICHAEL GOUGH is a biologist and expert on risk assessment who has participated in science policy issues at the Congressional Office of Technology Assessment, in Washington think tanks, and on various advisory panels. He edited *Politicizing Science: The Alchemy of Policymaking* (Hoover Institution Press, 2003).

ED HAMER is a freelance journalist.

JAYDEE HANSON is policy director at the International Center for Technology Assessment (http://www.icta.org).

ERIC HOFFMAN is Genetic Technology Policy Campaigner with Friends of the Earth (http://www.foe.org/healthy-people/biofuels-synthetic-biology).

CHRISTOPHER D. HOLMES is a PhD candidate at Harvard University's Department of Earth and Planetary Sciences.

KURT Z. HOUSE is the president of C12 Energy, based in Cambridge, MA.

GREGORY E. KAEBNICK is a research scholar at The Hastings Center and editor of the *Hastings Center Report*.

MARISSA N. KARPOFF is currently pursuing an MD at Tulane Medical School in New Orleans.

DIANE KATZ is director of Risk, Environment and Energy Policy for the Fraser Institute, an independent non-partisan research and educational organization based in Canada.

KEITH KLINE is a scientist in the Center for BioEnergy Sustainability at the Oak Ridge National Laboratory, Oak Ridge, Tennessee.

TIM KRUEGER is working for Policy Matters Ohio, an economic research institute.

JOANNA KYRIAZIS is currently enrolled in law school at the University of Toronto.

ANDREA LARSON is a professor in the Darden School of Business, University of Virginia. Her research has focused on innovation and entrepreneurship, strategy, and sustainability.

BENJAMIN G. LEE is a postdoctoral researcher at the National Renewable Energy Lab in Golden, Colorado.

RUSSELL LEE is a scientist in the Center for BioEnergy Sustainability at the Oak Ridge National Laboratory, Oak Ridge, Tennessee.

PAUL LEIBY is a scientist in the Center for BioEnergy Sustainability at the Oak Ridge National Laboratory, Oak Ridge, Tennessee.

STEPHEN LESTER is science director of the Center for Health, Environment & Justice.

BJØRN LOMBORG is an adjunct professor in the Copenhagen Business School and organizer of the Copenhagen Consensus, a conference of top economists who work to prioritize the best solutions for the world's greatest challenges. His latest books are *Cool It! The Skeptical Environmentalist's Guide to Global Warming* (Knopf, 2007) and Smart Solutions to Climate Change: Comparing Costs and Benefits (Cambridge University Press, 2010).

ROBERT MCCORMACK is currently completing a Masters in Environmental Law and Policy at Vermont Law School.

MIKE MCKEE is Commissioner of Uintah County, Utah. He also co-chairs the Western Legacy Homestead Alliance.

ALLISON MACFARLANE is an associate professor of Environmental Science and Policy at George Mason University and a member of the U.S. Energy Department's Blue Ribbon Commission on America's Nuclear Future. She is the coauthor, with Rod Ewing, of *Uncertainty Underground: Yucca Mountain and High-Level Nuclear Waste Disposal* (MIT Press, 2006).

ALISON MARKLEIN is a graduate student in Earth Systems Ecology and Biogeochemistry at University of California, Davis.

LEE T. MURRAY is a PhD candidate at Harvard University's School of Engineering and Applied Sciences.

JUSTIN PARRELLA is a PhD candidate at Harvard University's School of Engineering and Applied Sciences.

GILLIAN S. PAUL has a masters of forest science from the Yale School of Forestry and Environmental Studies and now works for the Environmental Leadership and Training Initiative (ELTI), a global training and capacity building organization that is a joint initiative of Yale University and the Smithsonian Tropical Research Institute (STRI).

DAVID PIMENTEL is professor emeritus in the departments of Entomology and Ecology and Evolutionary Biology at Cornell University in Ithaca, New York.

J. WINSTON PORTER is the president of the Waste Policy Institute in Leesburg, Virginia. From 1985 to 1989, he was the EPA's Assistant Administrator for Solid Waste and Emergency Response.

ANNE RABE is the campaign coordinator for the Center for Health, Environment & Justice's Be Safe Campaign for Precaution.

DAVID M. ROMPS is a research scientist at Harvard University's Department of Earth and Planetary Sciences.

MARY ANNETTE ROSE is an assistant professor in the Department of Technology at Ball State University in Muncie, Indiana.

DUSTIN R. RUBENSTEIN is a behavioral and evolutionary ecologist. In 2009, he joined the Department of Ecology, Evolution and Environmental Biology at Columbia University.

DANIEL I. RUBENSTEIN is professor and chair of Princeton's Department of Ecology and Evolutionary Biology and Director of the Program in African Studies.

JASON RUGOLO is a PhD candidate at Harvard University's School of Engineering and Applied Sciences.

CARL SAFINA is the founding president of the Blue Ocean Institute. His latest book is *Voyage of the Turtle* (Henry Holt and Company, 2006).

PAUL W. SHERMAN is a professor in the Department of Neurobiology and Behavior at Cornell University. His specialty is animal social behavior.

KRISTIN SHRADER-FRECHETTE is the O'Neill Family Professor, Department of Biological Sciences and Department of Philosophy, at the University of Notre Dame. She is the author of *Taking Action, Saving Lives: Our Duties to Protect Environmental and Public Health* (Oxford University Press, 2007).

AGNE SIRINSKIENE is an associate professor in the Faculty of Law, Department of Biolaw, Mykolas Romeris University, Vilnius, Lithuania. Her research interests include the regulation and ethical problems of the beginning of life and the precautionary principle in European Union law.

JIM THOMAS is program manager at ETC Group (Action Group on Erosion, Technology and Concentration) (http://eee.etcgroup.org/issues/synthetic_biology).

MICHELE L. TRANKINA is professor of biological sciences at St. Mary's University and adjunct associate professor of physiology at the University of Texas Health Science Center, both in San Antonio, Texas.

MEGAN A. TOTH is a graduate student in Environmental Studies at the University of Oregon in Eugene, Oregon.

JESSICA TSAI is an assistant editor at *CRM* magazine.

FOOD AND AGRICULTURE ORGANIZATION OF THE UNITED NATIONS is a United Nations agency that coordinates efforts to achieve global food security.

MARK T. WINKLER is a PhD candidate at Harvard University's Department of Physics.

DEBORAH WEISBERG is an award-winning journalist whose work appears in the *Pittsburgh Post-Gazette*, *New York Times*, and other publications.